# LIGHTWEIGHT ALLOYS FOR AEROSPACE APPLICATION

*Edited by:*
*Dr. Kumar Jata, Dr. Eui Whee Lee,*
*Dr. William Frazier and Dr. Nack J. Kim*

# LIGHTWEIGHT ALLOYS FOR AEROSPACE APPLICATION

*Edited by:*
*Dr. Kumar Jata, Dr. Eui Whee Lee,*
*Dr. William Frazier and Dr. Nack J. Kim*

Proceedings of Symposium Sponsored by
the Non-Ferrous Metals Committee of
the Structural Materials Division (SMD) of
TMS (The Minerals, Metals & Materials Society).

Held at the TMS Annual Meeting
in New Orleans, LA, USA
February 12-14, 2001.

A Publication of

Partial funding for this publication was provided by the Seeley W. Mudd Fund.

**A Publication of The Minerals, Metals & Materials Society**
184 Thorn Hill Road
Warrendale, Pennsylvania 15086-7528
(724) 776-9000

**Visit the TMS web site at**
http://www.tms.org

ISBN Number   978-0-87339-491-8

If you are interested in purchasing a copy of this book, or if you would like to receive the latest TMS publications catalog, please telephone 1-800-759-4867 (U.S. only) or 724-776-9000, EXT. 270.

# TABLE OF CONTENTS

# ALUMINUM -LITHIUM ALLOYS

# TITANIUM ALLOYS

# COMPOSITES

# FOREWORD

This manuscript contains a collection of 27 papers presented at the Lightweight Alloys for Aerospace Applications symposium at the TMS Annual Meeting in New Orleans, February 12-15, 2001. The manuscript contains outstanding papers on:

• Current understanding of coarsening of precipitates in novel aluminum alloys
• Deformation, fracture and fatigue and corrosion resistance of aluminum-lithium alloys
• Nano-crystalline aluminum alloys
• Processing, fatigue and fracture and relationships to microtexture in Ti
• Status of gamma titanium aluminides
• Microstructure and properties of metal matrix composites

# ACKNOWLEDGEMENTS

The organizers of the symposium would like to extend their sincere appreciation to Professors Erhard Hornborgen, Gary Shiflet, Thomas Sanders, Hugh McQueen, Arun Gokhale and to Drs. Jonathan Paul, Kamran Nikbin and Awadh Pandey for participating in the symposium as invited speakers and to Dr. Mary C. Juhas and Prof. O.S. Es-Said who helped in chairing the sessions. The organizers would like to thank all the participants who made this symposium a big success.

Kumar V. Jata
Air Force Research Laboratory, USA
June 2001

# LIGHTWEIGHT ALLOYS FOR AEROSPACE APPLICATION

*Edited by:*
*Dr. Kumar Jata, Dr. Eui Whee Lee,*
*Dr. William Frazier and Dr. Nack J. Kim*

## ALUMINUM ALLOYS

Precipitation Hardening –
The Oldest Nanotechnology

*Erhard Hornbogen*

Pgs. 1-11

184 Thorn Hill Road
Warrendale, PA 15086-7514
(724) 776-9000

# PRECIPITATION HARDENING - THE OLDEST NANOTECHNOLOGY

Erhard Hornbogen

Ruhr-University Bochum, Institute for Materials
44780 Bochum, Germany

## Abstract

Precipitation hardening of aluminum was discovered about 100 years ago by Dr. Alfred Wilm [1,2]. Using aluminum alloys as example a survey is given on mechanism and limits of precipitation hardening. It is discussed how hard, nanometer-size particles (nanos, greek, the dwarf) can form as an ultra fine dispersoid. A simple example for optimum conditions is provided by diamond cubic particles (Si, Ge) in the f.c.c. Al-Matrix. The role of a sequence from more to less metastable phases is discussed, as well as the effects of additional (and trace) alloying elements. The amount of precipitation hardening is limited, besides by the volume fraction of particles, by their strength. This in turn determines the critical diameter above which the transition from passing to by-passing by dislocations takes place. Simple models for the calculation of this microstructural parameter are discussed.
From combinations of precipitation hardening with other hardening mechanisms the limits for ultra high strengths are defined.

Lightweight Alloys for Aerospace Applications
Edited by Kumar Jata, Eui Whee Lee,
William Frazier and Nack J. Kim
TMS (The Minerals, Metals & Materials Society), 2001

# The story of a discovery

It is now about 100 years ago that Dr. Alfred Wilm started experiments with a wide range of aluminum alloys at the metallurgical department of the Central Institute for Scientific and Technological Studies in Neubabelsberg (close to Berlin). Major motivation for his work was to increase the insufficient strength of this then still relatively young material. By 1906 he had developed a new type of alloy with 3.5-5.5% Cu and less than 1% Mg and Mn with a strength of more than 400 MPa. It was soon well known by its trade name *Duralumin*. The prefix contains a possibly intended ambiguity: *durus* (latin, hard), but also *Dürener Metallwerke*, (Rhineland) the industrial firm, where the alloy was produced and shaped. [1,2]

There are the options for evolutionary, predictable progress in science and technology or revolutionary developments. The discovery of precipitation hardening was unpredicted.

Alfred Wilm knew all the physical metallurgy of his days. He knew that metals could be solid solution hardened and work hardened. He also knew that the carbon steels had to be quenched for hardening. Therefore he melted a large number of alloys which he investigated in an as-quenched, slowly cooled and worked state. To his great frustration he found out that – different from steels – some alloys became even softer by quenching.

However, one day, many quenching experiments were performed on saturday morning. Not all the hardness measurements were completed the same day. The sun was shining and Wilm went out for the week-end to go sailing. Next monday morning the hardness measurements were completed, and surprisingly hardness as well as all strength properties had increased considerably. At first, the technician was blamed for sloppy work, but then all the measurements were repeated carefully. In addition, duration and temperature of aging was varied systematically.

1. Wilm had to follow the wrong hypothesis of the analogy between steel and aluminum to expect hardening by rapid quenching.
2. He had to be lazy, not to complete his measurements immediately, to realize the importance of aging.
3. He had to recognize the completely unexpected result, reproduce, optimize and apply it.

He could, however, not understand that the production of a fine dispersoid of nm-size particles is the physical cause of precipitation hardening. So he did not know that he discovered the first nano-technology.

# Lattice correspondences

Considerable strength of Al-alloys is always due to precipitation hardening. This requires the formation of an even and ultra-fine dispersoid of hard particles. The optimum would be achieved by hard diamond cubic particles in the fcc Al-based matrix. Size and spacing should be in the nm-range. In fact, often either a fine microstructure of soft particles emerges or a coarser one of strong particles. Both do not lead to the desired hardening effect. (Fig. 1)

This paper will discuss the mechanisms of precipitation of second phases, using fcc solid solutions of Al as examples. They have been well explored in the past 50 years [3, 4]. The results can be interpreted by a generalized Ostwald step rule. In addition dislocation theory provides the principles on how to obtain optimum strengthening.

Starting condition is a supersaturated solid solution $\alpha_s$, which is often obtained by quenching from a temperature of high solubility. This implies that not only solute atoms are obtained in excess, but also lattice vacancies (Fig. 3). Higher super-supersaturations $\alpha_{ss}$ can be obtained by rapid solidification of liquids or vapour deposition - which may lead either to a glassy, or quasi-crystalline, or crystalline (massive crystallization) phases. Not only vacancies have to be considered for the analysis of precipitation reactions in Al-alloys, but also $0 < d \leq 3$-dimensional defects, such as dislocations (d = 1), grain boundaries, stacking faults (d = 2). The density of dislocations $\rho_d$, for example, may be varied by many orders of magnitude ($\rho_d \leq 10^{15}$ m$^{-2}$) by plastic deformation of supersaturated crystals $\alpha_{sd}$.

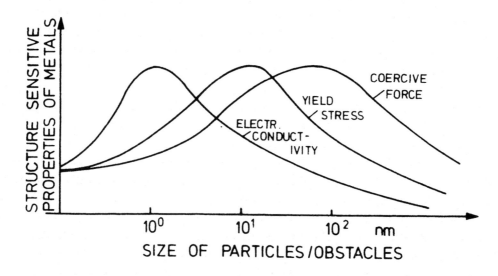

Figure 1: Anomalies of different physical properties due to nm-size particles

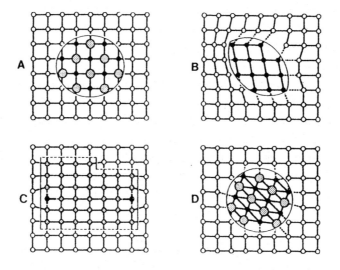

Figure 2: Types of coherency between matrix $\alpha$ and particle $\beta$, (a) coherent, ordered, (b) constrained coherent (shear), (c) partially coherent, (d) non coherent

Table I Nucleation mechanisms in Al-Cu- alloys

| | | | defects geom. dimension | 0-d | 1-d | 2-d |
|---|---|---|---|---|---|---|
| crystal structure | stability | coherency with fcc | no defect | vacancy v | dislocation d | grain boundary b |
| cluster, GP-zones | ↓ | c | + | + | + | + |
| $\Theta''$ | ↓ | c | + | + | ++ | +++ |
| $\Theta'$ | ↓ | pc | +++ | ++ | + | ++ |
| $\Theta$ | max | nc | +++ | +++ | ++ | + |

c coherent, pc partially c, nc non c, + low activation barrier for nucleation, +++ high activation barrier for nucleation

4

Finally a comparison must be made with respect to coherency between the structure of the matrix crystal $\alpha$ and the newly formed phases $\beta_i$. This relationship may imply:
a) non-coherency, b) partial coherency, c) constrained coherency, d) full coherency (Figs. 2, 3). These possibilities provide a decreasing amount of the structural term of the interfacial energy $\gamma_{\alpha\beta}$, which essentially controls the nucleation behavior of $\beta_i$ from $\alpha$ [5, 6, 7] (Table 1). The crystal structures of strong particles (termodynamically stable and high resistance to dislocation motion: $\Theta$-Al$_2$Cu, $\delta$-AlLi, Si) do not fulfill the prerequisites for coherency with fcc Al (Fig. 3).

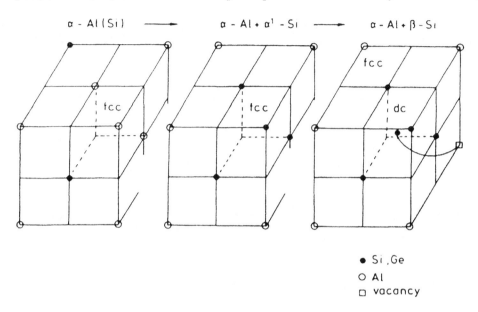

Figure 3: Formation of non-coherent Si from fcc Al (Si) solid solution

## Metastable phases

The sequence of the precipitate phases $\beta_i$ for the Al-Cu-alloys implies a decreasing coherency with increasing thermodynamical stability (Fig. 4): $\Theta'' \rightarrow \Theta' \rightarrow \Theta$.

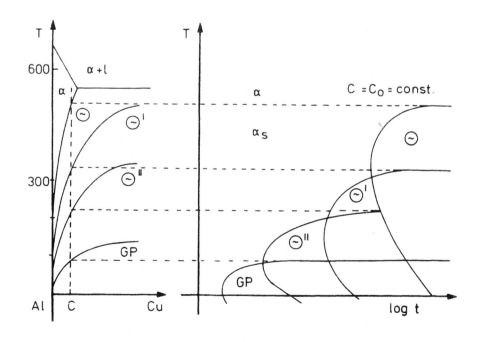

Figure 4(a): Sequence of metastable, coherent $\rightarrow$ non-coherent $\Theta$–Al$_2$Cu in AlCu-alloys

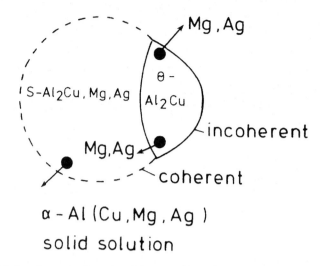

Figure 4(b): Effect of Mg, Ag, on the transition to $\Theta$–AlCu from metastable S-Al$_2$, Mg, Ag

At a certain amount of undercooling the relatively smaller driving force becomes sufficient for a quicker formation of the metastable phase, relative to the more stable one. Its nucleation is favoured by the formation of a low interfacial energy $\gamma_{\alpha\beta}$ with $G_N$ activation energy of nucleation, $g_\beta$ specific free energy change Jm$^{-3}$, i number of atoms in the nucleus, [6]:

$$G_N(T) = g_\beta(T)\ i + \gamma_{\alpha\beta}\ i^{2/3} \tag{1a}$$

$$\frac{dg_{\Theta''}}{di} < \frac{dg_{\Theta'}}{di} < \frac{dg_\Theta}{di} \tag{1b}$$

$$\gamma_{\alpha\Theta''} < \gamma_{\alpha\Theta'} < \gamma_{\alpha\Theta} \tag{1c}$$

Thus crystallographic coherency and classical nucleation theory explain why the less stable phase forms first and not the one which leads to the most stable state. There are two reasons why multi-stage reactions cannot be expected: 1. No metastable phases exist (Al-Si, Al-Ge, Al-Zn) [8], 2. The most stable phase can form coherently. Examples for the second case (as ($\gamma$ + $\gamma'$)-Ni alloys) do not exist for Al-alloys.

**Extrinsic defects**

Structural correspondences are also evident between lattice defects and types of interfaces: d = 0, lattice vacancy-coherency, d = 1, lattice dislocation-interfacial dislocation- partial coherency, d = 2, grain boundary - non-coherent interface.

This in turn leads to particular combinations of lattice defects and interfaces [5, 6, 7, 8], which provide minimum activation barriers and consequently maximum rates of formation (Table 1). Such favourable nucleation processes occur only at sites where lattice defects are preexisting. Consequently the bulk rate depends on grain size or dislocation density. This leads to the phenomenon that different phases form simultaneously, but at the various sites inside the alloy. During particle growth small particles dissolve in favour of large ones. But even more important, less stable phases are dissolved in the environment of more stable phases due to the differences in local solubilities. Consequently particle-free zones and uneven distribution of particles in the interior of grains develop (Fig. 5).

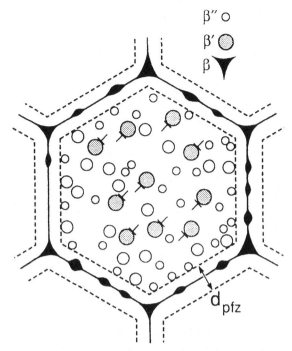

Figure 5: Defect-induced nucleation of different phases (compare Table 1), formation of particle free zones.

At any defect:
$$G_N(T) = g_\beta(T)i + (\gamma_{\alpha\beta} - \gamma_{def})i^{2/3} \qquad (2a)$$

At dislocation:
$$G_{N\,min} = g_{\Theta'}(T)i + (\gamma_{\alpha\Theta'} - \gamma_d)i^{2/3} \qquad (2b)$$

At boundary:
$$G_{N\,min} = g_\Theta(T)i + (\gamma_{\alpha\Theta} - \gamma_b)i^{2/3} \qquad (2c)$$

Besides lattice defects some trace elements are known to have favorable effects on the dispersoid microstructure. Wilm reported already the effect of Mg on the binary Al-Cu-alloy. An additional effect of Ag was found later on [4]. Such elements should favor the formation of a finer dispersoid. This can be explained by solubility of the trace element in the coherent metastable phase, while the stable non-coherent phase shows no solubility [9]. This in turn retards anomalous coarsening by the in-situ-transformation, coherent → non-coherent.

**Rules for the creation of nano-size dispersoids**

Unconstrained coherent nucleation comes closest to homogeneous nucleation and therefore to the formation of the finest possible dispersoid of the second phase $\beta_i$ [10]. However, a dispersoid of a incoherent, hard phase is desired.

For Al-alloys there is usually a multitude of options for reactions which lead closer to equilibrium at a wide range of rates (Equ. 1, 2). For Al-Cu the following possibilities exist: The stable phase $\Theta$ and three metastable phases, combined with three types of lattice defects and eventually a continuous and a discontinuous mode; i.e. there is a competition between 13 options. The fastest ones will win and produce the microstructure. The thermodynamical principle behind this phenomenon seems to be the maximization the initial rate of entropy production. The eventual creation of structural order ($-S_{str}$) is overcompensated by the fast production of thermal entropy $+S_{th}$. This in turn is a generalisation of Ostwalds step rule [10].

$$\frac{d(S_{th} - S_{str})}{dt} \rightarrow max.$$ (3)

The initial, microstructure can be interpreted in terms of Equ. 3. This relation is favored by the following set of sometimes conflicting properties:

1.   Enthalpy of the formation for precipitate phases                     $\rightarrow$   MAX.,
2.   Size of its elementary cell, i.e. min. nucleus size               $\rightarrow$   MIN.,
3.   Coherency with fcc lattice                                        $\rightarrow$   MAX.,
4.   Correspondence between structure of defect and interface          $\rightarrow$   MAX.,
5.   Size of prenucleation clusters*                                   $\rightarrow$   MAX.*,
6.   Diffusion path                                                    $\rightarrow$   MIN.

* this is favored in ternary alloys by the ±-effect, i.e. one solute larger, the other smaller than Al (Table 2).

Table II Atomic size ratio of solutes in Al

| Smaller | Larger |
|---------|--------|
| Cu | Ti |
| Zn | Sn |
| Si | Cd |
| Ni | Ge |
| Co | Li |
| Fe | Mg |
| Mn | Ga |
| Cr | |

**Hardening**

The formation of a fine dispersoid of particles is not sufficient for hardening [11,12]. Necessary for the validity of the OROWAN-equation (Figs. 6, 7, 8) are strong particles which sustain the stress excerted by moving dislocations, $d > d_C$:

$$\Delta\sigma_p = \frac{Gb}{S_{eff}} = C\frac{Gb}{S - d}$$ (4)

there is always a critical particle diameter $d_C$ below which the particles are sheared. This diameter should be as small as possible (Table 3). The critical diameter $d_C$ defines the maximum hardening effect by a certain volume fraction of particles (Fig. 6). It can be estimated by examining the force F which a single particle is able to excert on a looping dislocation. For $F < Gb^2$, shearing occurs and by-passing for $F \geq Gb^2$ (Figs. 7, 8).
Particle strengthened Al-alloys may contain a dispersoid of particles: ordered or disordered, coherent or non-coherent with the fcc-matrix. For coherent ordered particles the critical diameter $d_C$ follows from $Gb^2 = \gamma d = F$ where $\gamma$ is the antiphase domain boundary energy (APB) and d the particle diameter:

$$d_C = C\frac{Gb^2}{\gamma}$$ (5a)

and $C \approx 1$ is a geometrical factor depending on the shape of the particle. G and b are shear

8

modulus and Burgers vector of the fcc-matrix solid solution.

For disordered coherent particles no APB has to be created, but the amount of the difference in critical shear stress between matrix $\alpha$ and particle $\beta$ ($\tau_\alpha$-$\tau_\beta$) becomes relevant for modest hardening:

$$d_C = C \frac{Gb}{(\tau_\alpha - \tau_\beta)} \qquad (5b)$$

Pores or liquid inclusions are sheared at any size, inspite of their strong hardening effect (Equ. 4):

$$d_C = \infty \qquad (5c)$$

Particles $\beta$ with a different crystal structure from the matrix $\alpha$ require the theoretical shear stress to create a dislocation $b_\beta$. Consequently only very small incoherent particles are sheared:

$$d_C = C \frac{4\pi b G_\alpha}{G_\beta} \qquad (5d)$$

Low $d_C$-particles like Si will provide the highest hardening pro volume fraction.

These equations have to be modified if not one but a pair or more dislocations interact with a particle. It follows that the "art" of causing precipitation hardening implies the production of even and (S $\rightarrow$ min, Equ. 4) fine dispersoids of particles with small critical sizes $d_C$ (Table 3).

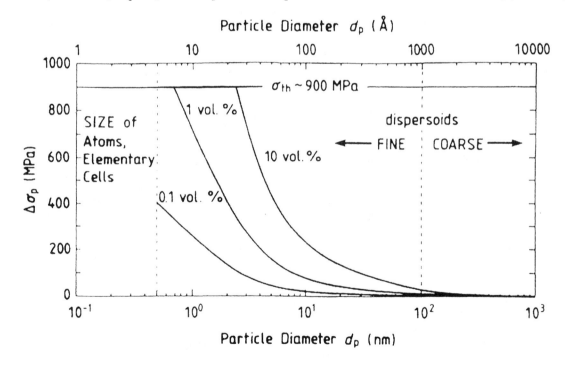

Figure 6: Calculated (Equ. 4) upper limit of hardening
by different volume fractions of particles in Al.

Usually, it is easy to form small subcritical particles d < $d_C$. As they are sheared they must cause less hardening than the by-passing mechanism. Also consequences on localization of strain have to be considered, for example on initiation and propagation of cracks under fatigue or stress corrosion conditions.

Finally it has to be considered that ultra high strength cannot be obtained by precipitation hardening $\Delta\sigma_p$ alone. A high yield stress $\sigma_y$ must be built up from contributions of additional hardening mechanisms. They can be systematically discussed by considering the 0- to 2-dimension of obstacles to the motion of dislocations [12].

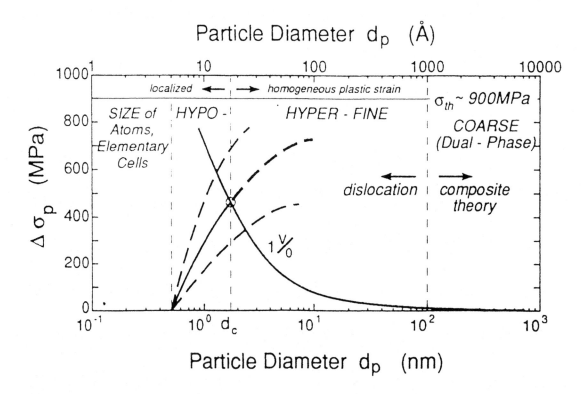

Figure 7: Definition of size ranges and critical particle diameters $d_C$, for 1% particles

$$1000MPa \approx \frac{G}{2\pi} \approx \sigma_y = \sigma_o + \Delta\sigma_s + \Delta\sigma_d + \Delta\sigma_b + \Delta\sigma_p \qquad (6)$$

Where $\sigma_o$ is the very low strength of pure Al, and $\Delta\sigma_s$ the contribution of solid solution hardening, $\Delta\sigma_d$ of a dislocation forest, $\Delta\sigma_b$ of grain boundaries. The different terms are not independent of each other. For our discussion it is important that for $d > d_C$, fine grain hardening $\Delta\sigma_b$ becomes irrelevant. $\Delta\sigma_s$ and $\Delta\sigma_d$ are used to built up high strength of Al-alloys. However, precipitation hardening $\Delta\sigma_p$ contributes usually the biggest share to high strength. This in turn is always due to strong particles in the size range between 1 and 10 nano-meters (Table 3).

Table III Examples for critical particle diameters $d_c$ in Al

| Examples | Coherency | $d_c$/nm | |
|---|---|---|---|
| Si, Ge | n, p | 2 | diamond structure |
| $\Theta$'-Al$_2$Cu<br>T-Al$_2$CuLi | n, p | 3-10 | intermetallic compounds |
| Zn | n, p | 20 | non coherent, solid solutions |
| Al$_3$Li | c | >50 | coherent ordered, depending on coherency stress, $\gamma_{APB}$ |

10

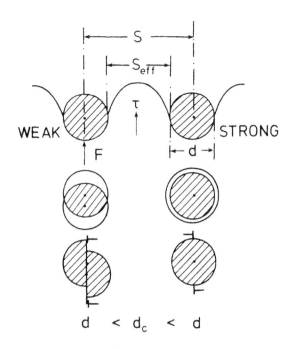

Figure 8: Geometrical features and particle-dislocation interactions for dispersoid microstructures

## References

1.  A. Wilm, Metallurgie, 8 (1911), 223
2.  C. Kammer, "Success for aluminium thanks to 75 years of materials research", Aluminium, 75 (1999) 753-775
3.  I. W. Martin, Micromechanisms in Particle-Hardened Alloys (Cambridge, UK: Cambridge Univ. Press, 1979)
4.  I. J. Polmear, Light Alloys, $3^{rd}$ ed. (London, UK: Edward Arnold, 1995), 41, 105
5.  E. Hornbogen, "Nucleation of precipitates in defect solid solutions", Nucleation, ed. A.C. Zettlemoyer (New York, NY: Marcel Dekker, 1969), 309-378
6.  E. Hornbogen, "Electronmicroscopy of Precipitation in Al-Cu Solid Solutions", Aluminium, 43 (1967), 41, 115, 163, 170
7.  E. Hornbogen, "Combined Reactions", Met. Trans., 10a (1979), 947-971
8.  E. Hornbogen, A. K. Mukhopadhyay and E.A. Starke, "An Exploratory Study of Hardening of Al-(Si, Ge) –alloys", Z. Metallkunde, 82 (1992), 577-580
9.  A. K. Mukhopadhyay, "Compositional Characterization of Cu-rich Phase Particles Present in As-Cast Al-Cu-Mg-alloys Containing Ag", Mat. Trans. A, 30A (1999), 1693-1704
10. E. Hornbogen, "Formation of nm-size dispersoids from Supersaturated Solid Solutions of Al", Materials Science Forum, Vol. 331-337 (2000), 879-888
11. C. P. Blankenship, E. Starke and E. Hornbogen, "Microstructures and properties of Al-alloys", Microstructure and properties of materials 1, ed. J.C.M. Li (Singapore: World Scientific 1996) 1-51
12. E. Hornbogen, E. Starke, "High Strength low alloy aluminum", Acta Mat., 41 (1993), 1-16

# LIGHTWEIGHT ALLOYS FOR AEROSPACE APPLICATION

*Edited by:*
*Dr. Kumar Jata, Dr. Eui Whee Lee,*
*Dr. William Frazier and Dr. Nack J. Kim*

## ALUMINUM ALLOYS

## The Role of Ledge Nucleation/ Migration in Ω Plate Thickening Behaviour in Al-Cu-Mg-Ag Alloys

*C.R. Hutchinson, X. Fan, S.J. Pennycook and G.J. Shiflet*

Pgs. 13-23

184 Thorn Hill Road
Warrendale, PA 15086-7514
(724) 776-9000

# THE ROLE OF LEDGE NUCLEATION/MIGRATION IN Ω PLATE THICKENING BEHAVIOUR IN Al-Cu-Mg-Ag ALLOYS

[1]C. R. Hutchinson, [2,3]X. Fan, [3]S. J. Pennycook and [1]G. J. Shiflet

[1]Dept. of Mat. Sci. and Eng., University of Virginia, Charlottesville, VA, 22903, USA.
[2]Dept. of Chem. and Mat. Eng., University of Kentucky, Lexington, KY, 40506, USA.
[3]Solid State Division, Oak Ridge National Laboratory, Oak Ridge, TN, 37831, USA.

## Abstract

The thickening kinetics of Ω plates in an Al-4Cu-0.3Mg-0.2Ag (wt. %) alloy have been measured at 200 °C, 250 °C and 300 °C using conventional transmission electron microscopy techniques. At all temperatures examined the thickening showed a linear dependence on time. At 200 °C the plates remained less than 6nm in thickness after 1000h exposure. At temperatures above 200 °C the thickening kinetics are greatly increased. Atomic resolution Z-contrast microscopy has been used to examine the structure and chemistry of the $(001)_\Omega \parallel (111)_\alpha$ interphase boundary in samples treated at each temperature. In all cases, two atomic layers of Ag and Mg segregation were found at the broad face of the plate. The risers of the growth ledges and the ends of the plates were free of segregation. No significant levels of Ag or Mg were detected inside the plate at any time. The necessary redistribution of Ag and Mg accompanying a migrating thickening ledge occurs at all temperatures and is not considered to play a decisive role in the excellent coarsening resistance exhibited by the Ω plates at temperatures up to 200 °C. Plates transformed at 200 °C rarely contained ledges and usually exhibited a strong vacancy misfit normal to the plate. A large increase in ledge density was observed on plates transformed at 300 °C, concomitant with accelerated plate thickening kinetics. The high resistance to plate coarsening exhibited by Ω plates at temperatures up to 200 °C, is due to limited ledge nucleation under these conditions. The prohibitively high barrier to coherent ledge nucleation on the broad faces of plates aged at 200 °C arises from the contribution to the total free energy change attending nucleation from elastic interactions between the misfitting coherent ledges and the significant strain field that can exist normal to the broad face of the Ω plate.

## Acknowledgments

CRH and GJS greatly acknowledge the support of the Southeastern Universities Research Association 1999 Summer Cooperative Research Program and the National Science Foundation under grant number DMR-9904034. The work at Oak Ridge National Laboratory was supported by the Division of Materials Sciences, US Department of Energy under contract No. DE-AC05-00OR22725 with UT-Battelle, LLC. Dr. Simon Ringer of the Electron Microscopy Unit at the University of Sydney, Australia is thanked for kind provision of materials used in this study.

Lightweight Alloys for Aerospace Applications
Edited by Kumar Jata, Eui Whee Lee,
William Frazier and Nack J. Kim
TMS (The Minerals, Metals & Materials Society), 2001

# 1. Introduction

The addition of trace amounts of Ag to Al-Cu-Mg alloys with high Cu:Mg ratios (eg., 10:1) significantly alters the precipitation sequence usually observed in these alloys [1-4]. The most notable change is the appearance of a thin, hexagonal-shaped phase, designated $\Omega$, that forms as platelets on the $\{111\}_\alpha$ slip planes of the matrix. Several structures for the $\Omega$ phase have been proposed [5-7] although the most widely accepted structure is orthorhombic (Fmmm, a=0.496nm, b=0.859nm, c=0.848nm) [8, 9]. The orientation relationship between $\Omega$ and the $\alpha$ matrix is $(111)_\alpha \parallel (001)_\Omega$ and $[\bar{1}10]_\alpha \parallel [010]_\Omega$. The appearance of the $\Omega$ phase promotes greater hardening, and alloys based on the Al-Cu-Mg-Ag system have shown promising creep properties at temperatures up to 200 °C because of the apparent resistance of the $\Omega$ phase to particle coarsening [10]. The reported coarsening resistance of $\Omega$ has been confirmed by Ringer et. al. [11]. Those researchers directly measured the changes in plate thickness using conventional transmission electron microscopy (CTEM) techniques as a function of time for temperatures between 200 °C and 300 °C. The plates examined in that study remained less than 6nm in thickness after 1000h exposure at 200 °C. At temperatures above 200 °C the thickening was greatly accelerated.

It is now generally accepted that rationally oriented plate-like precipitates thicken by a ledge mechanism [12] (Fig. 1). The thickening of plate-like precipitates therefore depends on the kinetics of the *nucleation and growth of thickening ledges* on the broad faces of the plates. Previous studies on precipitate thickening kinetics in Al-alloys [13-15] have concluded that the overall thickening kinetics are ultimately restricted by limited ledge nucleation.

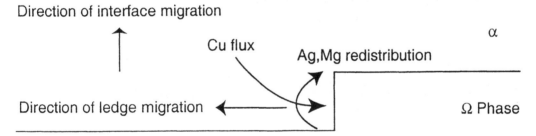

**Figure 1.    Schematic illustration of the necessary Ag and Mg redistribution and Cu flux accompanying the migration of a thickening ledge on an $\Omega$ plate.**

$\Omega$ plates differ from other $\{111\}_\alpha$ precipitate plates in at least two important aspects. The first concerns the relatively large lattice misfit between the precipitate and the matrix that exists normal to the $(111)_\alpha \parallel (001)_\Omega$ interphase boundary. Other examples of $\{111\}_\alpha$ plates include $T_1$ ($Al_2CuLi$) in Al-Cu-Li alloys, $\eta'$ ($MgZn_2$) in Al-Zn-Mg and $\gamma'(\gamma)$ ($AlAg_2$) in Al-Ag. Each of these phases has a hexagonal structure whereas $\Omega$ is usually assummed to be hexagonal. The misfit normal to the precipitate for each of these phases is 0.12% for $T_1$, 0.03% for $\eta'$, 1.46% for $\gamma'$ and 9.3% for $\Omega$. The misfit normal to the $\Omega$ plate is accordingly considered large. On the basis of this large misfit, Fonda et. al. [16] initially postulated that the source of the enhanced thermal stability of $\Omega$ may be the relationship between ledge nucleation and propagation and the elastic strain field. Fonda et. al. [16] investigated the accommodation of misfit strain surrounding $\Omega$ plates and found the plates consistently exhibit a vacancy type strain field normal to the habit plane (Fig. 2), independent of plate thickness. Two types of thickening ledges were observed, coherent $1/2\Omega$ unit cell high ledges and less commonly, larger ledges which contain a misfit compensating dislocation of the type $\mathbf{b}=1/3<111>_\alpha$. Similar dislocations were also observed at the ends of the plates with an average spacing of 2 1/2 or 3$\Omega$ unit cells, which produces a minimum strain normal to the plate.

The second characteristic that differentiates $\Omega$ from other $\{111\}_\alpha$ plates is the well reported segregation of Ag and Mg to the broad face of the $\Omega$ plate. This segregation was first detected by Muddle and Polmear [9] and most recently by Reich et. al. [17] using 3D-APFIM. The 3D-APFIM work of Reich et. al. has provided evidence to warrant careful consideration of the usual assumption that ledge nucleation controls the overall plate thickening rate in Al-based alloys. In Reich et. al.'s atom probe study, they captured an $\Omega$ plate thickening ledge in an Al-1.9Cu-0.3Mg-0.2Ag (at. %) alloy aged 10h at 180 °C. Their observations show the presence of a monoatomic layer of Ag and Mg at the $\Omega$ plate/matrix $(001)_\Omega \parallel (111)_\alpha$ interface but no Ag or Mg was detected within the plate itself or at the riser of the ledge. The motion of the thickening ledge must then involve the simultaneous flux of Cu from the matrix to the riser of the ledge *and* the redistribution of Ag and Mg from the original broad face of the $\Omega$ plate to the terrace of the migrating thickening ledge. Fig. 1 is a schematic illustration of this process[1]. This complicated diffusion geometry raises two interesting questions. (a) What interaction (if any) is there between the redistributing Ag and Mg and the incoming flux of Cu? and (b) If an interaction is expected, could it be sufficient to retard ledge migration to the point where it becomes the rate controlling process for plate thickening instead of ledge nucleation?

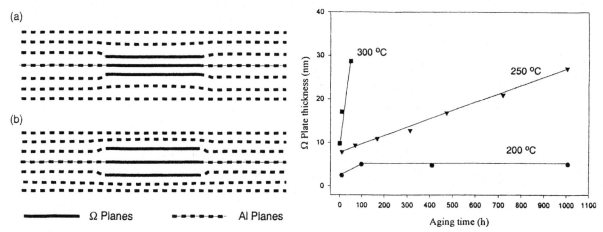

**Figure 2.** Schematic illustration of (a) vacancy and (b) interstitial strain fields normal to the broad face of a precipitate plate.

**Figure 3.** Mean $\Omega$ plate thickness (nm) as a function of time at 200 °C, 250 °C and 300 °C, in an Al-4Cu-0.3Mg-0.4Ag (wt. %) alloy.

The present work addresses the need for a systematic study of the structure and chemistry of the $\Omega$ plate/matrix $(001)_\Omega \parallel (111)_\alpha$ interface as a function of time and temperature to examine the respective roles of ledge nucleation and migration in accounting for the excellent coarsening resistance of $\Omega$ plates at temperatures up to 200 °C.

---

1. The diffusion path for Ag and Mg redistribution from the broad face of the plate to the terrace of the migrating thickening ledge shown in Fig. 1 is only one of several possible diffusion paths. This schematic is not intended to imply that this is the path of solute redistribution, only that some interaction between the flux of Cu and the Ag and Mg may be expected.

# 2. Experimental Procedure

The composition of the alloy used is Al-4.0Cu-0.3Mg-0.4Ag (wt. %). Strips of material 0.5-1mm thick were solution treated (ST) at 525 °C for 1h, water quenched (WQ) and aged in molten salt baths at 200 °C ± 2°C, 250 °C ± 2°C or 300 °C ± 2°C for various times up to 1000h.

Specimens for transmission electron microscopy (TEM) were punched mechanically from the strips and twin-jet electrolytically polished in a solution of 33 vol. % nitric acid and 67 vol. % methanol at -25 °C. The microstructural evolution was monitored using CTEM techniques with a 200 kV microscope. High resolution phase contrast microscopy was performed using a top entry HREM operating at 400 kV. The high resolution phase contrast simulation was performed using the Crystalkit and MacTempas software packages [18]. Atomic resolution Z-contrast microscopy [19, 20] was used for the systematic examination of the composition and structure of the $\Omega$ plate/matrix $(001)_{\Omega} \parallel (111)_{\alpha}$ interface as a function of time and temperature. This technique is capable of providing two-dimensional intuitively interpretable images of atomic structures with compositional sensitivity without the need for model structures and simulations associated with the phase contrast imaging techniques. A Z-contrast image is formed by scanning an electron probe of atomic dimensions across a specimen and collecting the high angle scattered electrons with an annular dark-field (HAADF) detector. Since the scattering is incoherent at high scattering angles, the image is essentially a map of the total scattering intensity of each atomic column, which is approximately proportional to the square of the atomic number (Z). This technique is especially well suited to the investigation of Ag in the Al-Cu-Mg-Ag system due to the relatively high atomic number of Ag. The microscopy was performed using a VG Microscope HB603U scanning transmission electron microscope operating at 300kV which is capable of forming an electron probe size of 0.126nm. The EDS analysis was carried out using a ATEM operating at 200 kV equipped with a field emission gun and an EDSX system.

Measurements of the thickness of $\Omega$ precipitate plates were made from CTEM micrographs recorded with the electron beam parallel to the precipitate habit plane (i.e. parallel to $<112>_{\alpha}$// $[100]_{\Omega}$ or $[110]_{\Omega}$). In each case, the "edge-on" thickness of between 70 and 100 precipitates was measured from the negatives magnified using a 4x graticule.

# 3. Results

## 3.1 Conventional Transmission Electron Microscopy (CTEM)

Observations of the $<100>_{\alpha}$, $<110>_{\alpha}$ and $<112>_{\alpha}$ zone axes of the Al matrix were made to ensure a true representation of the precipitate distribution was obtained. At all times and temperatures examined, the $\Omega$ phase was found to be present. At 200 °C and 250 °C, the $\Omega$ phase coexists with $\theta'$ (Al$_2$Cu) and S (Al$_2$CuMg) phases. At 300 °C, the $\Omega$ phase was the only precipitate found at all times observed. The thickening kinetics of $\Omega$ plates were measured at each temperature. A plot of plate thickness as a function of time is shown in Fig. 3. At all temperatures examined the precipitate thickness shows a linear dependence on time. At 200 °C, the $\Omega$ plates reach a thickness of approximately 5.5nm after 100h exposure, after which there is no detectable change in average thickness. At 300 °C the rate of thickening is rapid and thicknesses greater than 30nm are reached within 50h at 300 °C. At 250 °C, thicknesses of 25-30nm are obtained after 1000h exposure. These observations are qualitatively consistent with those of Ringer *et. al.* [11].

## 3.2  High Resolution Electron Microscopy

### 3.2.1  Z-Contrast Microscopy

Atomic resolution Z-contrast microscopy was used to examine the structure and chemistry of the $\Omega$ plate/matrix $(001)_\Omega \parallel (111)_\alpha$ interface in samples transformed at each temperature.

A low magnification Z-contrast image of an $\Omega$ plate (right) and a $\theta'$ plate (left) is presented in Fig. 4(a). The intensity in a Z-contrast image is approximately proportional to $Z^2$ and the bright bands bounding each side of the $\Omega$ plate are interpreted as preferential segregation of at least Ag to the $\Omega$ plate/matrix $(001)_\Omega \parallel (111)_\alpha$ interphase boundary, consistent with previous investigations [9, 17]. $\Omega$ plates in this orientation (Fig. 4(a)) were found to be very long, straight and typical of $\Omega$ plates observed in samples transformed at 200 °C. An atomic resolution Z-contrast image of a 4 unit cell thick $\Omega$ plate is shown in Fig. 4(b). Two atomic layers of enhanced intensity are seen at the $\Omega$ plate/matrix $(001)_\Omega \parallel (111)_\alpha$ interface. These correspond to two layers of segregation. This is in contrast to the monoatomic layer reported by Reich *et. al.* [17]. The layers of enhanced intensity within the plate parallel to the habit plane are separated by 0.424nm and correspond to layers enriched in Cu. This is qualitatively consistent with the projection of the proposed orthorhombic $\Omega$ structure down zone axes parallel to the habit plane. EDS was used to determine that the interfacial segregation contained both Ag and Mg, consistent with recent analytical investigations [17]. No significant quantities of Ag or Mg were detected within the $\Omega$ plate or the adjacent matrix.

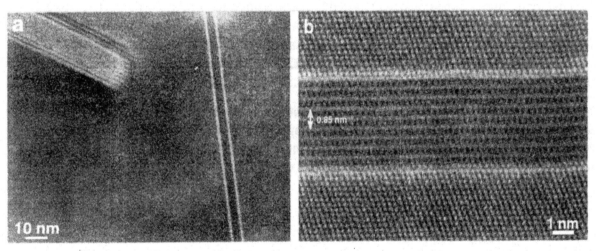

**Figure 4.  Z-contrast images of sample transformed 100h at 200 °C. (a) Low magnification image of $\Omega$ plate (right) and $\theta'$ plate (left). (b) Atomic resolution image of an $\Omega$ plate illustrating two atomic layers of interfacial segregation.**

Fig. 5(a) is a Z-contrast image of an $\Omega$ plate thickening ledge in a sample exposed for 70h at 250 °C. The ledge is 1/2 $\Omega$ unit cell high and coherent with the matrix. The image shows a double layer of interfacial segregation to the terraces of the growth ledge but not at the riser of the ledge. The lack of segregation at the riser of the thickening ledge is consistent with the APFIM observations of Reich *et. al.* [17]. Energy dispersive spectra were obtained from the matrix, the $(001)_\Omega \parallel (111)_\alpha$ interphase boundary and wholly within the $\Omega$ plate. They showed segregation of both Ag and Mg to the interface. As was the case for the sample treated at 200 °C, no significant quantities of Ag or Mg were detected within the $\Omega$ plate or the adjacent matrix. At 300 °C the $\Omega$ plates thickened at a greatly enhanced rate, and the Z-contrast image in Fig. 5(b) shows thickening ledges are plentiful on $\Omega$ plates transformed at this temperature. Two atomic layers of segregation to the $\Omega$ plate/matrix $(001)_\Omega \parallel (111)_\alpha$ interface were again observed. EDS analysis observed at the ends of the plates with an average spacing of 2 1/2 or 3Ω unit cells, which produces a minimum strain normal to the plate.

The second characteristic that differentiates Ω from other $\{111\}_\alpha$ plates is the well reported segregation of Ag and Mg to the broad face of the Ω plate. This segregation was first detected by Muddle and Polmear [9] and most recently by Reich *et. al.* [17] using 3D-APFIM. The 3D-APFIM work of Reich *et. al.* has provided evidence to warrant careful consideration of the usual assumption that ledge nucleation controls the overall plate thickening rate in Al-based alloys. In Reich *et. al.'s* atom probe study, they captured an Ω plate thickening ledge in an Al-1.9Cu-0.3Mg-0.2Ag (at. %) alloy aged 10h at 180 °C. Their observations show the presence of a monoatomic layer of Ag and Mg at the Ω plate/matrix $(001)_\Omega \| (111)_\alpha$ interface but no Ag or Mg was detected within the plate itself or at the riser of the ledge. The motion of the thickening ledge must then involve the simultaneous flux of Cu from the matrix to the riser of the ledge *and* the redistribution of Ag and Mg from the original broad face of the Ω plate to the terrace of the migrating thickening ledge. Fig. 1 is a schematic illustration of this process[1]. This complicated diffusion geometry raises two interesting questions. (a) What interaction (if any) is there between the redistributing Ag and Mg and the incoming flux of Cu? and (b) If an interaction is expected, could it be sufficient to retard ledge migration to the point where it becomes the rate controlling process for plate thickening instead of ledge nucleation?

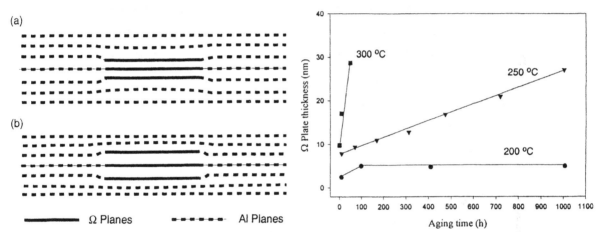

Figure 2. Schematic illustration of (a) vacancy and (b) interstitial strain fields normal to the broad face of a precipitate plate.

Figure 3. Mean Ω plate thickness (nm) as a function of time at 200 °C, 250 °C and 300 °C, in an Al-4Cu-0.3Mg-0.4Ag (wt. %) alloy.

The present work addresses the need for a systematic study of the structure and chemistry of the Ω plate/matrix $(001)_\Omega \| (111)_\alpha$ interface as a function of time and temperature to examine the respective roles of ledge nucleation and migration in accounting for the excellent coarsening resistance of Ω plates at temperatures up to 200 °C.

---

1. The diffusion path for Ag and Mg redistribution from the broad face of the plate to the terrace of the migrating thickening ledge shown in Fig. 1 is only one of several possible diffusion paths. This schematic is not intended to imply that this is the path of solute redistribution, only that some interaction between the flux of Cu and the Ag and Mg may be expected.

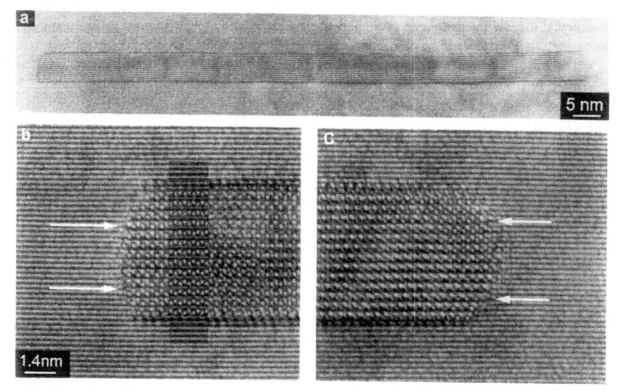

Figure 6. a) High resolution phase contrast image of an $\Omega$ plate after exposure for 1000h at 200 °C. The plate contains no ledges. b=$<112>_\alpha$//$[110]_\Omega$ (b) High magnification image of the left end and (c) right end of the plate in (a). The ends of the plate contain two dislocations (arrowed) of type 1/3$<111>_\alpha$, with the extra half plane in the precipitate phase.

Table I: Summary of $\Omega$ plate thickness and ledge density recorded from high resolution phase contrast images of samples treated at 200 °C.

| $\Omega$ Plate | Heat Treatment Condition | Plate Thickness ($\Omega$ unit cells) | Number of coherent 1/2 $\Omega$ unit cell high ledges | Inferred number of 1/3$<111>_\alpha$ dislocations at plate ends | Unrelaxed misfit (%) in $[001]_\Omega$ ‖ $[111]_\alpha$ direction. | |
|---|---|---|---|---|---|---|
| 1 | 200 °C, 10h | 4 | 0 | 1 | -24.6 | |
| 2 | 200 °C, 10h | 1 | 0 | 0 | -18.7 | |
| 3 | 200 °C, 10h | 2.5-3.5 | 2[a] | 1 | +3.38 | -15.3 |
| 4 | 200 °C, 10h | 2.5 | 0 | 1 | +3.38 | |
| 5 | 200 °C, 10h | 3 | 0 | 1 | -5.9 | |
| 6 | 200 °C, 1000h | 6 | 0 | 2 | -11.9 | |
| 7 | 200 °C, 1000h | 6 | 0 | 2 | -11.9 | |
| 8 | 200 °C, 1000h | 7 | 0 | 2 | -30.5 | |
| 9 | 200 °C, 1000h | 6.5 | 0 | 2 | -21.2 | |
| 10 | 200 °C, 1000h | 5.5 | 0 | 2 | -2.6 | |

a. The ledge on this plate is in fact a single 1 $\Omega$ unit cell high coherent ledge.

# 4. Discussion

This study has shown that despite considerable differences in $\Omega$ plate thickening kinetics between 200 °C and 300 °C, no significant changes in the segregation behavior of Ag and Mg are observed. Two atomic layers of Ag and Mg segregation are found at the coherent broad faces of the plates. No segregation is found at the risers of thickening ledges, consistent with the observations of Reich *et. al.* [17], nor is any significant segregation observed at the ends of the plates or within the plates themselves. Consequently, the necessary Ag and Mg redistribution from the broad face of the plate to the terrace of a migrating ledge (Fig. 1) must accompanying ledge migration at all temperatures. Any interaction between the redistributing Ag and Mg and the flux of Cu to the ledge riser is expected to occur at all temperatures and is not considered unique to temperatures below 200 °C. Therefore, if an interaction does exist it is not considered to play the decisive role in the excellent coarsening resistance exhibited by $\Omega$ plates at temperatures up to 200 °C. The characteristic feature that distinguishes $\Omega$ plates transformed at 200 °C from those at higher temperatures is the thickening ledge density. Samples transformed at 200 °C for times greater than 10h rarely contained ledges (Table I). A large increase in thickening ledge density was observed for samples treated at 300 °C, concomitant with greatly enhanced plate thickening kinetics. The excellent coarsening resistance of $\Omega$ plates at temperatures up to 200 °C can be ascribed to a limited supply of thickening ledges.

The most commonly observed thickening ledges on $\Omega$ plates are coherent and 1/2 $\Omega$ unit cell in height. Nucleation of these ledges replaces two matrix $\{111\}_\alpha$ planes and introduces a vacancy misfit of more than 0.04nm normal to the habit plane of the plate [16]. The nucleation of a coherent misfitting crystal in an elastically constrained matrix can be strongly influenced by elastic interactions with pre-existing strain fields in the matrix. Since the 1/2 $\Omega$ unit cell high coherent ledge exhibits a vacancy misfit normal to the plate, the nucleation probability is highest (activation barrier is lowest) when the plates exhibit a residual interstitial misfit and increasingly less probable for increasing vacancy strains normal to the broad face of the plate. A plot of the change in unrelaxed misfit (expressed in nm) normal to the $\Omega$ plate habit plane as a function of $\Omega$ plate thickness is shown in Fig. 7. As successive coherent vacancy misfitting 1/2 $\Omega$ unit cell ledges are added to the broad face of a plate, the overall vacancy misfit normal to the plate increases. This is represented by the solid line ($\overline{AB}$) in Fig. 7. Above some critical value of this misfit[1] it becomes energetically favorable for a dislocation of the type $\mathbf{b}=1/3[111]_\alpha$ to form with the extra half plane in the precipitate phase (e.g. Fig. 6(b)). The resulting misfit associated with the plate becomes interstitial in nature ($B \rightarrow C$, Fig. 7). Subsequent nucleation of coherent vacancy misfitting ledges reduces the interstitial misfit and the strain field normal to the broad face once again becomes vacancy in nature ($C \rightarrow D$, Fig. 7), with an associated increase in the activation barrier for coherent vacancy misfitting ledge nucleation. For those plates transformed at 200 °C and listed in Table I, the inferred number of Frank partial dislocations and the unrelaxed misfit normal to the plate have been evaluated on the basis of Fig. 7. The table shows that in all cases except plates 3 and 4, a vacancy misfit exists normal to the plate. In the case of plate 4, the misfit is interstitial, albeit only slightly. These plates contain no ledges because the nucleation barrier for a coherent vacancy misfitting ledge in a pre-existing vacancy field is prohibitively large. Plate 3 provides an interesting example. It was the only plate observed with a ledge, and the ledge is a single coherent full $\Omega$ unit cell in height. The plate was initially 2.5 cells thick, corresponding to an interstitial strain field ($E$, Fig. 7), but after propagation of the ledge, the plate is 3.5 cells thick with a strong vacancy misfit ($F$, Fig. 7). Nucleation of the coherent

---

1. The critical misfit normal to the plate corresponding with the formation of a Frank partial dislocation of the type $\mathbf{b}=1/3[111]_\alpha$ may depend on plate thickness. At very small plate thicknesses, the plate is expected to behave as an elastically constrained thin film, but as the plate thickens the elastic properties of the plate would be expected to more closely approach those of bulk $\Omega$.

vacancy misfitting ledge is assisted through mitigation of the interstitial strain field normal to the plate. In all cases observed, plates that contain ledges have a residual interstitial misfit (or only slightly vacancy) and those that contain no ledges exhibit a vacancy misfit[1]. Most of the plates listed in Table I exhibit strong vacancy misfits, conditions very unfavorable for the nucleation of coherent vacancy misfitting ledges. It is concluded that nucleation of coherent ledges during the coarsening stages of $\Omega$ plate thickening are dominated by strain energy considerations and that this effect is responsible for the excellent coarsening resistance of $\Omega$ plates at temperatures up to 200 $^{\circ}$C.

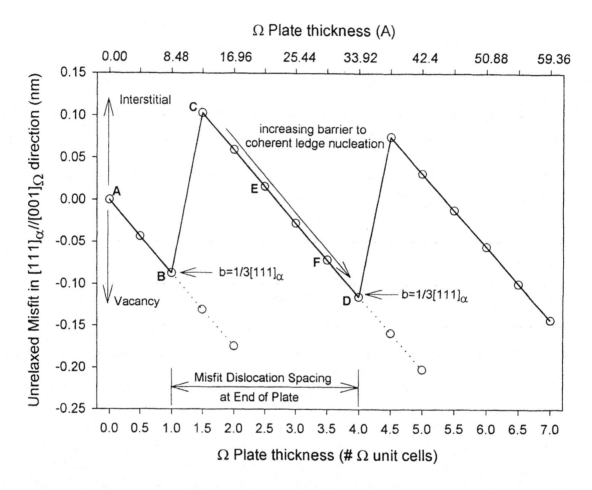

**Figure 7.** **Plot of the change in unrelaxed misfit (nm) normal to the $\Omega$ plate habit plane as a function of $\Omega$ plate thickness.**

The analysis presented above applies only at relatively small plate thicknesses. The matrix dislocation density greatly increased during the later stages of thickening of $\Omega$ plates at 250 $^{\circ}$C and 300 $^{\circ}$C and these dislocations interacted strongly with both the edges and the broad faces of the plates, presumably to aid in the accommodation of strain associated with the plate. The regime of thickening that includes the generation of matrix dislocations and their interaction with the strain fields of the plates is beyond the scope of this study.

---

1. Fonda *et. al* [16] observed that the average spacing of Frank partial dislocations at the ends of thick plates was 2.5-3.0 $\Omega$ unit cells. This is an *average* spacing and was used to construct Fig. 7. Deviations from this average spacing may reveal plate thicknesses that appear inconsistent with the arguments presented above on the basis of Fig. 7. In those cases, direct observation of the number of dislocations at the ends of the plates is necessary.

# 5. Conclusions

Z-contrast observations revealed two atomic layers of Ag and Mg segregation to the $(001)_\Omega \parallel (111)_\alpha$ interphase boundary at all times and temperatures examined. No segregation was found to the risers of thickening ledges or to the ends of the plates. No Ag or Mg was detected in any significant quantities within the plate at any time. The necessary Ag and Mg redistribution from the broad face of the plate to the terrace of the migrating thickening ledge must accompany ledge migration at all temperatures and is not considered to play a decisive role in accounting for the excellent coarsening resistance of $\Omega$ plates at temperatures up to 200 °C. Consistent with previous investigations of the thickening kinetics of precipitate plates in Al-alloys [13, 14, 15], the thickening of $\Omega$ plates is restricted by a limited supple of ledges. The density of thickening ledges on plates transformed for times longer than 10h at 200 °C was very low, usually zero. A large increase in ledge density is associated with the increase in thickening kinetics at 250 °C and 300 °C. The prohibitively high barrier to coherent ledge nucleation on the broad faces of plates aged at 200 °C arises from the contribution to the total free energy change attending nucleation from elastic interactions between the misfitting coherent ledges and the significant strain field that can exist normal to the broad face of the $\Omega$ plate.

# 6. References

1.  Auld, J. H. and Vietz, J. T., in *The Mechanism of Phase Transformations in Crystalline solids, Monograph and Report Series*, No. 33, pp. 77-79. Inst. Metals, London (1969).
2.  Taylor, J. A., Parker, B. A and Polmear, I. J., *Metals Sci.*, 1978, **12**, 478.
3.  Chester, R. J. and Polmear, I. J., *Micron*, 1980, **11**, 311.
4.  Chester, R. J. and Polmear, I. J., in *The Metallurgy of Light Alloys*, pp. 75-81, Inst. Metals, London (1983).
5.  Auld, J. H., *Mater. Sci. Technol.*, 1986, **2**, 784.
6.  Scott, V. D., Kerry, S. and Trumper, R. L., *Mater. Sci. Technol.*, 1987, **3**, 827.
7.  Garg, A. and Howe, J. M., *Acta. metall.*, 1991, **39**, 1939.
8.  Knowles, K. M. and Stobbs, W. M., *Acta crystallogr.*, 1988, **B44**, 207.
9.  Muddle, B. C. and Polmear, I. J., *Acta. metall.*, 1989, **37**, 777.
10. Polmear, I. J. and Couper, M. J., *Metall. Trans.*, 1988, **19A**, 1027.
11. Ringer, S. P., Yeung, W., Muddle, B. C. Polmear, I. J., *Acta. metall. mater.*, 1994, **42**, 1715.
12. Aaronson, H. I., in *Decomposition of Austentite by Diffusional Processes*, pp. 387-548, Eds. Zackay, V. F. and Aaronson, H. I., Interscience, (1962).
13. Aaronson, H. I. and Laird, C., *Trans. AIME*, 1968, **242**, 1437.
14. Sankaran, R. and Laird, C., *Acta Metallurgica*, 1974, **22**, 957.
15. Laird, C. and Aaronson, H. I., *Acta Metallurgica*, 1969, **17**, 505.
16. Fonda, R. W, Cassada, W. A and Shiflet, G. J., *Acta. metall. mater.*, 1992, **40**, 2539.
17. Reich, L., Murayama, M. and Hono, K., *Acta mater.*, 1998, **46**, 6053.
18. Software development by Roar Kilaas, National Center for Electron Microscopy (NCEM), Lawrence Berkeley National Laboratory, Berkeley, CA.
19. Pennycook, S. J. and Jesson, D. E., *Phys. Rev. Lett.*, 1990, **64**, 938-941.
20. Pennycook, S. J. and Nellist, P. D., in *Impact of Electron and Scanning Probe Microscopy on Materials Research*, pp. 161-207, Ed.'s Rickerby, D. G., Valdré, U. and Valdré, G., Kluwer Academic Publisers, The Netherlands, 1999.

# LIGHTWEIGHT ALLOYS FOR AEROSPACE APPLICATION

*Edited by:*
*Dr. Kumar Jata, Dr. Eui Whee Lee,*
*Dr. William Frazier and Dr. Nack J. Kim*

## ALUMINUM ALLOYS

Improving Recrystallization Resistance
in Wrought Aluminum Alloys with
Scandium Addition

*Y.W. Riddle, M. McIntosh and T.H. Sanders Jr.*

Pgs. 25-39

184 Thorn Hill Road
Warrendale, PA 15086-7514
(724) 776-9000

# Improving Recrystallization Resistance in Wrought Aluminum Alloys with Scandium Addition

Y. W. Riddle*, M. McIntosh, and T.H. Sanders Jr.

School of Materials Science and Engineering
Georgia Institute of Technology
Atlanta, GA 30332-0245  USA

*Currently at:* Institutt for Materialteknologi og Elektrokjemi,
Norges Teknisk-Naturvitenskapelige Universitet
7491 Trondheim, Norway

## Abstract

The 5xxx, 2xxx, and 7xxx aluminum alloys can all benefit from minor additions of Sc to control recrystallization.  The addition of Sc rapidly precipitates high volume fraction of homogeneously distributed dispersoids having coherent $Al_3Sc$ ($L1_2$) structure.  $Al_3Sc$ dispersoids impart some improvement in recrystallization resistance to wrought alloys, namely with higher annealing temperatures and volume fractions, compared to alloys employing $Al_3Zr$. However, the higher coarsening rate of $Al_3Sc$ compared to that of $Al_3Zr$ may limit its usability. When both scandium and zirconium are used in the same alloy $Al_3(Sc_{1-x}, Zr_x)$ dispersoids form, which are more effective than either $Al_3Sc$ or $Al_3Zr$ acting alone. The $Al_3(Sc_{1-x}, Zr_x)$ dispersoid benefits from the rapid precipitation characteristic of Sc in Al, the slow coarsening associated with Zr, and a high volume fraction as an effect of both Sc and Zr. Furthermore the distribution of $Al_3(Sc_{1-x}, Zr_x)$ is homogeneous. It was also established that alloys containing up to 3.5Mg showed improvement in recrystallization resistance with both Sc and Zr present.

Lightweight Alloys for Aerospace Applications
Edited by Kumar Jata, Eui Whee Lee,
William Frazier and Nack J. Kim
TMS (The Minerals, Metals & Materials Society), 2001

# Introduction

## Inhibiting Recrystallization with Dispersoids

The use of dispersoid phases in aluminum alloys is well established as a method for controlling grain growth, recovery, and recrystallization in wrought aluminum alloy products. Several reasons exist for the need to control recrystallization. Strengthening 5xxx alloys by strain hardening is effective if the unrecrystallized wrought state of the alloy can be maintained. In 2xxx alloys recrystallization negatively impacts corrosion resistance. Recrystallization in 7xxx alloys increases weld cracking, decreases fracture toughness, and decreases corrosion resistance.

Dispersoid phases precipitate via a solid state reaction to form a distribution of <0.5μm diameter particles when the alloy is preheated properly. Dispersoids are known to prevent the motion of subgrain boundaries during annealing, termed "Zener drag", which inhibits recrystallization [1]. Zener s theory shows: increasing volume fraction, decreasing coarsening kinetics, and maintaining a coherent interface between the dispersoid and matrix all increase the resistance to recrystallization. Also, a homogeneous distribution of dispersoids is essential for good recrystallization resistance. Currently the dispersoids $Al_{12}Mn_2Cr$, $Al_{20}Mn_3Cu_2$, and $Al_3Zr$ are employed as recrystallization inhibitors in aluminum alloys of the 2xxx and/or 7xxx family. Of these, $Al_3Zr$ is known to be the most potent recrystallization inhibitor [2-4].

The recent literature is replete with references to the commercial potential of adding scandium to wrought aluminum alloys. A recent review contains references to much of this information [5]. Scandium forms $Al_3Sc$ ($L1_2$) dispersoids that impart a high degree of recrystallization resistance in many wrought aluminum alloy systems. Al-Mg [6-9] alloys show a large increase in recrystallization resistance when $Al_3Sc$ is present. Al-Mg containing $Al_3Sc$ can sustain a large degree of deformation while retaining an unrecrystallized microstructure. Al-Zn-Mg-(Cu) [10] alloys benefit from $Al_3Sc$ by increased corrosion resistance, decreased weld cracking, and increased fracture toughness [11, 12]. Copper containing alloys, such as Al-Cu [10, 13, 14], and Al-Zn-Mg-Cu benefit from scandium addition although the scandium and copper contents must be closely controlled to avoid deleterious response to alloy performance by formation of W-phase.

## Dispersoid Phase Equilibria

Zirconium, a commonly used dispersoid forming element, is peritectic with Al having a maximum usable solid solution of 0.14wt% which is limited by the break in the liquidus curve. Scandium is known to be eutectic with aluminum having a maximum solid solubility of 0.38wt%Sc and eutectic isotherm at 655 C [15, 16]. Scandium forms coherent stable $Al_3Sc$ dispersoids with aluminum and has an equilibrium $L1_2$ structure.

The literature shows that elemental substitution exists in $Al_3Sc$ and $Al_3Zr$ dispersoids allowing alloys containing both zirconium and scandium to precipitate $Al_3(Sc_{1-x}, Zr_x)$ phase [10]. No reports showing co-existence of independent $Al_3Sc$ and $Al_3Zr$ phases are known. Figure 1 compares the solvus lines of scandium and zirconium in aluminum. Both solvi are in the same temperature and composition region making their respective dispersoids well suited for co-precipitation in aluminum alloys.

The $Al_3Sc$ phase is reported to have rapid precipitation kinetics at 400 C [10, 11] combined with a moderate coarsening rate at this same temperature [17, 18]. The precipitation temperature range of $Al_3Zr$ is similar to $Al_3Sc$, although nucleation of $Al_3Zr$ is much slower, as shown in

27

figure 2 [10, 11].

Scandium diffusion in Al has $D_O$=5.31 cm²/s and a high activation energy for diffusion of Q=173kJ/mol [19]. For comparison, zirconium has $D_O$=6800 cm²/s and Q=241.6 kJ/mol [20]. Figure 3 compares the diffusivity of Sc and Zr in Al between 350 -600 C. Diffusivity of Sc in Al can be several orders of magnitude greater than Zr, which may lead to rapid coarsening of $Al_3Sc$ compared to $Al_3Zr$.

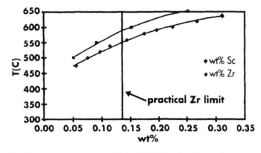

Figure 1 — Solvus lines of Sc and Zr in Al.

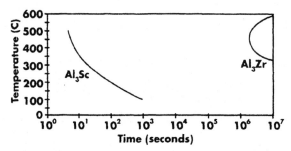

Figure 2 — Precipitation curves for $Al_3Sc$ and $Al_3Zr$.

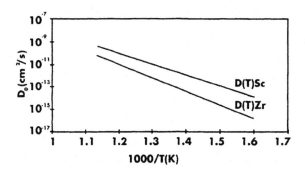

Figure 3 — Diffusivity of Sc and Zr in Al.

Thermodynamics and kinetics of Sc in Al are compatible with current preheating practices of aluminum alloys with and without Zr. Furthermore, rapid precipitation kinetics of $Al_3Sc$, high volume fraction, and homogeneous distribution of $Al_3Sc$ are favored over the slow precipitation, limited volume fraction, and often heterogeneous distribution of $Al_3Zr$ [21, 22]. Howver, faster coarsening kinetics of $Al_3Sc$ compared to $Al_3Zr$ limits the direct replacement of Zr by Sc in wrought Al alloys. This makes scandium an attractive addition to Al alloys.

**Experimental**

Ingots

High purity aluminum (99.99%) was provided by Alcoa, portions of an Al-2Sc master alloy from Ashurst Technology, Ltd., Al-6.3Zr from Alcoa, and Al-6Mg from Alcoa were used to cast ingots. The composition of all alloys are reported in weight percent (wt%). Portions of these master alloys were added to clay bound graphite crucibles coated with boron nitride and placed in a resistance furnace at 720 C until completely dissolved. The liquid was stirred and poured into room temperature steel book molds coated with boron nitride to produce ingots having dimensions of 15 x 7 x 2cm.

Ingots follow one of two thermo-mechanical treatments after casting. The following describes the two treatments.

**TMP 1:** The ingot was first preheated from room temperature to 450 C at 50 C/hr and held for 4 hours, followed by quenching in cold water. After preheating the ingot was hot rolled in three stages. The hot rolling temperature, 425 C, was reached within 15 minutes by heating the ingot in a preset resistance furnace. The passes through the rolling mill reduced the ingot to 40, 63, and finally 80% of the original thickness then quenched in cold water. Ingots were heated to the rolling temperature between each pass.

**TMP 2:** Following casting the ingots were immediately cold rolled at room temperature to 50% reduction of the original thickness in one pass through the mill. Next, the ingots were solution heat treated above the solvus temperature for the alloy s composition for one hour then quenched in cold water.

Table 1 describes alloy compositions and thermo-mechanical conditions for alloys used in this research.

Table 1 — Alloy compositions and thermo-mechanical treatments of alloys in this study.

| Alloy # | Alloy composition (wt%) | | | | TMP |
|---|---|---|---|---|---|
| | Zr | Sc | Mg | Al | |
| 1 | 0.12 | - | - | bal. | TMP 1 |
| 2 | 0.12 | - | 2 | bal. | TMP 1 |
| 3 | - | 0.2 | - | bal. | TMP 1 |
| 4 | - | 0.2 | - | bal. | TMP 2 |
| 5 | - | 0.3 | - | bal. | TMP 2 |
| 6 | - | 0.2 | 2 | bal. | TMP 1 |
| 7 | 0.12 | 0.2 | - | bal. | TMP 1 |
| 8 | 0.12 | 0.2 | - | bal. | TMP 2 |
| 9 | 0.12 | 0.1 | 2 | bal. | TMP 1 |
| 10 | 0.12 | 0.2 | 2 | bal. | TMP 1 |
| 11 | 0.12 | 0.2 | 3.5 | bal. | TMP 1 |

<u>Sample preparation, microscopy, and stereology</u>

Samples for optical microscopy were sectioned from the ingots and mounted in bakelite having the long transverse direction available for inspection. Polishing to a 0.04μm finish was performed in the usual manner. Samples were anodized using a solution of 0.5L $H_2O$, 2.3mL fluoboric acid, and 3.5g boric acid with an applied voltage of 12VDC at room temperature. A Reichert-Jung MeF3a with a polarized lens was used for optical microscopy. Recrystallization data were taken from midline in the sample.

TEM samples were prepared by electrothinning of pre-thinned 3mm diameter samples using a Struers Tenupol 3. The thinning solution was 75% methanol, 25% $HNO_3$ at —40 C using 13VDC. A JEOL 100C operated at 100keV was used for TEM work concerning general characterization, select area diffraction (SAD), and coarsening measurements. Measurements of particles using TEM were calibrated with a magnification calibration standard in the appropriate range. At least 500 dispersoids, photographed in dark field, were used for each datum point to ensure statistical significance of results. Figures that include TEM photographs were recorded in either bright field (BF) or dark field (DF) and indicated accordingly. EDS spectra were collected using a Hitachi HF-2000 TEM operated at 200keV equipped with a Noran EDS detector.

# Results

## Dispersoid Coarsening
### Coarsening Rates

Coarsening data for Al-0.2Sc, Al-0.3Sc, and Al-0.12Zr-0.2Sc alloys having been isothermally aged are presented in $r^3$ vs. t form in figure 4. Standard deviation, which was measured experimentally, however, is omitted in figure 4 to increase the clarity of the presented data. Standard deviation is considered in the next paragraph. Coarsening rate of $Al_3Sc$ in Al-0.2Sc is moderate at 450 C but increases dramatically as temperature is increased to 500 C. A strong dependence of coarsening on volume fraction of $Al_3Sc$ is also noted when comparing the trends of Al-0.2Sc and Al-0.3Sc at 450 C. An increase from 0.2Sc to 0.3Sc substantially increases the coarsening rate of $Al_3Sc$. In Al-0.12Zr-0.2Sc alloys, dispersoid coarsening is significantly slower than Al-Sc alloys, even though the volume fraction of dispersoids present is higher.

A functional relationship exists between average particle radius, $R$, and standard deviation, $s$, and is presented in figure 5 a,b,c. When plotted as standard deviation vs. average radius, regardless of thermo-mechanical treatment, each alloy composition shows strong linear behavior. A linear regression fit yielded a $s=mr+b$ relationship for each data set. When the particle has $r=0$ the standard deviation $s=0$ so the value of $b$ in all cases is forced to be zero in the linear regression solution. Thus, the linear regression is reduced to $s=mr$ form where $m$ is the slope of the line. The linear fit line, including the equation, for each alloy composition is presented in figure 5d. Note in figure 5a some Al-0.2Sc alloys were aged directly from the 50% cold rolled condition (CR50). Alloys of this study contain minor solute additions. Therefore, the slopes of $s$ vs. $r$ should vary only slightly between compositions. The data reflect this trend.

### Coherent to non-coherent transformation

Using SAD it was determined that the dispersoids of this study, $Al_3Sc$ and $Al_3(Sc_{1-x}, Zr_x)$, precipitate as coherent $L1_2$ spheres. In both cases the distribution of dispersoids is homogeneous as a result of preheating and homogeneous nucleation characteristic of scandium [23]. Figures 6 a, b show well distributed coherent dispersoids in Al-0.2Sc and Al-0.12Zr-0.2Sc in alloys thermo-mechanically treated using TMP 2, respectively. $Al_3Sc$ dispersoids remain coherent up to 25nm radius, at which point coherency to the matrix is lost. The dispersoids lose the Ashby-Brown contrast, indicative of coherent particles. $Al_3(Sc_{1-x}, Zr_x)$ dispersoids coarsen more slowly than $Al_3Sc$ at a given temperature. The size at which the coherent to non-coherent transition occurs for $Al_3(Sc_{1-x}, Zr_x)$, if it occurs at all, was not reached in the present coarsening study. $Al_3(Sc_{1-x}, Zr_x)$ dispersoids remain as coherent spheres throughout all the coarsening studies of this research. The largest dispersoids of $Al_3(Sc_{1-x}, Zr_x)$ produced in this study were 28nm radius after 100 hours at 500 C.

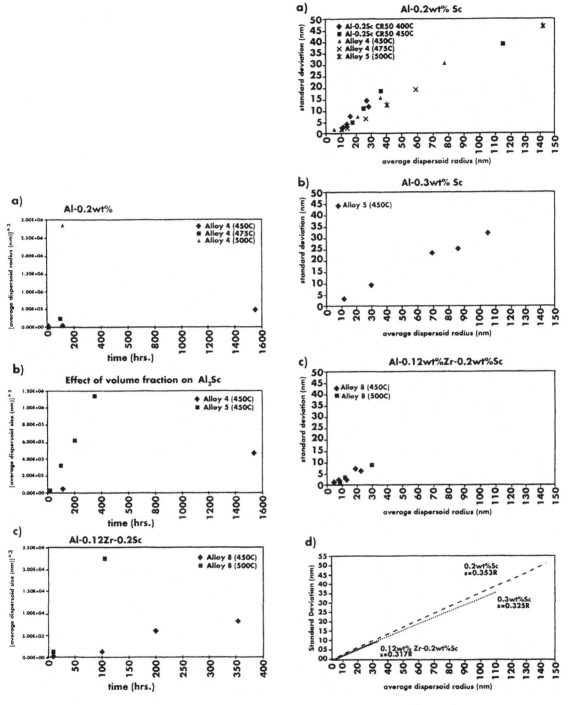

Figure 4 — $r^3$ vs t coarsening data. Aging temperature is in parenthesis.

Figure 5 — Standard Deviation vs. Radius. a),b),c) alloys grouped by composition. d) trend lines for a, b, c.

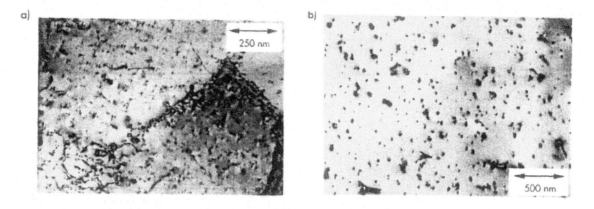

Figure 6 — Well distributed a) $Al_3Sc$ dispersoids in Al-0.2Sc,
b) $Al_3(Sc_{1-x}, Zr_x)$ dispersoids in Al-0.12Zr-0.2Sc.

Recrystallization

*Recrystallization in Mg-free wrought Al alloys*

Alloys containing 0.12Zr and/or 0.2Sc with the balance being Al have been preheated and hot deformed to 80% total reduction (TMP 1), as described in the experimental procedure. Following this, the alloys were annealed for 1 hour at 400, 450, 500, 550, or 590 C. Figure 7 presents volume fraction recrystallized for these alloys. Al-0.12Zr has no ability to resist recrystallization even at 400 C, the lowest temperature studied in this work. Al-0.2Sc is able to retard recrystallization up to 550 C, while Al-0.12Zr-0.2Sc completely suppressed recrystallization up to 590 C.

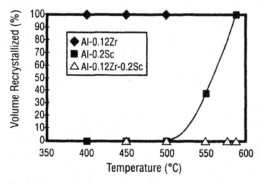

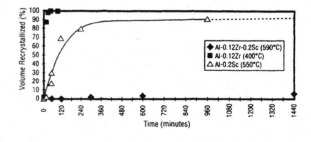

Figure 7 — Volume fraction recrystallized after 1 hour annealing for Al-0.12Zr, Al-0.2Sc, and Al-0.12Zr-0.2Sc. All alloys have been subjected to TMP 1.

Figure 8 — Volume fraction recrystallized during isothermal annealing up to 24 hours for Al-0.12Zr, Al-0.2Sc, Al-0.12Zr-0.2Sc. All alloys have been subjected to TMP 1. The annealing temperature for each alloy is indicated.

Annealing was extended to 24 hours for each alloy, namely; Al-0.12Zr at 400 C, Al-0.2Sc at 550 C, and Al-0.12Zr-0.2Sc at 590 C to determine the long term stability of unrecrystallized structures in the Sc containing alloys. Figure 8 presents the findings of the 24 hour annealing study. As presented in figure 7, Al-0.12Zr is fully recrystallized in 1 hour at 400 C. Al-0.2Sc is about 80% recrystallized after annealing at 550 C for 6 hours. However, Al-0.12Zr-0.2Sc remains essentially unrecrystallized throughout the study at 590 C.

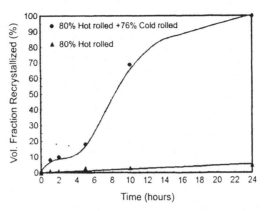

Figure 9 — Volume fraction recrystallized in Al-0.12Zr-0.2Sc after TMP 1 plus an additional 76% cold rolling during isothermal annealing at 590 C.

For the Al-0.12Zr-0.2Sc alloy temperature could no longer be raised in an attempt to induce recrystallization since 590 C is near the melting point of this alloy. An additional 76% cold deformation was added to the Al-0.12Zr-0.2Sc (TMP 1) alloy and recrystallization recorded during a 24 hour isothermal anneal at 590 C. Increasing the level of deformation increases the driving force to recrystallize and may reduce the temperature at which recrystallization begins. Figure 9 presents this data and shows that a full 24 hours at 590 C was necessary to completely recrystallize this highly deformed Al-0.12Zr-0.2Sc alloy.

*Recrystallization in Mg-containing wrought Al alloys*

Ternary Al-Sc-Mg Alloys     The effect of Sc on alloys containing up to 2Mg is illustrated in figure 10. These alloys were preheated and 80% hot rolled (TMP 1) then annealed for 1 hour at 400, 450, 500, or 590 C. Up to 500 C alloys with less than or equal to 2Mg remain unre-crystallized. However, partial recrystallization was observed in both alloys at 550 C, and complete recrystallization at 590 C. Addition of up to 2wt%Mg does not dramatically alter the behavior of Al-0.2Sc when in TMP 1 condition.

Quaternary Al-Zr-Sc-Mg
     Effect of Mg content     Adding Sc to Al-Zr-Mg has a similar effect on recrystallization as adding Sc to Al-Mg. Al-0.12Zr-0.2Sc alloys containing up to 3.5Mg were preheated, 80% hot rolled (TMP 1), and annealed for 1 hour at 400, 450, 500, 550, or 590 C. Volume fraction recrystallized was recorded and is presented in figure 11. No significant recrystallization occurs when Al-0.12Zr-0.2Sc contains 2Mg. At 3.5Mg partial recrystallization occurs between 400 and 500 C. At 550 C the Al-0.12Zr-0.2Sc-3.5Mg alloy was fully recrystallized within 1 hour.

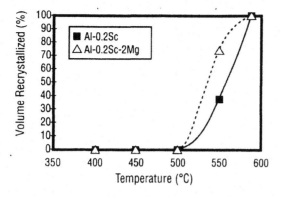

Figure 10 — Effect of 0.2wt%Sc on recrystallization behavior of Al-(0-2)wt%Mg. Alloys were subjected to TMP 1 followed by isothermal annealing for 1 hour.

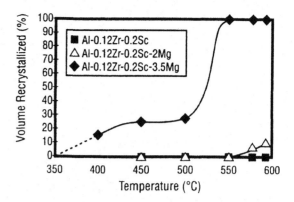

Figure 11—Effect of 0.12wt%Zr + 0.2wt%Sc on recrystallization behavior of Al-(0-3.5)wt%Mg. Alloys were subjected to TMP 1 followed by isothermal annealing for 1 hour.

Effect of Sc content    Scandium content less than 0.2wt% still contributed significantly to reducing recrystallization in wrought alloys. The change in recrystallization resistance after 1 hour annealing on Al-0.12Zr-2Mg alloys containing 0, 0.1, and 0.2wt%Sc are presented in figure 12. With no Sc the Al-0.12Zr-2Mg alloy fully recrystallized within 1 hour at 400 C. Adding 0.1Sc to Al-2Mg raised the recrystallization threshold to 550 C. Increasing the Sc content to 0.2wt% raised the temperature at which recrystallization begins to 590 C, near the melting temperature for this alloy.

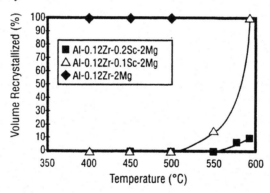

Figure 12 - Effect of Sc level on recrystallization in Al0.12Zr-2Mg alloys. All alloys are subjected to TMP1   followed by isothermal annealing for 1 hour.

*Recrystallized grain morphology*

The mechanism of recrystallization is different when comparing alloys containing Zr vs. Sc. Alloys that contain Zr (but no Sc) always recrystallize by nucleation and growth of new grains. When viewed in the optical microscope, recrystallized grains are equiaxed and exhibit no subgrain regions.  On the other hand, alloys with Sc (with or without Zr) tend to exhibit extensive recovery.  In the Sc containing alloys subgrain bands recover during annealing but do not nucleate new grains. After extensive recovery has taken place the final grains are similar in shape to the wrought grains, elongated in the rolling direction.  Figure 13 demonstrates the difference between recrystallization by nucleation and growth in Al-0.12Zr and extensive recovery in Al-0.2Sc.

## Recrystallization vs. Coarsening

A dispersoid s ability to counteract boundary pressure is inversely dependent on its size. At some dispersoid size the particle will become ineffective and recrystallization will occur. Knowing the size limit at which a particular dispersoid becomes ineffective is therefore important when designing recrystallization resistant alloys and their thermo-mechanical treatments. In this study, some alloys of Al-0.2Sc were cold rolled 50% then directly isothermally annealed (below the solvus temperature). Precipitation of $Al_3Sc$ from the wrought solid solution of Al-Sc occurred before recrystallization could take place. During isothermal annealing measurements of dispersoid size and volume fraction recrystallized were made. By comparing dispersoid size and volume fraction recrystallized determination of a size range above which $Al_3Sc$ was no longer effective was made. Figure 14 compares volume fraction recrystallized and dispersoid radius vs. time for annealing performed at 400 and 450 C on 50% cold rolled Al-0.2Sc. In both cases significant recrystallization begins to occur when dispersoids are in the 20-25nm radius range.

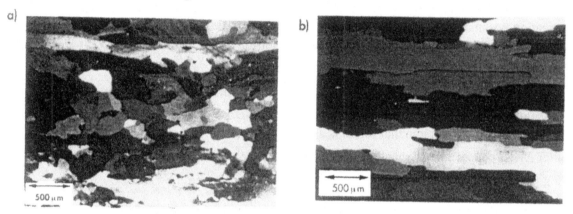

Figure 13 — Typical recrystallization behavior by a) nucleation and growth in Al-0.12Zr and b) extensive recovery in Al-0.2Sc.

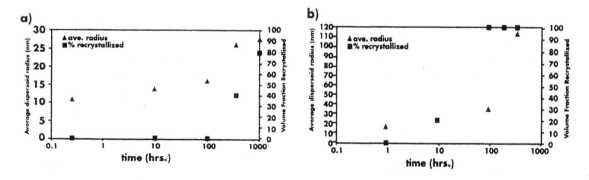

Figure 14 — Comparison of volume fraction recrystallized and dispersoid radius vs. time for 50% cold rolled Al-0.2wt%Sc isothermally annealed at a) 400 C, b) 450 C.

# Discussion

Dispersoid precipitation behavior

Precipitation of $Al_3Sc$ and $Al_3(Sc_{1-x}, Zr_x)$ from solid solution is rapid. The rapid decomposition of the solid solutions is an effect of Sc addition to the alloy. Cold rolled alloys containing solid solution Sc precipitate $Al_3Sc$ or $Al_3(Sc_{1-x}, Zr_x)$ rapidly enough to retard recrystallization. This observation is not typical of any other dispersoid forming elements. In comparison, the decomposition of Zr forming $Al_3Zr$ requires a significant incubation and requires a significant quantity of other elements to aid in its formation [22, 23]. Moreover, the distribution of $Al_3Sc$ and $Al_3(Sc_{1-x}, Zr_x)$ is always found to be homogeneous, regardless of preheating condition, unlike the heterogeneous nature of $Al_3Zr$. Assuming the solidus and liquidus arms of the Al-Sc phase diagram are straight then the value of the solute distribution coefficient, k, is constant and defined as Cs*/Cl*. When calculated, k has a value of 0.81 for Al-Sc. Given the moderate cooling rate from the melt and a high solute distribution coefficient (indicating possibility for significant coring effects) rejection of Sc to the grain boundary is not severe, as indicated by the homogeneous distribution of $Al_3Sc$ or $Al_3(Sc_{1-x}, Zr_x)$ from both cold rolled and solution heat treated microstructures.

Although $Al_3Sc$ has a very limited, nearly stoichiometric, compositional range it does have solubility for Zr [24]. In fact, when both Zr and Sc are present in the same alloys of this study the resultant dispersoid that forms is always $Al_3(Sc, Zr)$. In no cases were separate $Al_3Sc$ and/or $Al_3Zr$ phases found when both Sc and Zr were included in an alloy. This was confirmed using numerous EDS scans on dispersoids in alloys containing both Sc and Zr.

Coarsening Kinetics

Careful attention was given to measurements of dispersoid size for the coarsening study. Coarsening was measured in alloys that were first fully recrystallized and solution heat treated prior to isothermal annealing. This minimized competing and influential mechanisms that could have been acting on dispersoids outside of coarsening. It also ensured that dispersoid forming elements were well distributed before precipitation occurred.

Although not presented here, the coarsening data ($r^3$ vs. t) can also be re-plotted as ln(r) vs. ln(t). The slope of this line indicates the limit of the coarsening reaction. If the slope is 1/2, interface kinetics limit coarsening. A slope of 1/3 indicates diffusion limited coarsening. Given this criterion, alloys of this study agree satisfactorily with diffusion limited coarsening.

Several important coarsening trends prevail in this study. Coarsening rate of $Al_3Sc$ increases with increasing temperature from 450 to 500 C. However, comparatively slower coarsening rates were associated with $Al_3(Sc_{1-x}, Zr_x)$ dispersoids in Al-0.12Zr-0.2Sc at 450 and 500 C. Increasing the volume fraction of dispersoids in Al-Sc increases coarsening rate. For the data shown, at a coarsening temperature of 450 C the Al-0.2Sc and Al-0.3Sc equilibrium volume fraction of $Al_3Sc$ is 0.005 and 0.008, respectively.

Limit of Dispersoid Coherency

Coherency to the Al matrix is preserved with $Al_3Sc$ dispersoids up to 25nm radius. This result compares well with Drits, who showed $Al_3Sc$ maintains coherency up to 19nm radius in Al-0.3at%Sc (Al-0.5wt%Sc) [6]. However, this does not compare well with other data presented by Drits showing complete coherency of $Al_3Sc$ with Al up to 58nm radius [25]. However, that study used alloys of Al-(4.4-8.3)Mg-(0.25-0.3)Sc. A calculation of relative lattice parameters between Al and $Al_3Sc$ as a function of Mg content was included in previous work [26]. This calculation showed the difference between Al and $Al_3Sc$ lattice parameters decreases with increasing Mg content. This should stabilize $Al_3Sc$ against becoming incoherent by reducing misfit strain, prolonging its ability to remain coherent. Long incubation times necessary for recrystallization to occur in Al-Mg-Sc alloys of this study suggest this may be the case. Therefore, it is plausible that without the presence of magnesium $Al_3Sc$ dispersoids may lose coherency to the matrix at a smaller radius. Further investigation is necessary to confirm this suggestion. It is not known whether change in the lattice parameter of $Al_3Sc$, and thus dispersoid stability, is affected by substitution of Zr in the dispersoid.

The relatively slow coarsening rate of $Al_3(Sc_{1-x}, Zr_x)$, compared to $Al_3Sc$, suggests slower diffusion to the dispersoid interface and a lattice parameter closer to that of the Al matrix. Also, Sc addition resulted in a uniform distribution of dispersoids. Taking advantage of the slow diffusion, sustained coherency to the matrix, homogeneous distribution, and rapid precipitation kinetics associated with $Al_3(Sc_{1-x}, Zr_x)$ is a useful approach for alloy development when recrystallization control is concerned. Since Zener drag is inversely proportional to particle size, keeping small, well dispersed particles prolongs the effectiveness of the dispersoids to arrest migrating boundaries. Thus, $Al_3(Sc_{1-x}, Zr_x)$ is more effective than $Al_3Sc$ due to a slower coarsening rate. A greater potential volume fraction exists for $Al_3(Sc_{1-x}, Zr_x)$ compared to $Al_3Zr$ and $Al_3Sc$. Conventional wrought aluminum alloys may also benefit from the increased volume fraction effect not obtainable using Zr or Sc along, due to solubility limits of each in Al.

Recrystallized Grain Morphology

Aluminum alloys containing $Al_3Zr$ dispersoids tend to recrystallize by nucleation and growth of new grains. The recrystallized microstructure of Al-Zr alloys has equiaxed grains. This microstructure suggests recovery may be limited, especially in a heavily deformed matrix, leaving recrystallization the dominant path to reduce stored mechanical energy. However, aluminum alloys having $Al_3Sc$ dispersoids allow substantial recovery to occur, which significantly reduces the driving force available for the nucleation of new grains. Recovery continues in wrought Al-Sc until the final grains have a flat, uniform color when anodized and viewed under polarized light in the optical microscope. The well recovered grains contain no subgrain bands indicating all strain hardening has been eliminated. In effect, the microstructure is comparable to a recrystallized material. A notable difference with Al-Sc alloy "recrystallization" compared to Al-Zr recrystallization is that the final grain structure has grains with an aspect ratio elongated in the rolling direction, reminiscent of the prior rolled grains. When Sc is added to Al-Zr the tendency towards extensive recovery dominates. This may result from the contribution of Sc prolonging the recovery process and increased volume fraction of dispersoids when both Sc and Zr are present, allowing very highly recrystallization resistant microstructure to persist during high temperature annealing.

## Conclusion

Scandium addition to wrought aluminum is compatible with existing thermo-mechanical processes established for these alloys. The addition of Sc rapidly precipitates high volume fraction, homogeneously dispersed dispersoids having coherent $Al_3Sc$ ($L1_2$) structure. $Al_3Sc$ dispersoids impart some improvement in recrystallization resistance to wrought alloys, namely with higher annealing temperatures and volume fractions, compared to alloys employing $Al_3Zr$. However, the moderate coarsening rate of $Al_3Sc$ may limit its usability. When both scandium and zirconium are used in the same alloy $Al_3(Sc_{1-x}, Zr_x)$ dispersoids form, which are more effective than either $Al_3Sc$ or $Al_3Zr$ acting alone. The $Al_3(Sc_{1-x}, Zr_x)$ dispersoid benefits from the rapid precipitation characteristic of Sc in Al, the slow coarsening associated with Zr, and a high volume fraction as an effect of both Sc and Zr. Furthermore the distribution of $Al_3(Sc_{1-x}, Zr_x)$ is homogeneous.

It was also established that alloys containing up to 3.5Mg showed improvement in recrystallization resistance with both Sc and Zr present.

The use of scandium should be considered a potent and compatible solution for increasing recrystallization resistance in some wrought aluminum alloys. With scandium addition, thermal-mechanical processing of certain important industrial alloys can be extended to higher temperatures, longer times, and more severe deformation resulting in semi-finished product having better mechanical properties that remain stable. Scandium is currently an expensive consideration as an alloy additive. If combined with Zr, substantial benefits from Sc can still be realized when used at levels below the maximum solubility.

## References

1    C. Zener (quoted by C.S. Smith), Trans. Met. Soc. AIME, 175 (1948), pp. 345.

2    N. Ryum, J. Inst. Metals, 94 (1966), pp. 191.

3    N. Ryum, Acta Met., 17 (1969), pp 296.

4    E. Nes, Acta Met., 20 (1972), pp. 499.

5    H. Paris, T.H. Sanders Jr., Y.W. Riddle, "Assessment of Scandium Additions in Aluminum Alloy Design", Proceedings of ICAA-6, 1 (July 1998), pp. 499-504.

6    R.R. Sawtell, C. Jensen, "Mechanical Properties and Microstructures of Al-Mg-Sc Alloys", Met. Trans. A, 21A (Feb. 1990), pp. 421-430.

7    Y.W. Riddle, H.G. Paris, T.H. Sanders Jr., "Control of Recrystallization in Al-Mg-Sc-Zr Alloys", ICAA-6, 2 (1998), pp. 1179-1184.

8    M. Drits, L. Toropova, Y. Bykov, L. Ber, S. Pavlenko, "Recrystallisation of Al-Sc Alloys", Russian Metallurgy, 1 (1982), pp. 148-152.

9    J. Royset, N. Ryum, "Precipitation and Recrystallization of an Al-Mg-Sc Alloy", Mat. Sci.Forum ICAA-7, 331-337 (2000), pp. 194-201.

10   Zakharov, Rostova, "On the Possibility of Scandium Alloying of Copper-Containing Aluminum Alloys", Metal Science and Heat Treatment, 37, n1-2 (1995), pp. 65-69.

11    Elagin, Zakharov, Rostova, "Some Features of the Decomposition for the Solid Solution of Scandium in Aluminum", Metallovendenie I Termicheskaya Obrabotka Metallov, n7 (July 1983), pp. 57-60.

12    L.A. Willey, "Aluminum Scandium Alloy", United States Patent #361981, (Nov. 9, 1971).

13    M. Kharakterova, "Phase Composition of Al-Cu-Sc Alloys at Temperatures of 450 and 500C", Izvestiya Akademii Nauk SSSR, Metally, n4 (1991), pp. 191-194.

14    I.N. Fridlyander, Drits, Yeleseyev, "Effect of Scandium on the Structure and Properties of Al-Cu Alloys", Lightweight and High-Temperature Alloys and the Processing of Them, Moscow, (1986), pp126-130.

15    K.A. Gschneidner, F.W. Calderwood, "The Al-Sc (Aluminum-Scandium) System", Bulletin of Alloy Phase Diagrams, 10, n1 (1989), pp. 34-36.

16    S.I. Fujikawa, M. Sugaya, H. Takei, K.I. Hirano,"Solid Solubility and Residual Resistivity of Scandium in Aluminum", Journal of the Less-Common Metals, 63 (1979), pp. 9987-97.

17    Y.W. Riddle, T.H. Sanders Jr., "Contribution of $Al_3Sc$ to Recrystallization Resistance in Wrought Al-Sc Alloys", Materials Science Forum Proceedings of ICAA-7, 331-337 (2000),pp. 939-944.

18    M.Y. Drits, L.B. Ber, Y.G. Bykov, L.S. Toropova, G.K. Anastas Eva, "Ageing of Alloy Al-0.3at%Sc", Phys. Met. Metall., 57, n6 (1984), pp. 118-126.

19    S.I. Fujikawa, "Impurity Diffusion of Scandium in Aluminum", Defect and Diffusion Forum, 143-147 (1997), pp. 115-120.

20    T. Marumo, et al., Japan J. Light Metals, 23 (1973), pp. 17-25.

21    H.R. Last, Precipitation of Dispersoids in Aluminum Alloys , (Ph.D. Thesis, Georgia Institute of Technology, Atlanta, GA, USA, 1991).

22    K. Ranganathan, Recrystallization Resistance in Aluminum Alloys , (Ph.D. Thesis, Georgia Institute of Technology, Atlanta, GA, USA, 1991).

23    R. Hyland, "Homogeneous Nucleation Kinetics of $Al_3Sc$ in a Dilute Al-Sc Alloy", Met. Trans. A, 23A (July 1992), pp. 1947-1955.

24    N. Sano, Y. Hasegawa, K. Hono, H. Jo, H.W. Pickering, T. Sakurai, J. Phys. colloq. C6 Suppl. 11, (1987), pp. 337-342.

25    Drits, "Homogenization of Alloys in the System Al-Mg-Sc", Metals Sci. and Heat Treating, 25 (1983), pp. 550-554.

26    Y.W. Riddle, "Control of Recrystallization in Al-Mg Alloys Using Sc and Zr", (M.Sc. Thesis, Georgia Institute of Technology, Georgia, USA, August 1998).

-El Fin-

# LIGHTWEIGHT ALLOYS FOR AEROSPACE APPLICATION

*Edited by:*
*Dr. Kumar Jata, Dr. Eui Whee Lee,*
*Dr. William Frazier and Dr. Nack J. Kim*

## ALUMINUM ALLOYS

## Application of 3D Digital Image Processing to Quantify Fracture Micro-Mechanisms in Al 7050 Alloy

*Manish Dighe, Sunit Mukherjee and Arun Gokhale*

Pgs. 41-50

184 Thorn Hill Road
Warrendale, PA 15086-7514
(724) 776-9000

# APPLICATION OF 3D DIGITAL IMAGE PROCESSING TO QUANTIFY FRACTURE MICRO-MECHANISMS IN AL 7050 ALLOY

Manish Dighe[1], Sunit Mukherjee[2] and Arun Gokhale[2]
[1] Hi TecMetal Group, Cleveland, Ohio
[2] School of Materials Science and Engineering
Georgia Institute of Technology, Atlanta, Georgia .

## Abstract

7XXX series wrought aluminum alloys are extensively used for structural aerospace applications due to their high strength to weight ratio, excellent corrosion resistance, and high fracture toughness. 7050 is an important alloy of this group, which is widely used for the applications such as aircraft wing skin structures, landing gear parts, and fuselage frames. Therefore, it is of interest to investigate the fracture behavior of 7050 aluminum alloy. The global fracture of the alloy involves multiple fracture micro-mechanisms. It is the objective of this contribution to report development and application of a new digital image-based microscopy technique for the study of fracture micro-mechanisms. The technique permits simultaneous observation of three-dimensional fracture surface <u>and</u> the three-dimensional microstructure just beneath the fracture surface. The method has been utilized to quantify the relative contributions of different fracture micro-mechanisms to over all fracture path in plane strain fracture toughness test specimens of a partially recrystallized 7050 Al-alloy.

Lightweight Alloys for Aerospace Applications
Edited by Kumar Jata, Eui Whee Lee,
William Frazier and Nack J. Kim
TMS (The Minerals, Metals & Materials Society), 2001

# Introduction

7XXX series of wrought aluminum alloys are extensively used for aircraft structural components, where fracture toughness and strength are of prime concern. 7050 is a typical alloy of 7XXX series. Commercial 7050 alloy plates often have a partially recrystallized[1,2] microstructure. The degree of partial recrystallization depends on the alloy chemistry, cast ingot microstructure, and processing parameters such as hot-rolling temperature. The fracture toughness of partially recrystallized 7050 alloy depends on the microstructural attributes such as degree of recrystallization. Due to it complex microstructure, the fracture of partially recrystallised 7050 alloy involves multiple fracture micro-mechanisms. The microstructure determines the relative contributions of different fracture micro-mechanisms to the global fracture path, which in turn affect the fracture resistance of the alloy. Therefore, it is of interest to understand relationships among microstructure, fracture micro-mechanisms and the fracture path in this alloy. For this purpose it is essential to observe simultaneously both the fracture morphology and the microstructure immediately beneath it. It is the objective of this contribution to report the development and application of a new digital image analysis based technique to reconstruct the fracture surface and associate it with the exact three-dimensional microstructure beneath the fracture surface using serial sectioning. The technique permits unambiguous correlation of fracture morphologies to the corresponding fracture micro-mechanisms and the associated microstructural features. The technique involves serial sectioning of the microstructure just beneath the fracture surface, and association of this local microstructure with the fracture surface images of exactly the same locations. The technique has been applied to plane strain fracture toughness specimen of a partially recrystallized 7050 alloy to quantify the fraction of the fracture surface area generated due to different fracture micro-mechanisms. The next section of the paper gives a brief background on the chemistry, processing, and the microstructure of hot rolled Al 7050, and the dominant fracture mechanisms. The subsequent sections deal with the procedure and development of the technique, the correlation between fracture surface morphology and the underlying microstructure, and the identification of different fracture mechanisms on the fracture surface.

# Background

## Material and Microstructure

The nominal chemical composition of 7050 alloy is Al − 6.2 wt% Zn − 2.2wt% Cu − 2.2wt% Mg-0.1wt% Zr − 0.12wt% Si (max.) − 0.15wt% Fe (max). Si and Fe are restricted to 0.12 wt% and 0.15 wt%, respectively because they lead to formation of constituent particles, which are deleterious to the fracture toughness of the alloy. The alloy chemistry, solidification process during ingot casting, thermo-mechanical processing, and precipitation hardening treatment lead to the development of a very complex microstructure in hot rolled precipitation hardened 7050 alloy plates. The important microstructural features are as follows

1. The microstructure contains unrecrystallized regions consisting of ensembles of equiaxed sub-grains. The sub-grain sizes are in the range of 2 to 10 μm. The sub-grains are separated by low angle grain boundaries.

2. The unrecrystallized regions also contain prior high angle grain boundaries from the cast ingot. These grain boundaries have anisotropic orientations: they are mostly parallel to the rolling direction.

3. The recrystallized grains and regions are partitioned by high angle boundaries. The grain boundaries between the recrystallized grains, and the boundaries between the recrystallized grains and unrecrystallized regions are these high angle grain boundaries. The recrystallized grains are mostly oriented parallel to the rolling direction, and their sizes are typically in the range of 25 to 100μm.

4. The high angle grain boundaries created by the partial recrystallization as well as the prior high angle grain boundaries contain coarse incoherent and closely spaced η phase precipitates.
5. Metastable semi-coherent fine η' phase precipitates are present in the grain and sub-grain interiors. The η' precipitates[6,7,3] are responsible for increase in the strength of the alloy after precipitation hardening treatment.
5. Iron and silicon based coarse constituent particles (1 – 20 μm size range) are present in the microstructure. The volume fraction, average size, and anisotropy of these particles affect the fracture behavior of the alloy.
6. Uniformly distributed $Al_3Zr$ dispersoids are also present in the microstructure.

## Fracture Micro-Mechanisms

Microstructure of partially recrystallized 7050 Al-alloy contains features at length scales ranging from nanometers to millimeters. Many of these features are morphologically anisotropic, and their distributions are spatially non-uniform. All of these aspects affect the fracture processes and fracture toughness[4,8] of the alloy. Microstructural complexity leads to numerous fracture micro-mechanisms. Dominant fracture micro-mechanisms and their associated microstructural features are as follows.

- Micro-void induced transgranular fracture through the recrystallized grains
- Transgranular fracture through the sub-grains in the unrecrystallized regions
- Intergranular fracture associated with the high angle grain boundaries between the recrystallized grains, and, between unrecrystallized regions and recrystallized grains.
- Intergranular fracture associated with the prior high angle grain boundaries in the unrecrystallized regions of the microstructure.
- Fracture of constituent inter-metallic particles.

The fracture toughness of the alloy depends on the relative contributions of the various micro-mechanisms to the global fracture path, which in turn depend on the microstructure of the alloy.

## Experimental Work

The experiments were performed on partially recrystallized 7050 alloy plates donated by ALCOA. The thermo-mechanical processing and mechanical testing of these plates was carried out at ALCOA Tech. Center. The plate was rolled at different rolling temperatures to vary the degree of recrystallization. The fracture toughness tests were performed on the L-T orientation specimens as per the ASTM standard B399-83.

## Characterization of Fracture Surface Morphology

The fractographic measurements were performed in the plane strain region (center) of the plastic zone ahead of the fatigue pre-crack in the $K_{IC}$ test specimens. The proportion of the plastic zone fracture surface area generated by a given fracture mode is the quantitative measure of the contribution of the corresponding micro-mechanism to the fracture initiation process. Thus, it of interest to measure the area fractions of intergranular fracture regions, transgranular fracture regions, and fractured constituent particles in plastic zone area.

For the specimens under this investigation, the fracture morphologies are clearly resolved at the SEM magnification of 600X. Also, at this magnification, the size of the SEM field of view is very close to the plastic zone size. Thus by taking fractographs in the center of the $K_{IC}$ specimen, the plane strain region can be covered.

## Fracture Surface Montage Creation

As mentioned in the earlier sections, there is lot of ambiguity in identifying the different fracture micro-mechanisms on the fracture surface. One way to solve this problem is to observe the microstructure just above and below the fracture surface. But to accomplish this, creation of a fracture surface montage is necessary. The SEM was used to grab overlapping images of the fracture surface. The montage of the entire crack front was recreated. The edges of the fracture specimen were also grabbed in the montage. These act as a reference to correlate the exact location of a unique feature on the fracture surface to that on the optical sections of the underlying microstructure. Figure-1 shows a digitally compressed fracture surface montage created from six overlapping fractographs grabbed at 600X.

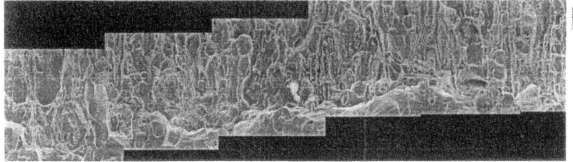

Figure-1: Digitally compressed fracture surface montage

## 3D Microstructural Reconstruction

The second step in solving the puzzle of fracture micro-mechanisms is creation of 3D microstructure just below and above the fracture surface. To achieve this, serial sections were taken perpendicular to the fracture surface. Figure-2 shows the schematic explaining the configuration. Note that to see microstructure both below and above the fracture surface, sectioning has to be done on both the fracture surfaces. The exact positioning of the montage and the depth of each section are monitored using a micro indent, the diameter of which has a simple mathematical relation to the depth of the indent.

## Digital Montage Creation

Figure-3 shows the locations of the optical montages of the microstructures grabbed. These are grabbed at a magnification of 50X objective (total magnification is 680X, which is close to the magnification at which SEM fractographs were grabbed). The number of microstructural fields in the montage is chosen so as to include at least one field of view below and one above the fracture profile (the fracture surface appears as a fracture profile on a 2D optical section). Thus for the plane strain region, 8X15 (i.e. 120 contiguous fields) montage is required. These images were then aligned using and reconstructed using a 3D image software VOXBLAST. Details of the technique have been given elsewhere[5].

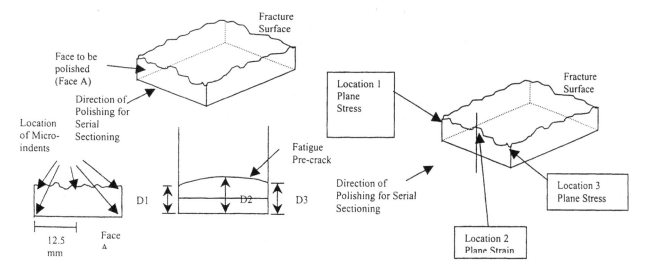

Figure-2:Schematic of Serial Sectioning          Figure-3:Location of montage

45

Associating the Fracture Surface and deciphering the Fracture Surface Morphology

Once the serial sectioning[1] and the fracture surface montage creation[1] were done, the next step is to associate these two structures to each other. The first step is to locate and identify prominent features on the fracture surface that can be clearly identified on the serial sections. The distance of this unique feature from the edge of the specimen also helps in this association of fracture surface and the microstructure.

To identify the fracture micro-mechanism and its source, first, particular fracture surface morphology is noted on the fracture surface, and then, the distance of this fracture morphology from the unique feature is measured. One can then go back to the serial section image that contains the microstructural feature just below the morphological feature of interest on the fracture surface.

The five predominant mechanisms and corresponding microstructural features identified in this manner are discussed below.

Intergranular Fracture Between Recrystallized Grains

Figure-4 shows this type of fracture as it appears on the two matching fracture surfaces and on the corresponding serial section montage. In this case, the opposite fracture surfaces (Figure-4) have a perfectly matching morphology consisting facets of recrystallized grains. One fracture surface shows a "crest" and the other one shows a matching "trough", and there is a negligible local plastic deformation at these features. This results in the fracture region to be preserved. The micrograph (Figure-4) shows a part of this section montage with both the fracture profiles. In the present case (Figure-4), the region below and above the feature of interest on the fracture of surface (intergranular fracture facet) is recrystallized grain, and therefore this intergranular fracture morphology is generated by fracture through the high angle grain boundary between recrystallized grains. Thus the fracture through the high angle grain boundary between two recrystallized regions has the following attribute

- The surface is very flat and highly faceted.
- There is absence of any ductility or dimples at the magnification of 600X. At higher magnifications, one can see some evidence of dimples.
- There is a perfect match between the features on both the fracture surfaces, due to absence of any plastic deformation.

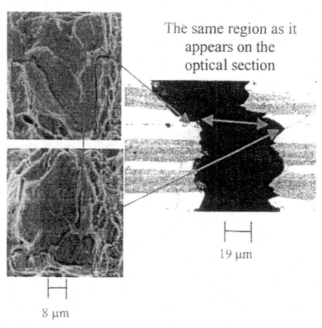

The same region as it appears on the optical section

19 μm

8 μm

Figure-4:Intergranular fracture through recrystallized grains

46

## Particle Induced Fracture

This type of fracture not only includes fractured particle(s), but also the region of the fracture surface generated due to void growth around the fractured constituent particle(s). This area mostly consists of some recrystallized regions as most of the particles are inside the recrystallized region. Figure-5 shows one of such particle induced fracture morphologies. One characteristic difference of this type of fracture morphology as compared to the fracture through the recrystallized grains is the presence of an elongated particle at the center of this region. Another striking feature is the smoothness of the area surrounding this elongated constituent particle, as compared to the recrystallized grain surface. But the two fracture surfaces do not match perfectly. Both the fracture surfaces have a concave region around the particle. This would mean that a void has nucleated around the particle and has grown considerable before final fracture. Figure-6 shows the 3D reconstructed image of a broken particle with a void around it. Observe that, although the void has grown significantly, the adjacent high angle grain boundaries have not fractured through intergranular fracture mechanism at this point. This support the hypothesis that in the sequence of events involving multiple micro-mechanisms based fracture, the constituent particle induced fracture occurs before intergranular fracture along high angle grain boundaries. In summary, the main distinguishing features of particle-induced fracture are as follows.

- These regions usually have a particle sitting in the center of the region.
- The region surrounding the particle is very smooth as compared to the recrystallized grain surface.
- The opposite fracture surfaces usually do not match as well as the recrystallized grain fracture, though there is striking similarity between the opposing surface

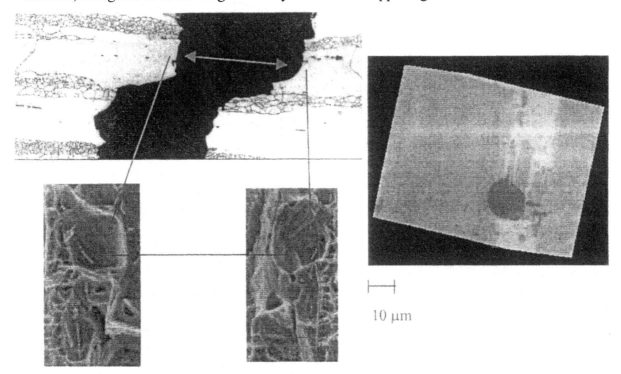

Figure-5:Particle induced fracture          Figure-6: 3D reconstructed void around broken particle

## Transgranular Fracture Through Recrystallized Region

This type of fracture, also known as micro-void induced ductile fracture, appears as collection of dimples on the fracture surface. The fracture mechanism leads to lot of energy dissipation, and hence is preferred for high toughness. Figure-7 shows the transgranular fracture through recrystallized grains. The corresponding images from a serial section that pass through this region are also shown in the same figure. As expected, in this case the two fracture profiles do not perfectly match due to local plastic deformation associated with this fracture mechanism. As it is evident in the figure, the recrystallized regions appear wavy in the area on the fracture profile (on the optical sections). This is vastly different from the morphology of the recrystallized regions as seen on the optical section, in case of particle fracture, and intergranular fracture through recrystallized regions. This type of fracture is the last to happen in the chain of events leading to final fracture. Also, since there are no constituent particles, causing the nucleation of these dimples, the nucleation energy and well as the growth energy for this type of dimples is expected to be very high. Consequently, not much of this type of fracture morphology is seen on the fracture surface.

Fracture Through Prior High Angle Grain Boundaries

The prior high angle grain boundaries are carried over from the ingot casting process. To detect the intergranular fracture along the prior high angle grain boundaries on the fracture, and to prove its existence unambiguously, is of significant interest in understanding the role of these boundaries in the over all fracture of the alloy. The fracture along prior high angle grain boundaries can be discerned by using the following criteria: (i) there must be unrecrystallized region both above and below the fracture segments generated by this mechanism, (ii) the morphology must be distinctly different from that of the sub-granular fracture, and therefore, these fracture segments should not contain the dimples characteristic of sub-granular fracture, and (iii) the fracture segments may or may not contain constituent particles. Figure-8 shows the typical morphology of the fracture along the prior high angle grain boundaries. The associated serial sections indicate that unrecrystallized regions are present above and below

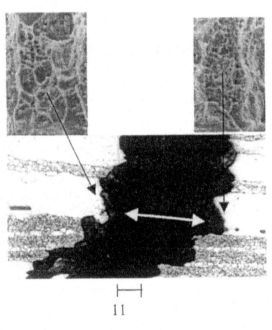

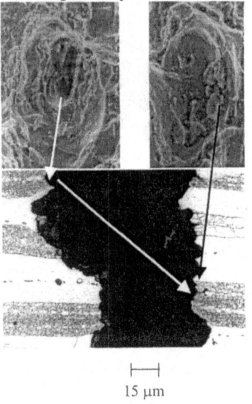

Figure 7. Transgranular fracture

15 μm

Figure-8: Fracture through prior high angle grain boundaries

48

## Sub-Granular Fracture

The final type of fracture morphology is the fracture through sub-grain regions. This type of fracture is termed as the sub-granular fracture. Figure-9 show this type of fracture morphology in more detail. The main distinguishing factor is the presence of the jagged small regions. The size of these regions is of the order of two to five microns, which is of the same length scale as the sub-grains. Also, by looking at the sections through this region, it is clear that such type of fracture is due to the fracture through the sub-grain regions. There is another type of morphology that also represents the crack through the sub-grain region. Figure 10 illustrates this other morphology of the fracture through the sub-grains. However, in both of these cases, it is not possible to determine if these morphologies are due to inter-sub-granular fracture or trans-sub-granular fracture, or a combination of these two modes. However, both of these fracture modes associated with sub-grains involve high plastic energy dissipation, and lead to higher fracture toughness. In summary, the following characteristics differentiate the sub-grain fracture from other types of fracture.

- The morphology has a very uneven type of surface devoid of any facets and particles.
- Very fine features are seen on this fracture morphology. These features are of the same size range as the sub-grains (1-5 μm). These features do not show any depth/depression, which is a characteristic feature of dimples.

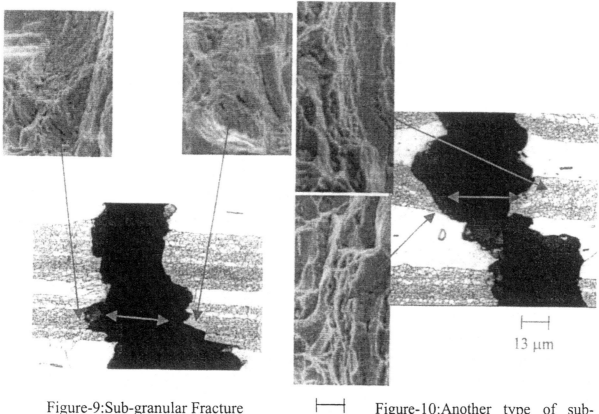

Figure-9:Sub-granular Fracture

⊢——⊣
5.7 μm

Figure-10:Another type of sub-granular fracture

49

## Conclusions

A digital image processing based methodology to reconstruct the three-dimensional (3D) fracture surface and associate it with the exact three-dimensional microstructure beneath the fracture surface, was developed. This was used to characterize the relationships among microstructure, fracture surface morphology, and fracture toughness of hot-rolled partially recrystallized precipitation hardened 7050 alloy. The fundamental fracture micro-mechanisms, which exist in a partially recrystallized 7050 alloy, have been identified and studied. This is accomplished by associating the fracture surface image with the three-dimensional reconstruction of the microstructure just below the fracture surface. The fracture mechanisms that are catalogued are

- Particle induced fracture.
- Fracture along the high angle grain boundaries between recrystallized grains.
- Fracture along the prior high angle grain boundaries.
- Fracture through the sub-grain regions.
- Transgranular fracture through recrystallized grains.

## Acknowledgements

This research has been supported by grants and plane strain fracture toughness samples from Alcoa and U.S. National Science Foundation (Grant No. DMR-9816618). The financial support is gratefully acknowledged.

## References

1. E.Hornbogen and M.Graf, "Fracture Toughness of Precipitation Hardened Alloys Containing Narrow Soft Zones at Grain Boundaries", Acta Metalurgica, Vol. 25, pp. 877-881, 1977.

2. J.W.Yeh and Kuo-Shung Liu, "The Correlation of Fracture Toughness and transgranular Fracture in Al-5.6%Zn-2.5%Mg Alloy with Small Additions of Fe", Trans. Jpn. Inst. Met., Vol. 27, No. 7, pp. 504-511, 1986.

3. T. Kawabata and O. Izumi, "Ductile Fracture in the Interior of Precipitate Free Zone in an Al-6%Zn-2.6%Mg Alloy", Acta Metallurgica, Vol. 24, pp. 817-825, 1976.

4. E.Hornbogen and M.Graf, "Fracture Toughness of Precipitation Hardened Alloys Containing Narrow Soft Zones at Grain Boundaries", Acta Metalurgica, Vol. 25, pp. 877-881, 1977.

5. Manish Dighe, Development of digital image processing based methodology to study, quantify and correlate the microstructure and three dimensional fracture surface morphology of aluminum alloy 7050",( 2000, Ph.D.Thesis Dissertation, Georgia Institute of Technology)

6. Hahn, and A.R.Rosenfield, "Metallurgical Factors Affecting Fracture Toughness of Aluminum Alloys", Met. Trans., Vol. 6A, pp. 653-670, 1975.

7. Sanders, Jr. and E.A.Starke, Jr., "The Relationship of Microstructure to Monotonic and Cyclic Straining of Two Age Hardening Aluminum Alloys", Met. Trans., Vol. 7A, pp. 1407-1418, 1976.

8. J.W.Yeh and Kuo-Shung Liu, "The Correlation of Fracture Toughness and transgranular Fracture in Al-5.6%Zn-2.5%Mg Alloy with Small Additions of Fe", Trans. Jpn. Inst. Met., Vol. 27, No. 7, pp. 504-511, 1986.

# LIGHTWEIGHT ALLOYS FOR AEROSPACE APPLICATION

*Edited by:*
*Dr. Kumar Jata, Dr. Eui Whee Lee,*
*Dr. William Frazier and Dr. Nack J. Kim*

## ALUMINUM ALLOYS

## On the Effect of Stress on Nucleation, Growth, and Coarsening of Precipitates in Age-Hardenable Aluminum Alloys

*B. Skrotzki and J. Murken*

Pgs. 51-61

184 Thorn Hill Road
Warrendale, PA 15086-7514
(724) 776-9000

# ON THE EFFECT OF STRESS ON NUCLEATION, GROWTH, AND COARSENING OF PRECIPITATES IN AGE-HARDENABLE ALUMINUM ALLOYS

B. Skrotzki, J. Murken

Ruhr-University Bochum, Institute for Materials
44780 Bochum, Germany

## Abstract

Light metals are the materials of choice for technical applications where low specific weight combined with high specific stiffness is demanded. Their mechanical properties at room temperature are usually well characterized. However, their high temperature behavior has not been studied in great detail. Generally, two approaches can be applied to improve high temperature strength: (i) nanodispersion of precipitates, and (ii) use of intermetallic compounds. In the present study, the high temperature behavior and the accompanying microstructural changes during aging with and without external tensile stress were studied for different age-hardenable Al-alloys. Aging with stress represents creep loading, which is expected to occur in technical aircraft and space application.

In the nucleation stage it was found that precipitates are preferentially oriented parallel to an external tensile stress in the solution heat-treated condition of a binary Al-Cu and a quaternary Al-Cu-Mg-Ag alloy. The nucleation of precipitates is strongly affected by an external applied stress and there is a critical value of stress above which preferential nucleation on habit plane variants occurs. The effect of an external stress on growth and coarsening of precipitates depends on the specific alloys and precipitates, respectively. Growth and coarsening of precipitates present in an Al-Cu-Mg-Ag and an Al-Si-Ge alloy were not affected by the creep parameters used in this study. However, $\delta'$ precipitates present in an Al-Mg-Li alloy grew faster with an external stress applied than under stress free conditions.

Lightweight Alloys for Aerospace Applications
Edited by Kumar Jata, Eui Whee Lee,
William Frazier and Nack J. Kim
TMS (The Minerals, Metals & Materials Society), 2001

# Introduction

High strength Al-alloys are hardened by finely dispersed second phase particles. These precipitates are usually coherent or semi-coherent to the Al-matrix, and they are metastable. At elevated temperatures, they not only begin to grow and coarsen, but they moreover transform into their thermodynamic stable forms if the temperature is high and given enough time. Under creep conditions, the effect of stress and strain has to be considered as an additional parameter. A thermodynamic effect (stress: around particles or at (sub)grain boundaries) and the kinetic effect (strain: dislocations represent additional diffusion paths between particle and matrix) can be distinguished. Only a few investigations are available on the growth and coarsening of precipitates under creep conditions of technical precipitation hardened Al-alloys. Model materials such as Al - 11 wt. % Zn and Al - 5 at. % Mg were extensively studied with respect to their macroscopic creep deformation behavior and accompanying evolution of the microstructure [1-5]. However, those results are only partly applicable to technical Al-alloys. More recent studies on creep of precipitation hardened Al-alloys focus on the creep behavior of alloys of type Al-Cu-Mg-Ag that are being considered for use in a new civil supersonic aircraft yet to be developed. However, these investigations focus on the measurement of mechanical creep data while the microstructural development of the precipitate structure has not been the center of interest [6-9].

Under normal aging conditions, i.e., when no applied or residual stresses are present, an even distribution of precipitates should form on all habit planes. If precipitation occurs preferentially on certain habit planes, an anisotropy of strength properties may result. Coherency strains are usually considered to stabilize a single-phase field, i.e., they shift the solvus line into the equilibrium two-phase field. Externally imposed strains may change the stability of a phase and may move the solvus line either into the single-phase or two-phase region. Consequently, externally applied stresses and internal stresses associated with second-phase particles can affect both nucleation and growth of precipitates and subsequently their coarsening behavior [10-11]. In several alloy systems, it has been found that an externally applied stress may result in preferential orientation of precipitates [12-18]. However, the reported results are contradictory. This is due to the fact that different alloy systems were investigated with precipitates having different morphologies and different amounts of positive or negative misfit. Even when similar alloy systems were studied, they were aged at different temperatures and stresses. In some alloys, nucleation and growth of precipitates was studied, while in others, the coarsening behavior was examined.

Almost no attention has been paid to the question of what kind of precipitate is under consideration, i.e. whether it is coherent, semi-coherent or incoherent. Another interesting aspect is the stability of the phase. Metastable phases such as $\Theta'$ and $\delta'$ coarsen faster than the more stable precipitates $\Omega$ or $T_1$. In addition, the misfit between precipitate and matrix should also affect the coarsening. In the present work, investigations with respect to nucleation, growth and coarsening of precipitates of technical Al-alloys under creep conditions will be summarized. Application of Al-alloys requires increased improvement of thermal stability at elevated temperature (which means temperatures up to 250 °C for Al-alloys). This is the case e.g. in aviation (supersonic civil transport) and space technology.

# Experimental Procedure

The compositions of all investigated sheet materials are given in Table I. At appropriate temperatures, the binary Al-Cu system has the precipitate sequence GP zones, $\Theta''$ (GP II), and $\Theta'$ as transition phases before the equilibrium $\Theta$ ($Al_2Cu$) [19]. The microstructure of the Al-Cu-

Mg-Ag alloy is more complicated. In addition to the aforementioned precipitates, the phases $\Omega$ (habit plane is $\{111\}_{Al}$) and S' ($Al_2CuMg$) are formed. Both alloys were used to study the effect of an externally applied stress on the nucleation of precipitates. Different initial heat treatments were investigated, such as solution heat-treated (SHT), peak-aged (T6) and peak-aged with prior straining (T8) with details given in [20]. Two Al-Cu-Mg-Ag alloys of similar composition were used for creep experiments to study the microstructural changes associated with creep (i.e. growth and coarsening). All three of the Al-Cu-Mg-Ag alloys were produced by ALCOA. Details on processing, heat treatment and testing are given in [20, 21].

Table I: Chemical composition of the investigated alloys in wt. %.

| Alloy | Cu | Mg | Mn | Ag | Zr | V | Fe | Si | Li | Zn | Sc | Ge | Al |
|-------|-----|---------|------|------|-----------|------|------|------|----------|---------|----------|------|------|
| Al-Cu | 5.00 | - | - | - | - | - | - | - | - | - | | - | bal. |
| Al-Cu-Mg-Ag | 5.75 | 0.52 | 0.30 | 0.49 | 0.16 | 0.09 | 0.06 | 0.05 | - | - | | - | bal. |
| C415 | 5.09 | 0.82 | 0.66 | 0.51 | 0.13 | n.a. | 0.07 | 0.06 | - | - | | - | bal. |
| C416 | 5.38 | 0.53 | 0.31 | 0.52 | 0.13 | n.a. | 0.07 | 0.05 | - | - | | - | bal. |
| Al-Mg-Li | - | 5.6-6.0 | - | - | 0.08-0.12 | - | - | - | 1.4-1.65 | 0.4-0.6 | 0.06-0.1 | - | bal. |
| Al-Si-Ge | - | - | - | - | - | - | - | 0.45 | - | - | | 1.33 | bal. |

A technical Al-Mg-Li-alloy was investigated, which was received as rolled sheet material of 4 mm thickness. The alloy is mainly hardened by the $\delta'$-phase ($Al_3Li$) which is coherent to the Al-matrix and its misfit is small. A superlattice with $L1_2$ crystal structure characterizes its ordered fcc unit cell. The $\delta'$-phase is metastable and precipitates as spherical particles. This advanced lightweight alloy, which has its potential application in civil aircraft, also contains small amounts of Sc and Zr to inhibit recrystallization. The two elements also strengthen the material by the precipitation of $Al_3(Sc, Zr)$ which is a modification of the well known $\beta'$-phase. It has to be noted that this alloy is not considered for high temperature applications. However, the system was chosen to study the coarsening behavior (under stress) of coherent spherical precipitates with small misfit.

Additions of Si and Ge to aluminum alloys do not form stable or metastable intermetallic compounds, but precipitates with diamond structure. A recent study by Hornbogen, et al. [22, 23] has shown that ternary Al-(Si, Ge) alloys exhibit a considerable precipitation hardening effect considering the small volume fraction of precipitates. This alloy was chosen to study the effect of an applied stress on the coarsening behavior of incoherent (spherical) precipitates. The alloy was cast in an induction furnace under argon atmosphere using high purity elements. The as-cast material was machined to remove the as-cast surface, followed by a homogenization treatment (30 h/500 °C) and water quenching. Hot rolling was subsequently carried out in several steps at 425 °C to a final thickness of 3 mm. Finally, the alloy was solutionized at 490 °C for 1 h and aged at 160 °C for 24 h into a peak aged condition.

All creep tests were carried out in tension. The extensometers were directly attached to the samples. In case of Al-Cu and Al-Cu-Mg-Ag, samples of constant area were aged at 160 °C under a tensile stress in an ATS creep machine for different times (10, 100, and 1000 h) and under various constant loads in the as-solution-heat-treated condition and in the peak-aged (T6) condition. Details are given in [20, 21]. The tests on the Al-Mg-Li alloy were conducted in the aged condition at 120 °C with stresses varying between 190 and 280 MPa. Tests were interrupted at nominal 2 % total strain. Different strain levels between 0.5 and 5 % were accommodated at 250 MPa to study the effect of strain. The crept conditions were compared to those aged at 120 °C *without* any external stress. For Al-Si-Ge, creep tests were carried out at 120 and 140 °C and stresses of 65, 75 and 85 MPa, respectively. Most specimens fractured before the envisaged strain of 2 % was reached. Again, the crept conditions were compared to isothermally aged samples.

Standard techniques were used to prepare TEM samples of all samples. Images of $\Omega$ and $\Theta'$ precipitates were taken in the bright field mode in $[011]_{Al}$ and $[001]_{Al}$ orientation. The dark field mode was employed to take images from the $\delta'$ precipitates. SiGe precipitates were imaged in bright field using the $[001]_{Al}$ zone axis. The images were transferred onto a transparency and digitized followed by a quantitative image analysis. Details are given in [24].

## Results

Nucleation and Growth

Al-Cu-Mg-Ag and Al-Cu: It was found that $\Omega$ and $\Theta'$ precipitates orient preferentially parallel to an externally applied tensile stress in the solution-heat-treated condition. The nucleation process is strongly affected by the applied stress and there is a threshold stress that must be exceeded before preferential nucleation on habit plane variants is observed. Both $\Omega$ and $\Theta'$ precipitate plates have negative misfit with the matrix when very thin, which leads to plate nucleation on variants under compression. These results were published earlier and a detailed description is given in [20].

Growth and Coarsening

Al-Cu-Mg-Ag: Quantitative stereology was carried out on crept samples to determine the volume fraction, number density and size of $\Omega$ and $\Theta'$ precipitates. In summary, the precipitate microstructure can be described as very stable under the conditions used in the tests (i.e. 107 and 135 °C). In contrast to nucleation of precipitates, growth is not affected by creep in the way that orientation of precipitates was observed. For a detailed representation of the results see [21].

Al-Mg-Li: Aging with and without external stress shows very similar trends. The as-received state is characterized by the smallest particle size. Further aging causes larger particle sizes. The particle radii were calculated from the experimentally measured diameters assuming that the particles are spherical. The cube of the mean particle radii after aging with and without stress are plotted in Figure 1 (a) as a function of aging and creep time, t, respectively. The straight lines represent the results of the regression analysis and they indicate that coarsening with stress is somewhat faster than under stress free conditions. This implies that the contribution of stress and/or strain to $\delta'$-precipitate coarsening cannot be neglected. An oriented coarsening was not observed and was not expected due to the spherical shape of the precipitates and their small misfit. The volume fraction of $\delta'$-precipitates was calculated from the area fraction determined by quantitative image analysis. Figure 1 (b) shows that the volume fraction $V_V$ of $\delta'$ precipitates increases with time for both aging procedures. However, aging under stress results in slightly higher volume fractions. The data for the calculated number densities of precipitates, $N_V$, (not shown) reveal that isothermal stress free aging results in a decreasing number density of $\delta'$ precipitates while $N_V$ remains almost constant if aging takes place under an external stress implying that further nucleation of precipitates takes place.

Al-Si-Ge: The TEM results revealed that the size, the shape and the distribution of the precipitates vary from grain to grain as well as locally within the grains which can be attributed to the strong dependence of nucleation of the SiGe precipitates on the quenched in excess vacancies and on the variation in local vacancy supersaturation [26].

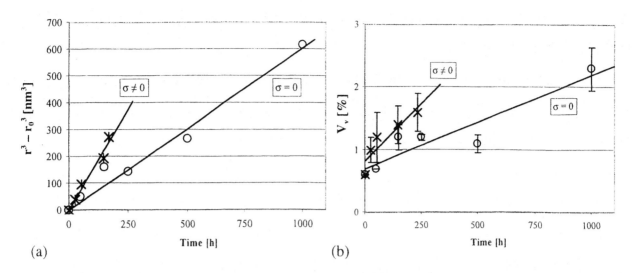

Figure 1: (a) The cube of the $\delta'$ particle radius, $r^3 - r_0^3$, vs. aging/creep time with and without stress. (b) Volume fraction, $V_V$, of $\delta'$ precipitates vs. aging/creep time with and without stress. [25]

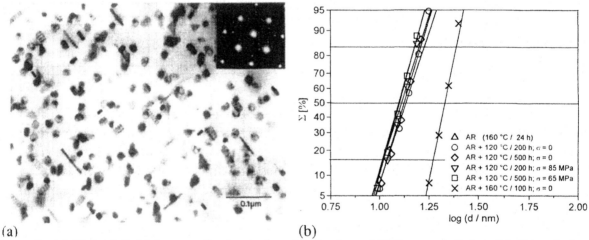

Figure 2: (a) Bright field image of spherical and lath-shaped SiGe precipitates (b) Cumulative frequency distribution of the logarithm of the diameter of equiaxed spherical precipitates [24, 25].

Due to the different morphologies of SiGe precipitates present in the peak-aged and crept conditions (see Figure 2 (a)), it was necessary to characterize the equiaxed and lath shaped particles separately. Only areas with a high precipitate population were taken into account. A complete treatment of the full quantitative characterization is given in [24]. Figure 2 (b) shows data obtained for the diameter, d, of the equiaxed precipitates plotted into a probability net. The logarithm of the data was used and mean size values can be read out at $\Sigma = 50\%$. Straight lines connecting the data points indicate that the data follow a logarithmic Gaussian distribution. It is clear from Figure (b) that aging at 120 °C with or without stress does not affect the precipitate size of both the equiaxed and the lath shaped precipitates. The curves of the aged (120 °C) conditions lie very close to that of the as received (i.e. peak aged) condition. However, aging at 160 °C results in a considerable shift to the right, i.e. to larger particle diameters or widths. The width and the height of the SiGe laths behave similarly. The volume fraction of SiGe precipitates remains constant within the experimental scatter; i.e. no further nucleation takes place. However, a slightly higher volume fraction of laths was observed after aging for 200 h at 85 MPa and 120 °C. This might be due to laths which have presumably formed during creep [26].

# Discussion

It is well known that the presence of an external stress in a two-phase crystalline solid may have a significant effect on the microstructure [10, 11, 27]. It can:

- Modify the morphology of isolated precipitates,
- Destabilize arrays of precipitates resulting in a rafted structure,
- Influence the precipitate coarsening kinetics, or
- Alter the relative stability of the two phases.

In precipitation hardened materials, the properties strongly depend on the shape, the size and the distribution of the second phase. Therefore, the effect of an external stress field on the stability of the microstructure and the morphology of precipitates is of high technological interest.

Martin et al. [28] pointed out that almost none of the useful structures in materials science are thermodynamically stable. The driving forces for microstructural changes are changes in chemical, strain and interfacial free energies, respectively. Precipitation hardened microstructures possess a considerable driving force for precipitate coarsening in the form of interfacial free energy [28]. Precipitate coarsening is generally described as Ostwald ripening. The physical mechanism emanates from the increased solubility of small particles, which is due to the increase in chemical potential due to the energy of the matrix/precipitate interface. The well known Lifshitz-Slyozov-Wagner equation [28-30]

$$\bar{r}^3 - \bar{r}_0^3 = \frac{8D\gamma c\Omega^2}{9RT}$$

gives four parameters which control coarsening of precipitates: (i) interfacial free energy, $\gamma$, of the precipitate/matrix interface, (ii) solubility, c, of the solute in the matrix, (iii) diffusion coefficient, D, of the solute, and (iv) the atomic volume, $\Omega$, of the precipitate phase. R is the universal gas constant and T the temperature. To keep coarsening rates small, one or more of the four parameters have to have small values.

The interfacial free energies of the precipitates studied in this work are given in Table II. The interfacial free energy depends on the type of the interface and on the misfit between the precipitate and the matrix, and data for these two parameters are given in Table II as well. The $\delta'$ precipitate shows the lowest interfacial free energy due to its coherent interface and the small misfit. The highest value is found for the incoherent SiGe precipitate while the semi-coherent $\Theta'$ and $\Omega$ precipitates show interfacial energies in between those of the coherent and the incoherent ones. These data imply according to the Lifshitz-Slyozov-Wagner equation that the precipitate with the smallest $\gamma$ should show the lowest coarsening rates. However, this is not in agreement with the experimental results.

Diffusion coefficients published in the literature [31, 32] for alloying elements used in the present work are also summarized in Table II. All of the elements relevant for the three alloys investigated diffuse faster in the Al-matrix than Al does in Al. Diffusion data for Cu are very close to the self-diffusion coefficient of Al; Cu is in fact the slowest diffusing element of the six elements considered. The data for Li and Mg are almost identical and both elements diffuse faster than Cu. The diffusion coefficients of Si and Ge are typical for normal solute diffusion in pure aluminum. Zumkley and Mehrer [31] have performed a study on Ge diffusion in binary and ternary Al-(Si,Ge) solid solution alloys. Si and Ge rather belong to the fast diffusing elements in aluminum. Therefore, diffusion data imply that the SiGe precipitate should coarsen faster than $\delta'$, $\Theta'$ and $\Omega$. But again, this contradicts the experimental results.

Table II: Parameters influencing precipitate coarsening rates for different phases ($\uparrow$ = high, $\downarrow$ = low, $\downarrow\downarrow$ = very low).

| | Al | $\delta'$ (Al₃Li) | $\Theta'$ (~Al₂Cu) | $\Omega$ | SiGe |
|---|---|---|---|---|---|
| Diffusion coefficient D [m²/s] at 120 °C [31, 32] | $2.0 \cdot 10^{-23}$ | Li: $6.3 \cdot 10^{-22}$ | Cu: $5.3 \cdot 10^{-23}$ | Cu: $5.3 \cdot 10^{-23}$ Mg: $5.8 \cdot 10^{-22}$ Ag: $3.4 \cdot 10^{-21}$ | Si: $1.2 \cdot 10^{-21}$ Ge: $8.0 \cdot 10^{-21}$ |
| Interface | - | coherent | semi-coherent | semi-coherent | incoherent |
| Interfacial free energy $\gamma$ [J/m²] [33-35] | - | 0.014 | 0.179 | n.a. ($\uparrow$) | 0.43 |
| Misfit [%] | - | -0.2 | -4.5 | -9.3 | - |
| Solubility $c_0$ [at.%] at 227°C [36] | - | 4.14 | Cu: 0.07 | Cu: 0.07 (Mg: 4.5) (Ag: 0.47) | Si: 0.01 Ge: 0.19 |
| Atomic volume $\Omega$ [nm³] | 0.0166 | 0.0164 | 0.0158 | 0.0941 | 0.0214 |
| Thermodynamic stability | - | metastable | metastable | stable? | stable |
| Coarsening rate | - | $\uparrow$ | $\downarrow$ | $\downarrow\downarrow$ | $\downarrow\downarrow$ |

The solubility of the alloying elements used in the investigated Al-alloys is given in Table II. As the solubility depends on temperature, those values are summarized in Table II, which were found for the lowest available temperature, i.e. for 227 °C (except for Ge: T = 177 °C). Although this temperature is higher than the one used for the aging experiments, the solubility data give at least a hint to the expected tendency. Table II reveals that the elements Si, Ge, and Cu have low or very low solubility in aluminum, while Li is easily dissolved in the Al-matrix. Mg and Ag perhaps play a role during coarsening of the $\Omega$ precipitate as well. However, it is more likely that the major importance of these two elements lies in the nucleation process of $\Omega$. The solubility data for Si, Ge, and Cu correspond well with the observed coarsening rates, implying that the low solubility of the participating alloying elements mainly controls the slow coarsening rates of the SiGe, the $\Omega$ and the $\Theta'$ precipitates.

The atomic volumes of the $\delta'$ and $\Theta'$ precipitates are very similar and comparable to that of the Al-matrix as is summarized in Table II. However, for the SiGe and the $\Omega$ precipitates the value differs considerably from that of the other precipitates predicting again high coarsening rates for these particles.

Table II summarizes data for the parameters discussed above and also includes information on the thermodynamic stability of the precipitates and on the experimentally observed coarsening rates. Concluding from the discussion on the parameters affecting precipitate coarsening and taking into account the stability of the precipitates studied here, it is proposed that the low solubility of the alloying element combined with a thermodynamically stable nature of the particle results in lowest coarsening rates. This is the case for SiGe and for $\Omega$. On the other hand, high solubility and metastability results in high coarsening rates of $\delta'$ although the interfacial energy is very small. When a more stable precipitate has formed according to Ostwald's step rule, it dissolves the less stable particles within its diffusion field, which results in coarsening of the dispersoids combined with decreasing coherency.

Effect of an External Stress

The previous discussion focused on isothermal aging *without* stress. Aging *with* stress, i.e. under creep conditions, has additional effects on the parameters described in the Lifshitz-Slyozov-Wagner equation. An applied stress affects the solubility of the solute as was shown by Johnson [27]. He carried out thermodynamic calculations and showed that coherency strains, which are almost always present around precipitates in Al-alloys, always move the interfacial concentration further into the two-phase region of the equilibrium phase diagram. An external stress applied to strain free precipitates shifts the solvus line in either direction depending on the difference in elastic constants between precipitates and matrix. However, if the shear modulus of the precipitate is higher than that of the matrix, the applied stress shifts the interfacial concentration line into the two-phase region. Johnson concludes that the average interfacial concentration of the solute influences the average precipitate growth rate and indicates the effect of an applied stress field on the relative stability of the precipitate and the matrix. His calculations revealed that for elastically soft precipitates, applying an external stress to a system initially in equilibrium induces a flux of atoms from the matrix to the precipitate, which results in an increase of the volume fraction of precipitates. On the other hand, the volume fraction of elastically hard precipitates decreases if an external stress is applied. This could explain the increasing volume fraction of $\delta'$ precipitates during creep (cf. Figure 1(b)). Another explanation is that due to the high solubility of Li in Al, the matrix is still supersaturated and able to precipitate new $\delta'$-particles. This corresponds well with the number density measurements.

If stresses are applied to precipitates, which exhibit coherency strains, then the sign of the stress, $\sigma$, and the strain, $\varepsilon$, must be considered [27]. When $\sigma$ and $\varepsilon$ are of the same sign, the effective supersaturation and therefore the growth rate increase. On the other hand, if $\sigma$ and $\varepsilon$ are of opposite sign, then the effective supersaturation and consequently the growth rate slow down. An applied stress also changes the interfacial free energy due to the fact that the external stress deforms the interface, which results in changes of the elastic strain energy of the interface [27]. Externally applied stress and interfacial stress are coupled. If both stresses are of opposite sign, then the interfacial free energy is decreased. Finally, diffusion processes are required for precipitate growth and coarsening. And it is well known that the diffusion coefficient not only depends on the temperature and the solute concentration but also on the applied stress $\sigma$ [27].

For the Al-alloys systems studied so far, the Al-Mg-Li alloy was the only one which exhibited an accelerated coarsening rate of precipitates under creep conditions. All other systems behaved either quite coarsening resistant or coarsening was not affected by creep stress.

Whether the thermodynamic effect due to hydrostatic tensile and compression stresses around precipitates enhances diffusion under creep conditions is not yet clear. Preferential orientation in certain directions with respect to the external stress was not observed during coarsening (although during nucleation). However, there might exist a critical stress and/or critical temperature to initiate such effects as was found for oriented nucleation of precipitates.

Acknowledgement

We gratefully acknowledge funding by the "Deutsche Forschungsgemeinschaft" (DFG Sk 47/1-1 and Sk 47/1-2). Thanks are due to Dipl.-Ing. G. Kausträter for experimental help with creep tests, and to Mr. R. Höhner and Dipl.-Ing. O. Girard for carrying out the quantitative image analysis on the Al-Mg-Li and the Al-Si-Ge alloy. We wish to thank Prof. G. Eggeler and Dr. A. K. Mukhopadhyay for stimulating discussion.

References

1   W. Blum, B. Reppich, "Creep of Particle-Strengthened Alloys", in: Creep Behavior of Crystalline Solids, B. Wilshire, R. W. Evans (eds.), Pineridge Press, Swansea, UK, 1985, 83 – 135

2   M. A. Morris, J. L. Martin, "Evolution of Internal Stresses and Substructure during Creep at Intermediate Temperatures", Acta metall. 32 (1984) 549 - 561

3   M. A. Morris, J. L. Martin, "Microstructural Dependence of Effective Stresses and Activation Volumes during Creep", Acta metall. 32 (1984) 1609 - 1623

4   M. J. Mills, J. C. Gibeling, W. D. Nix, "A Dislocation Loop Model for Creep of Solid Solutions Based on the Steady State and Transient Creep Properties of Al-5.5at.% Mg", Acta metall. 33 (1985) 1503 - 1514

5   M. J. Mills, J. C. Gibeling, W. D. Nix, "Measurement of Anelastic Creep Strains in Al-5.5at.% Mg using a new Technique:  Implications for the Mechanism of Class I Creep", Acta metall. 34 (1986) 915 - 925

6   S. M. Kazanjian, N. Wang, E. A. Starke, Jr., "Creep Behavior and Microstructural Stability of Al-Cu-Mg-Ag and Al-Cu-Li-Mg-Ag Alloys", Mater. Sci. Eng. A234 – 236 (1997) 571 – 574

7   J. Wang, X. Wu, K. Xia, "Creep Behavior at Elevated Temperatures of an Al-Cu-Mg-Ag Alloy", Mater. Sci. Eng. A234 – 236 (1997) 287 – 290

8   R. Mächler, et al., "High Strength Damage Tolerant AlCuMgAg Forging Alloy for Use at Elevated Temperatures", Materials Science Forum, Vols. 217 – 222 (1996) 1771 – 1776

9   I. J. Polmear, et al., "After Concorde: Evaluation of an Al-Cu-Mg-Ag Alloy for Use in the Proposed European SST", Materials Science Forum, Vols. 217 – 222 (1996) 1759 - 1764

10  W. C. Johnson, C. S. Chiang, "Phase Equilibrium and Stability of Elastically Stressed Heteroepitaxial Thin Films", J. Appl. Phys. 64 (1988) 1155 – 1165

11  C. S. Chiang, W. C. Johnson, "Coherent Phase Equilibria in Systems Possessing a Consolute Critical Point", J. Mater. Res. 4 (1989) 678 – 687

12  M. R. Louthan, Jr., "Stress Orientation of Titanium Hydride in Titanium", Trans AIME 227 (1963) 1166 – 1170

13  Y. Nakada, W. C. Leslie, T. P. Churay, "Stress-Orienting of $Fe_{16}N_2$ Precipitates in an Fe-N Alloy", Trans. ASM 60 (1967) 223 – 227

14  G. Sauthoff, "Orienting of Precipitating Au Particles in a Fe-Mo-Au Alloy by External Elastic Stress", Z. Metallkde. 68 (1977) 500 – 505

15  W. F. Hosford, S. P. Agrawal, "Effect of Stress During Aging on the Precipitation of Θ' in Al-4 wt % Cu", Metall. Trans. 6A (1975) 487 – 491

16  T. Eto, A. Sato, T. Mori, "Stress Oriented Precipitation of G.P. Zones and Θ' in an Al-Cu Alloy", Acta metall. 26 (1978) 499 – 508

17  J. K. Tien, S. M. Copley, "The Effect of Uniaxial Stress on the Periodic Morphology of Coherent Gamma Prime Precipitates in Nickel-Base Superalloy Crystals", Metall. Trans. 2 (1971) 215 – 219

18    T. Miyazaki, K. Nakamura, H. Mori, "Experimental and Theoretical Investigation on Morphological Changes of $\gamma'$ Precipitates in Ni-Al Single Crystals during Uniaxial Stress Annealing", J. Mater. Sci. 14 (1979) 1827 - 1837

19    G. Lorimer, "Precipitation Processes in Solids", K. C. Russel and I. Aaronson, eds., TMS-AIME, Warrendale, PA, USA, 1978, 87 – 160

20    B. Skrotzki, G. J. Shiflet, E. A. Starke, Jr., "On the Effect of Stress on Nucleation and Growth of Precipitates in an Al-Cu-Mg-Ag Alloy", Metall. Mater. Trans. 27A (1996) 3431 – 3444

21    B. Skrotzki, H. Hargarter, E. A. Starke, Jr., "Microstructural Stability under Creep Conditions of two Al-Cu-Mg-Ag Alloys", Materials Science Forum, Vols. 217 – 222 (1996) 1245 – 1250

22    E. Hornbogen, A. K. Mukhopadhyay, E. A. Starke, Jr., "Precipitation Hardening Al-(Si,Ge) Alloys", Scripta metall. mater. 27 (1992) 733 – 738

23    E. Hornbogen, A. K. Mukhopadhyay, E. A. Starke, Jr., "Nucleation of the Diamond Phase in Aluminum-Solid Solutions", J. Mater. Sci. 28 (1993) 3670 - 3674

24    J. Murken, et al., "Quantitative Characterization of Microstructure of a Al-Si-Ge Alloy after Isothermal Aging with and without Stress", Practical Metallography (submitted)

25    J. Murken, et al., "On the Influence of Stress and Strain on the Coarsening of Precipitates in Al-Alloys under Creep Conditions", Materials Science Forum, Vol. 331 - 337 (2000) 1507 – 1512

26    A. K. Mukhopadhyay, et al., "Nature of Precipitates in Peak Aged and in Subsequently Crept Al-Ge-Si Alloy", Materials Science Forum, Vol. 331 - 337 (2000) 1555 - 1560

27    W. C. Johnson, "Precipitate Shape Evolution under Applied Stress - Thermodynamics and Kinetics", Metall. Trans. 18A (1987) 233 - 247

28    J. W. Martin, R. D. Doherty, B. Cantor: Stability of Microstructures in Metallic Systems, 2nd ed., Cambridge University Press, Cambridge, UK, 1997

29    I. M. Lifshitz, V. V. Slyozov, "Kinetics of Diffusive Decomposition of Supersaturated Solid Solutions", J. Phys. Chem. Solids 19 (1961) 33

30    C. Wagner, "Theorie der Alterung von Niederschlägen durch Umlösen (Ostwald Reifung)", Z. Elektrochemie 65 (1961) 581 – 591

31    T. Zumkley, H. Mehrer, "Diffusion of Ge in Binary and Ternary Al-(Si,Ge) Solid-Solution Alloys", Z. Metallkde 89 (1998) 454 - 463

32    Landolt-Börnstein, Numerical Data and Functional Relationships, Vol. 26, Diffusion in Solid Metals and Alloys, H. Mehrer (ed.), Springer-Verlag, Berlin, Heidelberg, 1990

33    D. Y. Li, L. Q. Chen, "Computer Simulation of Stress-Oriented Nucleation and Growth of $\Theta'$ Precipitates in Al-Cu Alloys", Acta mater. 46 (1998) 2573 - 2585

34    H. J. Königsmann, "Microstructural Stability and Fracture Behavior in Al-Si-Ge Alloys", Dissertation thesis, University of Virginia, School of Engineering and Applied Science, Dept. of Materials Science and Engineering, Charlottesville, VA, USA, 1996

35    S. F. Baumann, D. B. Williams, "A New Method for the Determination of the Precipitate-Matrix Interfacial Energy", Scripta metall. 18 (1984) 611 - 616

36    L. F. Mondolfo, Aluminium Alloys: Structure and Properties, Butterworth, London, UK, 1976

# LIGHTWEIGHT ALLOYS FOR AEROSPACE APPLICATION

*Edited by:*
*Dr. Kumar Jata, Dr. Eui Whee Lee,*
*Dr. William Frazier and Dr. Nack J. Kim*

## ALUMINUM ALLOYS

## Hot Deformation Mode and TMP in Aluminum Alloys

*H.J. McQueen and M.E. Kassner*

Pgs. 63-75

184 Thorn Hill Road
Warrendale, PA 15086-7514
(724) 776-9000

# HOT DEFORMATION MODE AND TMP IN ALUMINUM ALLOYS

*H.J. McQueen and M.E. Kassner*
*Dept. of Mechanical Engineering, Concordia University, Montreal, Canada H3G 1M8*
*Dept. of Mechanical Engineering, Oregon State University, Corvallis, OR 97331, USA*

Hot forging and extrusion traditionally produce Al alloy components, providing suitable shaping capabilities combined with creation of beneficial dislocation, grain, preferred orientation and fiber microstructures. Likewise suitable structures can be produced by hot rolling to plate for direct milling and to strip for accurate finishing on cold mills followed by sheet-forming processes. While dependence of subtructure on T, $\dot{\varepsilon}$ and $\varepsilon$ is independent of deformation mode, both grain shape and orientation are strongly dependent on both strain path and schedule of passes and intervals. The hot shaping can be controlled and integrated into thermomechanical processing (TMP) to produce microstructures with improved service properties or capable of superplastic behaviour. Discussion of the above processes includes consideration of hot strength and ductility, constitutive equations, microstructural evolution, restoration mechanisms and product properties. The analysis also considers feedstock structure and segregation, effects of solute, dispersoid and precipitates and alterations by static mechanisms.

## INTRODUCTION

The hot workabilities of Al alloys and composites at any temperature T, strain $\varepsilon$ and strain rate $\dot{\varepsilon}$ are expected to be properties of the material dependent on texture, initial grain structure (segregation, etc.), solute, precipitates and dispersed particles [1-6]. The hot workability includes the flow stress, ductility, microstructural evolution and product properties. These should be independent of the mode of processing for similar T and $\dot{\varepsilon}$ with exception of texture development grain shape and fibrous character which create varied anisotropies. This assumption is vitiated to varying degrees by the effects that are dependent on different schedules with one or several stages (i = 1,2,3...) of $T_i$, $\varepsilon_i, \dot{\varepsilon}_i$ and $t_i$ (time) and on variation of T, $\varepsilon$, and $\dot{\varepsilon}$ with time and space paths during one stage. While it is theoretically possible to specify the complete sequence in detail and estimate every variation in substructure, the difficulty of doing it can be compared with genome mapping in biological research. It is necessary to make approximations based on simplified schedules and the variations within them. In this way it has been possible to organize the hot working behavior of different alloys in the process modes of extrusion, rolling and forging. The preheating and feed stock preparation will be classed as a stage; so will the final cooling; there is some consideration of interactions with expected secondary processing. When the sequence and condition of the stages are designed to produce specific mechanical properties, they comprise thermomechanical processing (TMP). [3-8].

Lightweight Alloys for Aerospace Applications
Edited by Kumar Jata, Eui Whee Lee,
William Frazier and Nack J. Kim
TMS (The Minerals, Metals & Materials Society), 2001

## TABLE 1: PROCESS VARIATIONS IN MICROSTRUCTURES AND TMP

| ROLLING | EXTRUSION | FORGING |
|---|---|---|
| Very long, flat (usually coiled) (rod for wire drawing not discussed in table) | Long, straight complex sections (tubes require rewelding) | Discrete contoured components ## defined by die shape, friction, chilling## |
| pancaked grains, thinned-elongated in plane strain $\varepsilon$ | needle-like grains diameter diminish in radially rising $\varepsilon$ | fiber evolution from initial shape due to ## |
| relatively uniform $\varepsilon$ but rising gradient of reversed shear near surface affects $\varepsilon$ energy | radial gradients of rising $\varepsilon$ and $\dot{\varepsilon}$ partially offset by rising T lead to denser substructure | wide variations in T, $\dot{\varepsilon}$, $\varepsilon$ and substructure due to ## |
| roll chilling near surface, which reheats by conduction from center | intense hot zone develop near die opening. | usually die chilling, die friction partial adiabatic heating, |
| | T, $\dot{\varepsilon}$, $\varepsilon$ maximum near surface | |
| MULTIPLE STAGE INTERVALS | SINGLE STAGE ON COOLING | MULTIPLE STAGES INTERVALS |
| static recovery SRV partial SRX near surface static recrystallization SRX | Spontaneous SRX only in dense substructure large grains due to gradients | SRX more variable than in rolling |
| SUBSTRUCTURE refined low finishing T, defined by $\Sigma \varepsilon_i$ $t_i$ for SRV final pass $\dot{\varepsilon}_i$ $T_i$ | SUBSTRUCTURE coarse high exit T defined by ram V billet T, cooling rate | Refined, low finishing T, die chilling spatially non-uniform Isothermal, coarse |
| RECRYSTALLIZATION Grain size defined by $\Sigma \varepsilon_n (< \varepsilon_s)$ $\dot{\varepsilon}_n$, $T_n$ since previous SRX | controlled by annealing T and heating rate Range of $\varepsilon$ defined by R, of T, $\dot{\varepsilon}$ defined by R, $V_R$ | Similar to rolling but number of stages near 3 instead of 6 to 17 for rolling |

The general hot working behavior of Al materials which has been described in recent reviews [9-17], is briefly summarized. The universal hot strength reduction mechanism is dynamic recovery (DRV) which leads to larger subgrain size $d_s$ with less dislocations $\rho_s$ at lower Z. The Zener-Hollomon parameter can be defined and related to stress $\sigma$ and substructure [3-6, 18]:

$$Z = \dot{\varepsilon} \exp (Q_{HW}/8.31 \, T) = A (\sinh \alpha \, \sigma_s)^n \tag{1}$$

$$D_s = a + b \log Z \tag{2}$$

$$\sigma_s = c + e \, d_s^{-1} \tag{3}$$

$$T_s = f + g \, \rho_s^{-1/2} \tag{4}$$

where $Q_{HW}$ is activation energy and A, $\alpha$, n, R, a, b, c, e, f and g are constants. In continuous straining at constant T and $\dot{\varepsilon}$, the flow stress rises to reach a steady state regime $\sigma_s$ at a strain $\varepsilon_s$ above which substructure also remains equiaxed and constant in character as the grains elongate [9-17]. The substructure character is very complex being dependent on the size of the subgrains, the density of dislocations within them and on the density in the walls (the misorientation) [11-14,19-24]. Rearrangement of the substructure upon change of T or $\dot{\varepsilon}$ is discussed under forging [3,4,10-14]. In terms of TMP, the hot work substructure is controlled primarily by T and $\dot{\varepsilon}$ but that in cold working by the strain [19-24]. As T decreases, the substructure alters in a continuous gradation of rising density and misorientation through warm into cold working [13-14]. Dynamic recrystallization (DRX) occurs in a few cases, such as extremely pure Al 99.999+ (discontinuous DRX), particle stimulated nucleation (PSN-DRX) and extreme grain thinning (geometric DRX) [11-16]. In periods following deformation steps, static recovery (SRV) or

# TABLE 1 (part 2): PROCESS VARIATIONS IN MICROSTRUCTURES AND TMP

| ROLLING | EXTRUSION | FORGING |
|---|---|---|
| **TEXTURE** | | |
| Deformation, 45° ears | Deformation causes 20% | Variable or symmetrical deform |
| No SRX in schedule, SRX only | strenthening, weak at 45° | textures |
| at end, 0-90 ears | Recrystallization weakens | Variable or symmetrical SRX |
| Spatially mixed; Blended, | Spatially mixed surface SRX | Spatially mixed |
| SRX intermediate, no ears | Blended not likely | Possible blending |
| **PRECIPITATION HARDENING** | | |
| Dissolved in preheat | | |
| Gradual precipitation as T | Solutioned from homogeni- | Ordinarily similar to rolling |
| Declines | zation high breakout pressure | |
| | Precipitated if preheated from | |
| | cold lower pressure | |
| Solution treatment usually | Fine particles dissolved in hot | |
| causes SRX and loss of | zone can serve as solution; | Isothermal forging could serve |
| substructure and | Quench at die, air or water jets | as solution |
| deformation texture | Age to T5 temper | Warm after solution and natural |
| | Suitable for 6000 series | aging could enhance strength |
| **DEFECTS** | | |
| Edge cracking due to bulging | Surface cracking | Cracking, tension $\sigma$ where |
| cooling, poor low T ductility | a) limited High T ductility | bulging without die support |
| Surface defects from casting | due to solute, particles | At parting line, flash due to high |
| imperfections | b) incipient melting | $\varepsilon$, $\dot{\varepsilon}$ and T |

static recrystallization (SRX) proceed with effects on the following stages [3,4]. If a deformation stage and following interval are at the same T, the rate of SRX increases with rising T and $\dot{\varepsilon}$; it also rises with $\varepsilon$ up to $\varepsilon_S$ where it becomes constant in response to constant substructure [3,4,25-28]. For annealing at a fixed T, the time and the grain size of SRX are greater for higher T and lower $\dot{\varepsilon}$ of deformation with $\varepsilon > \varepsilon_s$ having little effect thus quite different from cold working.

Solute and fine (<0.5μm) particles hinder dislocation glide and climb reducing DRV (SRX and DRX); large rigid particles (constituent > 2μm or ceramic reinforcements > 10μm) increase local dislocation activity and enhance heterogeneity [5,28-35]. In precipitation hardening alloys, the particles coalesce and decrease in volume fraction with rising T, thus raising solute content; the size may vary from dynamically precipitated (DPN) coherent zones to overaged particles with little effect on dislocation activity [6,36-38]. The subtle variations in this behavior for different forming modes will be described to assist in process optimization.

For DC-cast feed stock, homogenization is extremely important to overcome the problems of segregation [1,2,5,6]. There are always larger than equilibrium interdendritic volumes of eutectic and large constituent particles containing impurities combined with alloying phases. During heating near the melting point for 6 to 24 hours, the alloying elements diffuse into the primary dendrites. The properties of wrought products are usually superior to cast ones because of the welding shut internal voids and the aid to homogenization by the reduction of diffusion path lengths from elongation of the grains or dendrites [5,6,39]. Surface defects in cast feedstock, usually oxidized so they cannot weld, are extended so they become easily visible; however, if found by inspection and gouged out, the irregularities can be smoothed out by the deformation processing. Large particles or agglomerations of particles are broken up, separated from each other and redistributed so that their ability to initiate cracks is reduced. In powder feedstock that is usually homogeneous, the deformation must be great enough to close all voids and rearrange the prior particle boundaries; powder consolidation must be preceded by suitable vacuum degassing to remove all absorbed H or $H_2O$ to prevent formation of voids and blisters during heat treatment [40,41]. Powder produced by atomization, rapid solidification as ribbon and mechanical alloying are canned prior to extrusion, rolling and forging in the normal manner.

## HOT EXTRUSION

Hot extrusion of large billets is commonly employed to produce long straight products with complex cross sections, having multiple cavities as needed; many other features are summarized in Table X1. Direct extrusion with sticking friction against the chamber wall and a flat die with a dead zone at the die-chamber corner, retains the original surface layer as skull and butt, and produces a product with good accuracy and surface finish [42-44]. Extrusion is the only hot working process that can do this and Al is the only metal because of its low shear stress and hard abrasive oxide. The nonuniform flow results in a zone of intense deformation which heats up by 50 to 100°C [5,45-52]. The radial gradient of increasing $\varepsilon$ and $\dot{\varepsilon}$ has been investigated by studying flow patterns in macroetched extrusion butts [5,44,47-50] or markers inserted into billets [44,53] and by simulations employing plasticine with a multicolored pattern [42]. Shear line calculations showed that such flow patterns were related to the minimization of work and provided reasonable estimates of the necessary pressure [45-48,54]. Recently finite element analysis provides calculated distributions of grid distortion, effective strain, effective strain rate and temperature (Fig. 1) [55-57]. The behaviors of different alloys are compared by assigning them extrudability factors that decrease with rising strength because it requires increased pressure or billet temperature [43,45]. The higher work performed on stronger alloys causes greater heating so that the maximum temperature near the die land is higher, coming closer to the incipient melting temperature at low melting constituents associated with high alloy content [55-59]. Tearing at liquated boundaries or cracking at large, hard particles leads to defects or poor surface finish. These various factors narrow the operating range for extrusion and put limits on the permitted extrudate complexity, section thinness and projection length. Aluminum and Al-Mn alloy (3000 series) have extrudabilities over 120. The 6000 series with a relatively low content of Mg and Si are clustered near 100. The 2000 series and the 5000 series with close to

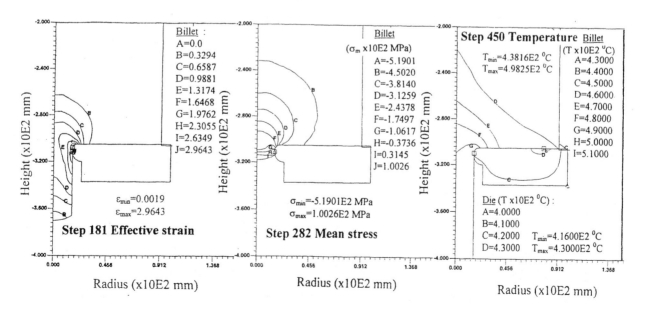

Figure 1. Distributions of strain, mean stress and temperature in extrusion calculated by finite element analysis (Deform $t_m$). While the $\varepsilon$ and $T_m$ contours develop very early, the temperature profile requires longer to reach a stable maximum, causing an initial sharp drop in the load [57].

5% Mg and 0.7% Mn are reduced to 40 or below as alloy content increases [43, 45]. The 7000 series have the highest alloy content and coarse precipitates lower ductility above 450°C to reduce extrudability to near 10 [55,56].

In a 1959 review, Chadwick [44] explained the macrostructure of Al-Cu-Mg extrusions. The outer zone consisted of extremely elongated grains from the cast columnar zone whereas the inner zone showed less elongated grains from the equiaxed region of the cast billet. On solution treatment at 500°C, the outermost part of the extrusion recrystallized to large elongated grains whereas the inner region underwent little change. Preferential etching and higher magnification showed that the inner region was a mixture of grains with either [001] or [111] directions parallel to extrusion axis whereas the outer region was almost pure [111] before SRX [44,60]. This texture confers a strength increase of 15-20% which can be retained after T5 aging and even after T6 solution as long as there has been no recrystallization; this is known as the press effect. Through TEM in the following decade [5,61,62], Sheppard and colleagues [47-52] studied the formation of subgrains in different regions from dead zone to center line at various distances from the die orifice in partially extruded billets as well as at different radii in the extrudate. These results for different alloys were compared with torsion subgrain sizes at various T, $\dot{\varepsilon}$ and $\varepsilon$ (Eqns. 2,3); such results were consistent with T and $\dot{\varepsilon}$ distributions from shear line analysis. The product strength $\sigma_y$ at 20°C or hardness could be tied to the subgrain size $d_s$ [9,49-51,63,64]:

$$\sigma_y = \sigma_o + K\, d_s^{-1} \qquad (5)$$

where $\sigma_o$ is strength without subgrains and K a coefficient. Of course, the subgrain size and dislocation density are related through Equations 3 and 4. Finite element distributions of T, $\dot{\varepsilon}$ and $\varepsilon$ in the deformation zone and in the extrudate can guide the choice of $T_B$ and $V_R$ to produce the substructure and texture with avoidance of incipient melting and cracking at the surface (Fig. 1) [55-57]. Forced cooling at the die exit may be required to prevent static recrystallization near the surface which results in a few grains growing inwards attaining large size and weakening the product [5,44]. Retention of the substructure provides 15-25% hardening (Fig. 2) partly from the [111], [110] fiber textures; strength is reduced and ductility enhanced at 45° to the axes [44]. In complex extrusions, the gradient of $\varepsilon$ and $\dot{\varepsilon}$ rises from thick regions to projections of increasing length and diminishing thickness. A substructure aimed at developing a refined grain structure during solution treatment can be induced by lowering $T_B$ or raising $V_R$, but this raises breakout pressure and is counteracted by increased heating [5,6,48,49,51].

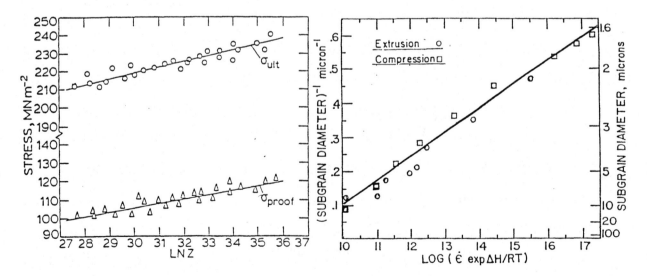

Figure 2. Increasing $V_R$ or decreasing $T_B$ to raise Z (Eqn 1) leads to improved extrudate strength and lower ductility (Q = 230 KJ/mol after Sheppard et al [48]) as a result of diminished subgrain size and dislocation spacings (Q = 156 KJ/mol after McQueen et al [61,108]).

The temperature rise in extrusion as a result of the intense straining, in a small zone near the die provides an opportunity for press heat treatment [6,37,38,44,65]. The temperature and duration of heating is sufficient to achieve solution treatment, if the particles are fairly fine as produced by rapid cooling from the homogenization treatment. Unfortunately direct cooling from homogenization to insert temperature causes problems of very high breakout pressures because of dynamic precipitation (DPN) which raises the initial flow stress tremendously followed by work softening as coalescence takes place [6,37,38,66,67]. The second step is suitable cooling at the die exit; this could be only air jets in less sensitive 6000 series alloys but water sprays may not be sufficient for very sensitive alloys of the 2000 and 7000 series [6,44,65]. This TMP produces T5 condition following aging treatment and seems fully successful in Al-Mg-Si and Al-Zn-Mg (7004) alloys and provides economy in eliminating both solution and quenching treatments. Such TMP was employed in the second world war where T5 treatment for 2000 series could guarantee a strength higher than 80% of T6 values [44]. For 7000 series containing Cu, such TMP is not suitable because the substructure adversely affects the precipitation sequence [6]. Moreover, achievement of optimized failure resistance by complex aging of 2000, 7000 and 8000 series requires solution treatment which eliminates the strengthening due to substructure and [111] fiber texture [5,44,48,49,51,65] while avoiding large grains. Deformation after solution, before or between aging, could be applied by stretching, drawing or bending to optimize strength and toughness [8,44,51,69].

## HOT ROLLING

Rolling of Al alloys creates very long products but, unlike extrusion, usually in the form of coils to facilitate handling in further processing. The preponderance is sheet and plate produced from DC-cast slabs. Since the cast surface is carried through the rolling, the billets are often scarfed beforehand to remove oxide, defects and segregation and afterwards the product is pickled in readiness for cold forming [1,2]. Simple shapes can be produced most commonly as rod in long coils for wire drawing usually from wheel casting in which the much faster solidification rates reduce the scale of grains, particles and segregation [8,31]. Rolling schedules consist of many passes on cold rolls with intervening intervals leading to a considerable decline in temperature (Fig. 3); the preheat is usually above 500°C so that the finishing passes near 300°C have suitable ductility and flow stress [1-4,70-74]. Because of the wide rolls and narrow bite (arc of contact) the deformation is almost pure plane strain; nevertheless there is some bulging at the edges (possibly corrected by vertical edger rolls) that leads to longitudinal surface stresses and possibly cracking, partly accerbated by more rapid cooling there.

Because friction pulls the workpiece into the roll gap, the surfaces tend to lead relative to the center of any section; this effect while being the opposite of extrusion is also much less [3,4,70-74]. The entry surface shear which diminishes gradually with depth is also partly reversed as the plate passes the neutral axes and moves faster than the rolls. The net shape change of a vertical slice is ultimately fairly uniform elongation in the rolling direction. The gradient of net shear is sufficient to alter the slip behavior so that Taylor theory predicts a different texture from the center of the plate [70-74]. While the shear gradient contributes to a strain energy gradient there is a much stronger variation due to straining at much lower T from chilling by the rolls during a pass; during interpass intervals the surface reheats due to conduction from the center which is heated by the deformation [72]. The combination of roll chilling and interpass equalization leads to augmented static recovery (SRV) near the surface which helps to uniformize the substructure across the section [71-73]. In more severe T changes or in alloys with lower degree of DRV or SRV, the outer layers of the plate undergo SRX at one or several interstage holds leading to layers of different substructive and texture [3,4,70-74].

69

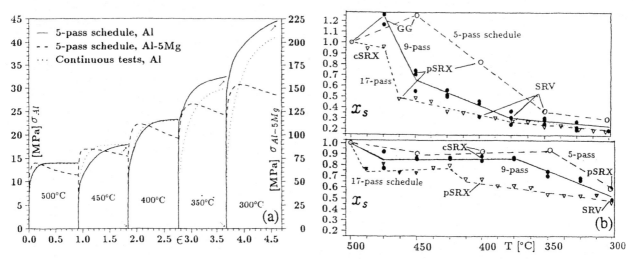

Figure 3. Torsional rolling simulation of Al and Al-5Mg for 3 schedules with same total strain 3.4, namely A) 17 pass, 0.23, 93s B) 9 pass, 0.46, 187s C) 5 pass, 0.92, 375s: a) stress strain curves plotted against cumulative strain case C and b) interval softening fraction $X_s$ for cases A, B and C with indication of complete static recrystallzation (SRX), partial (pSRX) or none, only static recovery (SRV) (after Poschmann and McQueen [25-27]).

The general physical metallurgy of rolling has many differences from extrusion although there are fundamental similarities. The initial temperature is much higher (to ensure finishing in the hot range) which initially develops coarser substructure. The gradual fall in T across the schedule results in substructure refinement through insertion of new walls while the old ones undergo rearrangement [3,4,70-72]. The intervals with SRV lead to loss of dislocations and some subgrain enlargement [24-26,71]. The substructures, textures and flowstresses have been modeled by Nes et al. [23,24]. In commercial production the individual pass strains are usually insufficient to attain steady state in recrystallized starting material but they may approach it as a result of the carried-over substructure. Recrystallization after the early stages would be beneficial for grain refinement (as industrially attained in rolling of steel [75,76]) but may not come about because of the small reductions to avoid cracking of as-cast material. The behavior may differ considerably for alloys in which solute or large particles enhance SRX (Fig. 3).

In production of a selected substructure, the important features are the finishing temperature, the rate of rolling, the retained accumulated strain since the last static recrystallization and the static recovery between passes. This behavior has been confirmed and clarified by simulation [25,26,32,77-82]. The ability to produce such material is quite feasible in pure Al where recovery is very high and in Al-Mn, Al-Fe and age hardenable alloys where small dispersoid and equilibrium precipitate particles inhibit nucleation [3,4,17,28,31,32,71-73,83-88]. For production of SRX in the latter alloys, prolonged holding at any rolling stage temperature is necessary or even heating up by 50 to 100°C to reduce the time below an hour [28,88]. Plate or strip produced without recrystallization develop a strong texture which is common to other pure fcc metals like Cu and Ni [87,89]. In many alloys, warm or cold rolling could be pursued without recrystallization; the substructures became more dense and the texture intensified. Annealing of sheet to recrystallize would produce refined grains and an intense cube texture. In rod of Al-0.65 Fe rolling in 13 passes from 8 cm billets to $\varepsilon$ = 4.3 produces subgrains of 1.0 μm; without annealing the rod is drawn through 12 dies to $\varepsilon$ = 2.6 resulting in cells of 0.6 μm [31]. Recovery annealing at 290°C for 3hrs produces subgrains of 1.4 μm which have superior stability, suitable ductility and conductivity for electrical wire.

Intermediate recrystallization, to cause grain refinement can be produced after the early stages. This can be enhanced by raising the pass strains and intervals or by accumulating strain in several passes in association with a suitably lengthy interval (Fig. 3) [25-28,70,71,90,91]. It is

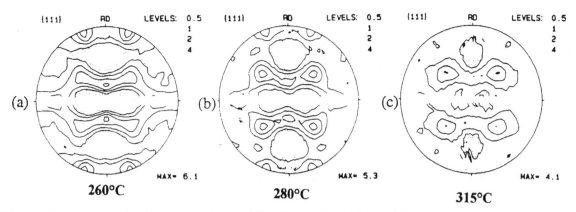

Figure 4. Textures in 3 pass tandem mill rolling of Al-1Mg-1Mn to total strain 60% with different finishing T; on relative earing (1.02-1.06) with additional cold rolling: (After J. Hirsch [73]).

| a) 260° unrecrystallized | b) 280° (strongly SRX) | c) 315°C complete SRX |
|---|---|---|
| 45° ears, 1.03 | 0-90° ears, 1.03 | 0-90° ears, 1.04 |
| cold rolled 92% | 75%, 0-45°, 1.02 | 50%, 0-90°, 1.02 |
| 45° ears, 1.06 | 92%, 45°, 1.04 | 92%, 45°, 1.04 |

possible in pure Al, lean age-hardenable alloys, Al-Mg and Al-Mg-Mn alloys. In the Mg solute alloys, the much more dense substructure from reduced DRV accelerates SRX (Fig. 3) and large $Al_6Mn$ particles stimulate nucleation [3,4,25-27,2,80,81,91-93]. Raising the preheat temperature can enhance SRX, but the facility for recrystallization with grain refinement after every pass never becomes similar to the behavior of carbon steel austenite [75,76]. The textures produced are quite variable depending on the strains at which the SRX repeats (Fig. 4) and whether it initiates by cube band nucleation, particle stimulated nucleation (PSN) (randomization due to intense cell formation around particles) or by stress induced boundary migration (SIBM) (retention of the rolling texture) [72-74,94]. Recrystallization is likely in the outer layers of the slab because of the more intense substructure due to the reversed shear strain and lower temperature from roll chilling; this occurs when the surface temperature rises due to conduction from the central layers. Recrystallization is usually wanted in the coil to prepare the material for cold rolling. A special texture may be wanted so that the cold rolling and possibly annealing produce sheet with planar anisotropy that enhances drawability (Fig. 4) [73]. In some cases a retained substructure increases draw deepth since it prevents wrinkling [95]. Superplastic sheet is produced in 7000 and 5000 alloys by warm rolling to develop a very dense substructure followed by a rapid high T anneal to induce SRX to very fine grains [96-98]. In contrast in materials such as supral, Al-10Mg-1Zr and 8000 series with a high density of fine particles that deter discontinuous SRX, a dense well-pinned substructure is produced by warm rolling and inter-pass recovery and is sent directly to superplastic forming where the new grains from by continuous dynamic recrystallization in the initial straining at low $\dot{\varepsilon}$ [36,84-86,99].

## FORGING

In hot forging, the selection of feedstock temperature for specific alloys is very important to restrain the forces within the press capacity and provide sufficient ductility through several stages with declining T. The forgeability decreases as strength increases and is similar to the extrudability [100]. The present discussion is directed toward the fiber structure which stems from the drawing out of minor phases whether impurity or intentional additions, as the grains elongate in response to die shape, friction and chilling (regions of die lock). In forging, the strain path is controlled to make the fibers follow the outer shape so that they are normal to potential cracks from stress concentrations arising from the shape and applied stress [100-101]. The fiber structure is strongly influenced by the preform shape and fiber pattern but is relatively independent of T and $\dot{\varepsilon}$; recrystallization can even take place without altering the fiber structure. The fiber continuity and thickness depends on second phase volume fraction; an excess of brittle or low melting phases may be seriously deleterious both in processing itself and in the product. Fiber structure anisotropy in extrusions and rolled products cannot be manipulated to any degree except by selection of the feedstock volume fraction and distribution. Although the flow patterns in some axi-symmetric and plane strain open die forging had been estimated, the analysis of multistage, three-dimensional, closed die forming required finite element analysis and great computing power [102-106]. Complexity lies not only in the design shape of the object but in the volume-proportioned initial shape and the need to allow excess material to escape at the parting line into the flash gutter. Fortunately methods of simplification have been found and distributions of T, $\dot{\varepsilon}$ and $\varepsilon$ calculated so that estimates can be made of the microstructural development.

In forging, the substructure density is much more variable due to non homogeneity in temperature, strain rate and strain; the number of intervals and the regularity of their duration is much less than in rolling. In symmetrical forgings, types of regions may develop consistent strain hardening and deformation textures. With suitable annealing, there can be distributions of recrystallized grain sizes and textures. With greater rates of change of T, $\dot{\varepsilon}$ and $\varepsilon$ than in rolling or in extrusion, the substructure development may not be able to follow them. Generally, a change in strain rate requires a substructure and $\sigma$-$\varepsilon$ transient proportional to the initial one. The establishment of the new substructure characteristics is facilitated by the continual rearrangement of the dislocation walls and internal network which maintains the subgrains equiaxed; the strain for the complete rearrangement in the steady state regime is estimated to equal that to reach steady state initially [12-15]. Moreover, the second phase that constitutes the fiber structure does affect the grain and subgrain evolution since particles are able either at small sizes to pin them or at large to generate both.

The forging process can be adapted to precipitation hardening in a number of ways. The ordinary hot forging schedule with declining T as in rolling leads to equilibrium precipitates with a range of sizes. Single blow hammer forging (aided by computer modeling) causes very little drop in T because of short dwell time in the cold dies and close to adiabatic heating from the high rate ($100s^{-1}$); this could be adapted as solution treatment followed by quenching [37,38]. Isothermal forging, which provides a substructure coarser and more uniform then the traditional process could be utilized as the solution treatment. However, this process is seldom applied to Al alloys but has proven excellent for Ti and Ni base superplastic alloys. Warm forging could be applied after solution, quenching and natural or partial artificial aging [107]. A rapid preheat and rapid forging would reduce particle coalescence and the substructure would cause considerable strengthening. Even with no net increase in strength there could be an improvement in resistance to fatigue and stress corrosion. Solution treatment would likely lead to SRX so forging conditions should aim at grain refinement.

## CONCLUSION

Extrusion is ideal for TMP since the product is produced by a single stage with a selected substructure, a strengthening texture and possibly solution treated, ready for aging (T5 for 6000 series). Need for better quenching and more complex aging for 2000 and 7000 series usually means sacrificing simple substructure and texture strengthening for optimized failure resistance. Rolling is noted for multistage processing so recrystallization during intervals can be employed to refine grain size and scheduled at the point where the final straining sequence produces the desired degree of substructure and texture. However it is difficult to finish the sheet in the solutioned condition. Forging is the most inhomogeneous but is also the most flexible so stages of heating or straining can be scheduled to attain final shape with desired TMP.

# REFERENCES

1. G.M. Raynaud, B. Grange and C. Sigli Al Alloys, Physical Mechanical Properties, ICAA3, L. Arnberg et al. eds. NTH, SINTEFF, Trondheim, (1992), pp 169-214.
2. J. Marshall (ICAA 5), Mat.Sci.Forum, 217-222 (1996), 19-30.
3. H.J. McQueen, JOM, (TMS) 50, [6], (1998), 28-33.
4. H.J. McQueen, Hot Deformation of Al Alloys, T.R. Bieler et al. eds., TMS-AIME, Warrendale PA, 1998, pp. 383-396.
5. H.J. McQueen and O.C. Celliers, Can. Metal. Quart, 35 (1996), 305-319.
6. H.J. McQueen and O.C. Celliers, Can. Metal. Quart, 36 (1997), 73-86.
7. H.J. McQueen, Thermomechanical Processing of Al Alloys, J.G. Morris, ed. Met. Soc. AIME, (1979), pp. 1-24; J. Met., 32 [2], (1980), 17-36.
8. H.J. McQueen and J.J. Jonas, Aluminum Alloys '90, ICAA2, C.Q. Chen, ed.Beijing, (1990), pp. 727-742.
9. H.J. McQueen, Hot Deformaton of Aluminum Alloys, T.G. Langdon and H.D. Merchant, eds. TMS-AIME, Warrendale, PA (1991) pp. 31-54.
10. W. Blum and H.J. McQueen, Aluminum Alloys, Physical and Mechanical Properties, (ICAA5) J.H. Driver et al., eds. Mat.Sci. Forum, 217-222 (1996) 31-42.
11. H.J. McQueen, "The Hot Worked State", (Section 9 "Current Issues in Recrystallization: a Review" R.D. Doherty, D.A. Hughes, F.J. Humphreys, J.J. Jonas, D. Juul-Jansen, M.E. Kassner (editor) W.E. King, T.R. McNelley, H.J. McQueen and A.D. Rollett).Mat. Sci. Eng., 238 (1998) 219-274
12. H.J. McQueen and W. Blum, Recrystallization and Related Topics (Proc. 3rd Intnl. Conf., ReX '96). T.R. McNelley,ed., Monterey Inst. Advanced Studies,CA,(1997),pp.123-136.
13. H.J. McQueen and W. Blum, Alloys, Physical Mechanical Properties ICAA6, T.Sato, ed., Japan Inst. Metals (1998) pp. 99-112
14. H.J. McQueen and W. Blum, Mat. Sci. Eng. A290 (2000) 95-107.
15. H.J. McQueen, Light Metals 2000, J. Kazadi et al. eds., Met Soc. CIM, Montreal (2000), in press.
16. H.J. McQueen, E. Evangelista and M.E. Kassner, Z. Metalkde, 82 (1991), 336-345.
17. H.J. McQueen, E. Evangelista, N. Jin, and M.E. Kassner, Metal. Trans., 26A (1995) 1757-1766.
18. H.J. McQueen and N.D. Ryan, Rate Processes in Plastic Deformation II, S.V. Raj et al. eds., TMS-AIME, Warrendale PA (2000) in press.
19. D.A. Hughes and N. Hansen, Advances in Hot Deformation Textures and Microstructures, J.J. Jonas et al. eds. TMS-AIME, Warrendale, PA (1995) pp 427-444.
20. N.Hansen and D. Juul-Jensen, Hot Working of Al Alloys, T.G. Langdon et al. eds. TMS AIME Warrendale PA, (1991) pp. 3-20.
21. D.A. Hughes and Y.L. Liu, ibid, pp. 21-30.
22. M. Richert and H.J. McQueen, Hot Workability of Steels and Light Alloys-Composites, H.J.McQueen,E.V.Konopleva,N.D.Ryan,eds.Met.Soc.CIM, Montreal (1996) pp. 15-26.
23. H.E. Vatne, R.K. Bolingbroke and E. Nes, Al Alloys,Physical Mechanical Properties ICAA4,T.H.Sanders and E.A.Starke. eds.,Georgia Inst. Tech Atlanta (1994), pp.251-258
24. E. Nes, H.E. Vatne, O. Daaland, T. Furu, R. Orsund, K. Marthinsen, Aluminum Alloys Physical and Mechanical Properties (ICAA4), T. Sanders and E. A. Starke, eds. (Georgia Inst. Tech., Atlanta 1994) Vol.2, pp. 250-257.
25. I. Poschmann and H.J. McQueen, Z. Metallkde.,87 (1996) 349-356.
26. I. Poschmann and H.J. McQueen, Z. Metallkde.,88 (1997) 14-22.
27. H.J. McQueen and I. Poschman, Thermec '97, T. Chandra and T. Sakai, eds.,TMS-AIME, Warrendale, PA, (1997), pp. 951-957.
28. B. Crawford, J. Belling, H.J. McQueen and A.S. Malin, Recrystallization '90 (Intnl. Conf. Recrystallization in Metallic Materials, T. Chandra, ed., TMS-AIME, Warrendale, PA,, 1990, pp. 655-660.
29. W. Blum, Q. Zhu, R. Merkel and H.J. McQueen, Z. Metalkde.,87 (1996), 341-348.
30. I. Poschmann and H.J. McQueen, Scripta Metal. Mat.,35 (1996), 1123-1128.
31. H.J. McQueen, H. Chia and E.A. Starke, Microstructural Control in Al Alloy Processing, H. Chia and H.J. McQueen, eds., TMS-AIME, Warrendale, PA, (1985), pp. 1-18.
32. H.J.McQueen, K.Conrod and G.Avramovic-Cingara,Can. Metal.Q.,32 (1993), 375-386
33. X. Xia, P. Sakaris and H.J. McQueen, Mat. Sci. Tech., 10 (1994), 487-494.

34. H.J. McQueen, E.V. Konopleva, M. Myshlyaev and Q. Qin, Proc. Tenth International Conf. on Composite Materials, ICCM-10, (Whistler, B.C.) A. Poursartip and D. Street, eds, Woodhead Pub., Abington, UK, (1995) vol 12, pp 423-430

35. M.M. Myshlyaev, E. Konopleva and H.J. McQueen, Mat. Sci. Tech., 14, (1998), 939-948.

36. G. Avramovic-Cingara, D.D. Perovic and H.J. McQueen, Aluminium, 76 (2000), 313-316, 402-406.

37. H.J. McQueen, Hot Deformation of Aluminum Alloys, T.G. Langdon and H.D. Merchant, eds., TMS-AIME, Warrendale, PA, USA, (1991), pp. 105-120.

38. E. Evangelista, E. Cerri, and H.J. McQueen, Modelling of Plastic Deformation and its Engineering Applications, (edited by S.I. Anderson, et al.), Riso National Lab., Roskilde, DK. (1992) pp 355-360

39. H.J. McQueen, Defects, Fracture and Fatigue, G. Sih and W. Provan, eds., Martinus Nihoff Pub., The Hague, NE, (1983), pp. 439-471

40. W.M. Griffith, Y-W, Kim and F.H. Froes, Rapidly Solidified Powder Al Alloys, M.E. Fine and E.A. Starke, eds., ASTM Philadelphia PA (1986) pp.283-303

41. S.W. Ping, ibid, pp 369-380

42. C.E. Pearson and R.N. Parkins, The Extrusion of Metals, 2nd edition, J. Wiley Inc., New York, NY, 1960.

43. K. Laue and H. Stenger, Extrusion Processes, Machinery, Tooling, Am. Soc. for Metals, Metals Park OH, 1981, pp. 1-62, pp. 124-152

44. R. Chadwick, Metal. Rev., 4 (1959), 189-255.

45. H.J. McQueen and O. Celliers, Materials Forum, 17, (1993), 1-3.

46. A.F. Castle and T. Sheppard, Met. Tech., 2, (1976), 454-464.

47. T. Sheppard, M.G. Tutcher, and H.M. Flower, Met. Sci., 13, (1979), 473-481.

48. T. Sheppard: Met. Tech., 8, (1981), 130-141.

49. T. Sheppard, Proc. 8th Light Metal Congress, J.Jeglitsch et al., eds., The University, Leoben, Austria (1987), pp. 301-311.

50. T. Sheppard and M.G. Tutcher, Met. Sci., 14, (1980), 579-589.

51. T. Sheppard, S.J. Paterson and M.G. Tutcher, Microstructural Control in Al Alloys, E.H. Chia and H.J. McQueen, eds., Met. Soc. AIME, Warrendale, PA, (1986), pp. 123-154.

52. T. Sheppard, M.A. Zaidi, M.G. Tutcher and N.C. Parson, Microstructural Control in Al Alloy Processing, H. Chia and H.J. McQueen, eds., TMS-AIME, Warrendale, PA, (1985), pp. 155-178.

53. H. Valberg and T. Malvik, Intl. J. Prod. Tech., 9, (1994), 428-463.

54. R. Akeret, J. Inst. Met., 95 (1967), 204-211.

55. H.J. McQueen and E.V. Konopleva, Al Alloys, Physical and Mechanical Properties (ICAA7), E.A. Starke and T. Sanders eds, TransTech Pub., Zurich (2000), pp. 1187-1192.

56. H.J. McQueen and E.V. Konopleva, Mathematical Modeling of Metal Processing and Manufacturing, Y. Verreman et al., Met Soc. CIM, Montreal (2000) (published electronically)

57. M. Sauerborn and H.J.McQueen Mat Sci Tech, 14 , (1998), 29-38.

58. O. Reiso, Proc. 4th Intnl Al Extrusion Tech. Sem., Aluminum Association, Washington, , (1988) vol. 2, 287-295.

59. T. Sheppard and M.P. Chode, Proc. 4th Intl. Al Extrusion Tech. Sem, Aluminum Association, Washington (1988) vol. 2, pp. 329-341

60. K.V. Gow and R.W. Cahn, Acta Met., 1, (1953), 238-241.

61. H.J. McQueen, W.A. Wong and J.J. Jonas, Can. J. Phys., 45, (1967), 1225-1239.

62. A. Plumtree and G. Deep, Met. Tech., 4, (1977), 1-5.

63. D.J. Abson and J.J. Jonas, Met. Sci., 4, (1970), 24-28.

64. H.J. McQueen, and W.B. Hutchinson, Deformation of Polycrystals, N. Hansen et al., eds., Riso Natl. Lab., Roskilde, Denmark, (1981), pp. 335-342.

65. T. Sheppard, Mat.Sci. Tech., 4, (1988), 635-643

66. J. Langerweger, Proc. 4th Intl. Al Extrusion Tech. Sem., Aluminum. Association, Washington (1988) vol 2, pp. 381-384

67. E. Evangelista, A. Forcellese, F. Gabrielli and P. Mengucci, Hot Deformation of Al Alloys, T.G. Langdon et al eds, TMS-AIME, Warrendale, PA, (1991) pp.121-129

68. R. Akeret, Z. Metallkd., 61, (1970), 3-10

69. M. Conserva, M. Buratti, E. Di Russo and F. Gatto, Mat. Sci. Eng., 11 (1973), 103-112.

70. N. Ragunathan, H.B.McShane C.P. Lee. and T.S. Sheppard, Hot Deformation of Al Alloys, T.G. Langdon et al, eds., TMS-AIME, Warendale, PA (1991) pp. 389-416.

71. M.A. Zaidi, T. Sheppard, Metal Sci, 16, (1982), 229-238.
72. J. Hirsch, K.Karhausen and R. Kopp, Al Alloys, Physical Mechanical Properties (ICAA4), T Sanders and E.A. Starke, eds., Georgia Inst. Tech., Atlanta, (1994) pp. 476-483.
73. J. Hirsch, Thermec'97, T. Chandra and T. Sakai, eds., TMS-AIME, Warrendale PA (1998), pp. 1083-1094.
74. J. Hirsch and O. Engler, Microstructural & Crystallographic Aspects of Recrystallization, N. Hansen et al., eds., Riso Natl.Lab., Roskilde DK (1995), pp. 49-62.
75. C.M. Sellars (Proc ReX'96) Recrystallization and Related Phenomena, Monterey Inst. Advanced Study (1997) pp. 81-94; (with Q.Zhu) pp 195-202.
76. H.J. McQueen, Can.Metal.Quart,. 21 (1982), 445-460.
77. H.J. McQueen, G. Avramovic-Cingara and P. Sakaris and A. Cingara, Proc. 3rd Intnl. SAMPE Metals Conf., (Toronto), (1992), pp. M192-M206.
78. G. Avramovic-Cingara, K. Conrod, A. Cingara and H.J. McQueen, Intl. Symp. on Light Metals Processing and Applications, Met. Soc. CIM, Montreal, (1993), pp. 495-510.
79. R.W. Evans and G.R. Dunstan, J.Inst. Metals, 99 (1971), 4-14.
80. M.M. Farag, C.M Sellars and W.J. McG. Tegart, Deformation Under Hot Working Conditions Iron Steel Inst, London (1968), pp. 60-67
81. T.B. Vaughan, ibid pp. 68-77
82. J.R. Cotner and W.J. McG. Tegart, J. Inst. Metals, 97 (1969), 73-79
83. J.P. Immarigeon and H.J. McQueen, Can. Metal. Quart, 8 (1969) 25-34.
84. S.J. Hales, T.R. McNelley and H.J. McQueen, Metal. Trans., 22A (1991) 1037-1047.
85. D.B. Brooks, H. Gudmundsen and J.A. Wert, (see reference 9) pp. 55-58.
86. B.M. Watts, M.J. Stowell, B.L. Baikie and D.G.E. Owen, Met.Sci. 10, (1976), 189- 197,198-205.
87. H.J. McQueen and H. Mecking, Z. Metallkde, 78 (1987), 387-395.
88. J. Schey, Acta Tech. Acad. Sci.Hung., 16 (1957), 131-152.
89. S.R.Goodman and H.Hu,Trans.Met.Soc.AIME, 230 (1964) 1413-19,233 (1965) 103-110.
90. C.M. Sellars, A.M. Irisarri, and E.S. Puchi, Microstructural Control of Al Alloys, E.H. Chia and H.J. McQueen, eds., Met. Soc. AIME, Warrendale, PA, (1987), pp. 179-196.
91. H.J. McQueen and N. Ryum, Scand. J. Met.14 (1985) 183-194.
92. F.J. Humphreys and P. Kalu, Acta Metal, 35, (1987), 2815-2829.
93. G.J. Baxter, T. Furu, J.A. Whiteman and C.M. Sellars (ICAA5), Mat.Sci.Forum, 217-222 (1996), 459-464.
94. J.R. Hirsch, Hot Deformation of Al Alloys, T.G. Langdon and H.D. Merchant, eds., TMS-AIME Warrendale PA (1991),p. 379-389.
95. R.Grimes, personal communication
96. J.A. Wert, N.E. Paton, C.H. Hamilton, M.W. Mahoney, Met. Trans., 12A (1981), 1267-1276.
97. R. Crooks, S.J. Hales and T.R. McNelley, Superplasticity and Superplastic Forming, C.H. Hamilton and N.E. Paton, eds. TMS-AIME, Warrendale, PA (1988) pp. 389-394.
98. H.J. McQueen and M.E. Kassner, Superplasticity in Aerospace II, T.R. McNelley and H.C. Heik-kenen, eds., Met.Soc. AIME, Warrendale, PA, (1988), pp. 189-206.
99. G. Avramovic-Cingara, K.T. Aust, D.D. Perovic, G. Palumbo and H.J. McQueen, Can. Met. Q, 34, pp. 265-273, Best Materials Science Paper CIM, 1995
100. Source Book on Selection and Fabrication of Al Alloys, ASM Intnl, Materials Park, OH, (1978), pp 84-143.
101. H.J. McQueen and E. Evangelistu, Materials in the Automotive Industry, E. Essadiqi ed., Met.Soc CIM, Montreal (2001) (in press)
102. C.J. Van Tyne, R.B. Focht, T.D. Nelson and W. Reese, JOM, 46, [9] (1994), 24-26.
103. R. Duggirala, JOM, 42, [2], 1990, 24-27.
104. H.E. Delgado, R.I. Ramakrishnan and T.E. Howson, JOM, 46, [9], 1994, 21-23.
105. D. Furrer, Advanded Mat. Proc., 155, [3], (1999), 33-36.
106. S.C. Jain and B.P. Bardes, JOM, 46, [5], (1994), 49-53.
107. A. Forcellese and F. Gabrielli, J. Machine Tools Man., 40 (2000), 1287-1297.
108. H.J. McQueen and J.E. Hockett, Met. Trans., 1, (1970), 2997-3004.

# LIGHTWEIGHT ALLOYS FOR AEROSPACE APPLICATION

*Edited by:*
*Dr. Kumar Jata, Dr. Eui Whee Lee,*
*Dr. William Frazier and Dr. Nack J. Kim*

## ALUMINUM ALLOYS

## Microstructural Characterization of Friction Stir Welds

*Mary C. Juhas, Peter C. Collins,*
*G.B. Viswanathan and Hamish L. Fraser*

Pgs. 77-85

184 Thorn Hill Road
Warrendale, PA 15086-7514
(724) 776-9000

# MICROSTRUCTURAL CHARACTERIZATION OF FRICTION STIR WELDS

Mary C. Juhas, Peter C. Collins, G.B. Viswanathan and Hamish L. Fraser

The Ohio State University
Department of Materials Science & Engineering
2041 College Road
Columbus, OH 43210

## Abstract

Microstructural characterization of friction stir welds (FSW) using advanced techniques such as transmission electron microscopy, TEM, is limited due to the wide range of microstructures produced across a short distance. Thus, the preparation of thin foils using conventional methods typically confines the observable microstructure to the area immediately surrounding the perforated hole, somewhere near the center of a 3mm diameter disk. This paper introduces the focused ion beam (FIB), a novel instrument for obtaining high quality thin foils from predetermined locations in bulk materials. Previously used in the semiconductor industry for the building and repair of integrated circuits, the FIB has been identified as an ideal tool for precisely locating and extracting TEM foils from structural materials and other engineering alloys. A longitudinal cross-section of an interrupted FSW, i.e., where the tool is retracted before it reaches the end of the workpiece, provided a means of capturing the microstructural evolution at various stages of processing. A microhardness contour map was generated for the area surrounding the site where the pin was retracted. The results of the microhardness map were used to determine the exact location of thin foils to be extracted using the FIB for TEM characterization. Samples from such locations were previously either unobtainable or only obtainable with tedious, lengthy techniques using conventional methods. The observed microstructures were compared with microtexture analysis obtaining using orientation imaging microscopy, OIM.

Lightweight Alloys for Aerospace Applications
Edited by Kumar Jata, Eui Whee Lee,
William Frazier and Nack J. Kim
TMS (The Minerals, Metals & Materials Society), 2001

## Introduction

Friction stir welding (FSW) is a recently developed solid state joining method that has received significant attention in a variety of industry sectors, notably aerospace, shipbuilding and automotive. The attractive features such as very low residual stress and distortion, and high reproducibility with exceptionally low defect rates have allowed this technology to emerge into a successful and often preferred alternative to fusion welding of aluminum alloy plate and sheet.

There has been quite a bit of research and development activity during the past decade to establish a basic understanding of the evolving microstructure as related to mechanical properties and hence the performance of friction stir welded components (1-7). The microstructural characterization of friction stir welds is typically carried out using optical metallography and scanning and transmission electron microscopy, SEM and TEM, respectively. While these methods are quite useful, each has limitations and thus the combination of all three can offer the complementary information necessary to provide the most complete picture possible of the deformation mechanisms that lead to the microstructures observed. For example while optical metallography and SEM are ideal techniques for revealing the various microstructural zones in a friction stir weld, a more detailed examination using TEM may inadvertently exclude important information due to the limitations associated with sample preparation. A wide range of microstructures occurs over a very short distance in friction stir welds making the task of TEM characterization using conventional sample preparation methods quite tedious if not impossible. In the current study, a new technique for obtaining TEM foils was used to better reveal the microstructural characteristics of FSW in the aluminum alloy 6061. In addition, Orientation Imaging Microscopy (OIM) was performed to assess microtexture as a complement to the TEM observations.

## Experimental Procedure

A 12 cm-long AA 6061 friction stir weld, produced at Edison Welding Institute, was provided in the as-welded condition. An interrupted weld was made such that the pin tool was retracted at the end of the plate with the hole left intact for metallurgical examination. Visual inspection indicated that the weld had no surface-connected defects. The weld was sectioned longitudinally along its centerline to reveal the range of microstructures produced both ahead of the pin tool and behind it, beginning at the initial plunge through to the point of pin tool retraction. The longitudinal section was prepared for metallographic examination using standard practices. All of the analyses presented in this paper were confined to the plane along the weld centerline.

A two-dimensional Vickers microhardness contour map was generated at the region around the hole where the pin tool was retracted. A 500 g load was used for 10 seconds at intervals of 500μm. In regions where discrepancies appeared to have occurred or dramatic hardness changes were observed, the interval was reduced to 250μm. In order to generate a complete hardness image map, an averaging system was used to fill in and predict hardness readings every 100μm. Spot testing of the predicted hardness was performed, and the predicted values matched those obtained experimentally.

The mating surface of the aforementioned sample along the weld centerline was prepared using standard metallographic techniques to produce a highly polished surface. A focused ion beam (FIB) was then used to very precisely extract TEM foils from predetermined locations in the sample based on the hardness image map. The ion beam is contained in a dual beam workstation that also contains an electron beam used for imaging the microstructure as in a conventional SEM. Once the area of interest is located in the SEM mode, the ion beam mills out a trench on either side of a thin foil as shown in Figure 1. A foil approximately 10μm x 8μm with 100% thin area is produced, thus removing the uncertainty of locating a specific area of the microstructure associated with conven-

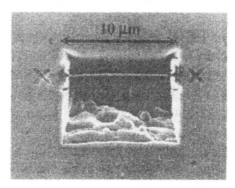

Figure 1: TEM foil shown on-edge, between "Xs", prior to complete extraction from the bulk using a FIB.

tional TEM foil preparation methods. To illustrate the value of the FIB method of foil preparation, consider how microstructural observations confined to the center of a 3 mm disk (conventional method of TEM foil preparation) would be severely limited in a 6 mm-thick plate such as in this study.

Orientation Imaging Microscopy was used to determine the microtexture as a function of location in the weld, thermomechanically affected zone (TMAZ), and base material (BM).

## Results

Figure 2 is a macrograph of the longitudinal cross-section of the FSW. The Roman numerals correspond to the areas of interest where TEM foils were extracted using the FIB. The area labeled "A" at the finish of the weld reveals a slight depression and incline on the surface which corresponds to the angle (~3°) between the axis of the pin tool and the axis perpendicular to the welding direction. The flash created behind the tool is visible at the start of the weld. This is due partially to the tool being plunged at a slight angle and also due to the fact that material is being forged locally outward and upward from the rotating pin. The region at the start of the weld is characterized macroscopically by a light, irregularly-shaped area where the fully developed stir zone microstructure (dynamically recrystallized), obtained during "steady state welding", had not yet been reached. Figure 3 is an optical micrograph of the same region showing the clear delineation between the TMAZ and the stir zone, which contains the signature onion skin pattern often observed in aluminum alloy FSW. This pattern, further revealed in Figure 4, is generated by a cyclic distribution of second phase particles including $Mg_2Si$ and $SiO_2$.

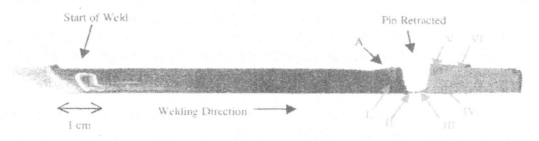

Figure 2: Longitudinal cross-section of an interrupted FSW in 6061 Al

The microhardness contour map, shown in Figure 5, indicates local softening at the surface of the plate below the shoulder in the region where the pin already passed, labeled "X". Similarly, softening occurs ahead of the tool in the region labeled "Y". This region corresponds to the

500 μm

500 μm

Figure 3: Optical micrograph of boundary between stir zone (right) and TMAZ (left).

Figure 4: Optical micrograph of stir zone showing "onion skin" pattern

1mm

Figure 5; Microhardness contour map of longitudinal cross-section showing location where pin was retracted

TMAZ/HAZ region just ahead of the rotating tool as it was retracted. It should be mentioned that the hardness of the base material is ~60 VHN, which corresponds to the darkest regions near the "X" and "Y" and also the three dark regions near the base of the plate. The hardness of the weld, behind the tool, is greater than that of the base material, as is the region immediately ahead of the tool and below the shoulder.

A TEM micrograph of the base material microstructure is shown in Figure 6(a). The second phase particles lying along grain boundaries and in the grain interiors are mainly $SiO_2$, while the $Mg_2Si$ precipitates are much larger as shown in Figure 6(b). A TEM foil, extracted from Region I (see Figure 2), is shown in Figure 7. It can be seen that the $SiO_2$ particles are pinned by dislocations and the overall dislocation density is low relative to a cold worked microstructure. Similarly, the microstructure in Region V, below the shoulder and ahead of the weld, Figure 8, also shows dislocation/particle interactions and a low dislocation density. In contrast, Region VI, in the TMAZ ahead of the tool, Figure 9, reveals dislocation forests.

The OIM results for the base material are shown in Figures 10(a) -(d). The pole figures for the base material, Figure 10(a), reveal what appears to be significant microtexture however it should be noted that these results are based on very limited data. The pole figures for the TMAZ/base material boundary area ahead of the tool (to the right of Region IV in Figure 2) are shown in Figure 10(b). There is some microtexture, probably derived from the base material which has a micro-

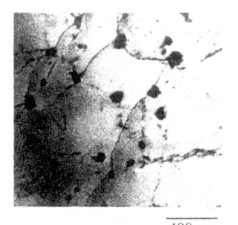

Figure 6(a): Base material microstructure showing SiO$_2$ impurities.

Figure 6(b): Base material microstructure showing a large Mg$_2$Si precipitate.

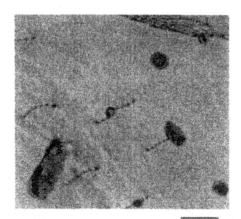

200 nm

Figure 7: TEM micrograph of Region I, see Figure 2.

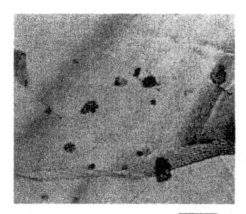

200 nm

Figure 8: TEM micrograph of Region V, see Figure 2.

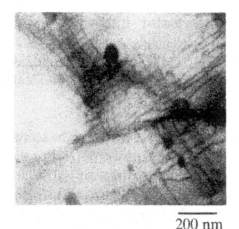

200 nm

Figure 9: TEM micrograph of RegionVI, see Figure 2.

structure that corresponds more closely to the OIM image map, Figure 10(c), than does the TMAZ microstructure. The pole figures for the area behind the pin, between Regions I and II, are shown in Figure 10(d). There is a higher degree of microtexture than observed in Figure 10(b) and less than the base material. This region of sample represents the weld stir zone.

## Discussion

A study of the longitudinal cross-section of an "interrupted" weld has shown to be an effective way of isolating the evolution of microstructural regions that may otherwise be lost in the more conventional transverse cross-section. Figure 2 shows how the microstructure in the region of the initial tool plunge evolved into the steady-state, stable stir zone microstructure. Similarly, the region ahead of the retracted tool, i.e., the stir zone, TMAZ, HAZ, and base material, can be isolated and characterized with respect to the relative proximity to the tool shoulder and pin. All of these observations may then be compared and contrasted with the microstructures observed in transverse cross-sections. The microhardness image map, generated from

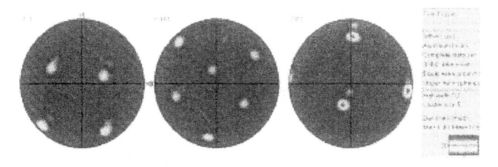

Figure 10(a): Pole figure, obtained using OIM, of base material

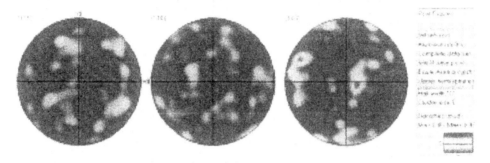

Figure 10(b): Pole figure, obtained using OIM, of TMAZ/base material boundary ahead of the tool.

50 μm

Figure 10(c): OIM image map of microstructure corresponding to the pole figures in Figure 10(b).

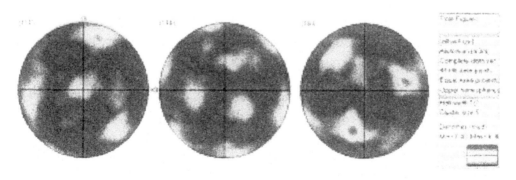

Figure 10(d): Pole figure, obtained using OIM, of stir zone between Regions I and II of Figure 2.

the longitudinal cross-section, serves as a means of basic screening of the various regions of the weld that warrent a more detailed microstructural analysis. It also provides a rough guide to the expected strength variations across the zones. The changes in microhardness values were location-specific. For example, the top and bottom surfaces of the plate behind (to the left of) the tool were softer than the mid-thickness. This was probably due to the highest local heating, provided by the shoulder and the bottom of the pin as the tool passed through the plate. The narrow region immediately ahead of the pin has hardness values equivalent to those in the stir zone behind the tool. This represents the evolving stir zone, the extent of which can be assessed as a function of tool rpm, travel speed or other process parameters.

The dual beam FIB was highly effective for the preparation of high quality TEM foils from predetermined locations in various regions of the FSW plate. The microstructural characterization of Regions II – VI, using conventional methods, would have required long, tedious sample preparation with no guaranteed outcome. Instead, the microstructures in these areas immediately adjacent to the tool, and throughout the thickness of the plate, were readily characterized with no evidence of damage due to sample preparation. The precise location and thinning of TEM foils across the wide range of microstructures produced in FSW is a powerful aid to the characterization process. For example, the use of the dual beam FIB for foil preparation in dissimilar alloy FSW will allow a detailed analysis of both the microstructure and composition across the stir zone and TMAZ.

The pole figure analyses revealed microtexture in the stir zone. Microtexture in FSW has been shown by previous investigators in aluminum alloys (8) and more recently in titanium alloys (9). It is interesting to note that, although the base material showed local microtexture in the area analyzed, the dynamic recrystallization that occurred in the stir zone yielded moderate microtexture. A local forging action occurs under the tool shoulder as the pin displaces the material across the weld centerline in a rotational fashion and also upward toward the shoulder (10, 11). The material flow resulting from this combination of forces produced some preferred grain orientation in the stir zone.

## Conclusions

1. The observation of microstructure in longitudinal cross-sections of interrupted FSW reveals the transitions from base material to HAZ, TMAZ and stir zone. Although the same microstructural regions can be observed in transverse cross-sections, the longitudinal view affords the opportunity to observe the evolution of these zones both in advance of the tool and behind it.
2. A microhardness image map of the longitudinal cross-section of interrupted FSW provides an effective rough screening method of the tensile properties across the various regions of the microstructure. The map can also be used to locate specific areas of the microstructure for more detailed analysis using TEM.
3. The dual beam FIB is a powerful tool for the precise location and extraction of high quality TEM foils from regions of the microstructure that would be at best very difficult to obtain using conventional methods.
4. Pole figure analyses performed using Orientation Imaging Microscopy revealed microtexture in the base material and moderate microtexture in the stir zone.

## References

1. P.J.Ditzel, "Microstructure-Property Relationships in Alumimum Friction Stir Welds", (M.S. Thesis, The Ohio State University, 1997).

2. O.T. Midling, "Material Flow Behavioral and Microstructural Integrity of Friction Stir Butt Weldments,(Paper presented at 4th International Conference on Aluminum Alloys, Atlanta, USA, ICAA4I, Sept., 1994).

3. M. W. Mahoney, C.G. Rhodes, J.G. Flintoff, R.A. Spurling and W.H. Bingel, "Properties of Friction Stir Welded 7075-T651 Aluminum, Metallurgical and Materials Transactions, (29A), (1998), 955-1964.

4. C.G.Rhodes, M.W. Mahoney, W.H.Bingel,R.A. Spurling,C.C. Bampton,, "Effects of Friction Stir Welding on Microstructure of 7075 Aluminum", Scripta Materialia, (36), No. 1, (January 1, 1997), 69-75.

5. R.S.Mishra, M.W. Mahoney, S.X. McFadde, N.A.Mara, A.K., Mukherjee, "High Strain Rate Superplasticity in a Friction Stir Processed 7075 Al Alloy" Scripta Materialia, (42), No. 2, (December 31, 1999),163-168.

6. K.V. Jata, and S.L.Semiatin, "Continuous Dynamic Recrystallization During Friction Stir Welding of High Strength Aluminum Alloys", Scripta Materialia, (43), No. 8, (September 29, 2000), 743-749.

7. L.E. Murr, G. Liu, and J.C. McClure, "Dynamic Recrystallization in Friction Stir Welding of Aluminum Alloy 1100", J. Mat. Sci. Let., (16), No. 22, 1801-1803.

8. T.W. Nelson, B. Hunsaker and D.P. Field, "Local Texture Characterization of Friction Stir Welds in 1100 Aluminum", (Proc. 1st Internat. Symp. of Friction Stir Welding, Thousand Oaks, CA, 1999).

9. M.C. Juhas, G.B. Viswanathan, N.I. Karogal, J.Fritz, and H.L.Fraser, "Characterization of Microstructural Evolution in a Ti-6Al-4V Friction Stir Weld", (Proc. of Symposium of Friction Stir Welding, TMS, New Orleans, 2001).

10. A.P. Reynolds, T.U. Seidel and M. Simonsen, "Visualization of Material Flow in an Autogenous Friction Stir Weld", (Proc. 1st Internat. Symp. of Friction Stir Welding, Thousand Oaks, CA, 1999).

11. P. Colegrove, "3 Dimensional Flow and Thermal Modeling of the Friction Stir Welding Process", (Proc. 2nd Internat. Symp. of Friction Stir Welding, Gothenburg,, 2000).

## Acknowledgements

This work was performed under the auspices of Edison Welding Institute. The authors are grateful to Ms. Su Meng for TEM foil preparation.

# LIGHTWEIGHT ALLOYS FOR AEROSPACE APPLICATION

*Edited by:*
*Dr. Kumar Jata, Dr. Eui Whee Lee,*
*Dr. William Frazier and Dr. Nack J. Kim*

## ALUMINUM ALLOYS

## Deformation, Fracture and Fatigue in a Dispersion Strengthened Aluminum Alloy

*Anthony P. Reynolds, Bob Wheeler and Kumar V. Jata*

Pgs. 87-97

184 Thorn Hill Road
Warrendale, PA 15086-7514
(724) 776-9000

# DEFORMATION, FRACTURE AND FATIGUE IN A DISPERSION STRENGTHENED ALUMINUM ALLOY

Anthony P. Reynolds[1], Bob Wheeler and Kumar V. Jata[2]

[1]University of South Carolina, Department of Mechanical Engineering, Columbia, SC
[2] AFRL/ML, WPAFB, Dayton, Ohio

## Abstract

Numerous attempts to produce aluminum alloys that retain useful property levels after extended exposure to elevated temperatures have been made in the past two decades. The typical approach has been to create a fine grain material that is stabilized against grain growth by the presence of numerous dispersoids. The dispersoid composition is generally tailored so as to inhibit coarsening (either oxide dispersion or transition metal bearing intermetallics). This paper reports the fatigue, fracture, and tensile behavior of a dispersion strengthened alloy consisting of a nearly pure aluminum matrix reinforced by a few volume percent of $Al_2O_3$. The behavior of this alloy is compared to that of previously developed dispersion strengthened aluminum alloys. While the properties of the new material are reasonably good, it suffers from drawbacks similar to those exhibited by other DS-Al alloys, notably reduced ductility with increasing temperature, low uniform elongation, minimal work hardening, and relatively low fatigue crack growth threshold at low stress ratio.

Lightweight Alloys for Aerospace Applications
Edited by Kumar Jata, Eui Whee Lee,
William Frazier and Nack J. Kim
TMS (The Minerals, Metals & Materials Society), 2001

## Introduction

The low density and good specific properties of aluminum alloys make them the primary material choice for many aerospace applications. However, typical precipitation hardened, ingot metallurgy (IM) aluminum alloys are limited to use at temperatures substantially below their precipitation heat treatment temperatures. Most IM alloys cannot be used for long periods at temperatures that approach 100°C. It has long been desired to develop high strength aluminum alloys that could compete with Titanium alloys in intermediate temperature applications (100-300°C).[1]

The common approach to development of so-called high-temperature aluminum alloys is to use dispersion rather than precipitation hardening. The dispersoids are normally either aluminum oxide (sometimes nitrides and carbides) or intermetallic compounds containing slow diffusing transition metals.[2] The key feature of the dispersoids is that unlike precipitates, they do not dissolve at any likely service temperature and they are highly resistant to coarsening because their components have very low equilibrium solubilities in the aluminum matrix. In order to obtain the dispersoid distributions necessary for effective grain size control and strengthening, rapid solidification and powder processing are usually required. For example, production of AA8009 (developed by Allied- Signal in the 1980's, originally FVS 0812) is a multi-step process that begins with ribbon casting to freeze in the desired Al-Fe-V-Si particles. The resulting ribbon is comminuted and then compacted by various thermo-mechanical techniques.[3] For any aluminum based powder metallurgy product, a substantial amount of mechanical work is required to break up the oxide layers that form on the cast particles so that good bonding between particles may be effected. Insufficient work results in planes of low ductility and toughness that result in easy delamination and highly anisotropic properties. Because of the relative complexity of processing required for production of these dispersion strengthened alloys, they are typically expensive compared to ingot metallurgy materials, hence, they must achieve a substantial delta in performance in order to be economically justifiable.

In this paper, the properties of a powder metallurgy, $Al_2O_3$ dispersion strengthened, aluminum alloy are examined. Tensile properties, da/dN-$\Delta$K behavior, R-curves, and fractographic features are reported and, where appropriate, will be compared to those of other dispersion strengthened alloys.

## Materials and Experimental Procedures

### Material
The dispersion strengthened alloy described in this paper was provided as sheet with a thickness of 1.2 mm and as a 63.5 mm diameter, round extrusion. Most of the testing was performed on the sheet material. The material was produced by powder milling of pure aluminum in an atmosphere containing inert components and a controlled level of oxygen. Powders were then compacted and extruded. Extruded material was rolled to make sheet. Controlled oxygen levels during milling gave re-producible oxide content in the final product. No carbides were detected although some might be present due to the stearic acid additions used during powder milling. TEM was performed to elucidate grain structure and to characterize the alumina particles. Optical microscopy was attempted but provided no useful information.

### Mechanical Testing

Tensile tests were performed on the sheet and extruded material at room temperature. Additional tensile tests were performed at 100°C, 125°C, 150°C, 200°C, 225°C, and 250°C on the sheet material. The initial strain rate was $0.0015s^{-1}$ for all tests.

Fatigue crack growth testing on the sheet was performed on center crack panels at 10 Hz at room temperature and 150°C. Stress ratios (R) of 0.05 and 0.8 were tested in both decreasing and increasing $\Delta K$ modes. Crack length was monitored using the direct current electric potential drop method. Crack closure was not measured.

R-curves were produced for sheet material at room temperature. Specimens were compact tension (CT) with W=50.8 mm and B=1.2 mm. Both L-T and T-L orientations were tested. Because of the thinness of the sheet, anti-buckling guides were used. The fracture toughness of the extrusion was measured as well. The specimens were disk shaped with W=50.8 mm and B=10 mm. The crack growth direction was radial.

Fractography was performed using scanning electron microscopy (SEM) for selected tensile and fatigue conditions.

## Results

**Microstructural Characterization** Figures 1a and 1b are respectively dark and bright-field TEM images of the microstructure. The dark-field image (1a) shows the grain structure of the DS alloy. The grains are somewhat elongated in the rolling direction and have typical dimensions of 150 nm in the thickness direction and 300 nm in the rolling direction. In figure 1b, the alumina particles show up faintly as the bright white features positioned predominantly along the grain boundaries of the aluminum matrix. The particles are of variable size but are generally less then several tens of nanometers in diameter and often much smaller. The $Al_2O_3$ volume fraction was roughly estimated at 2-4% by area fraction measurements on TEM images.

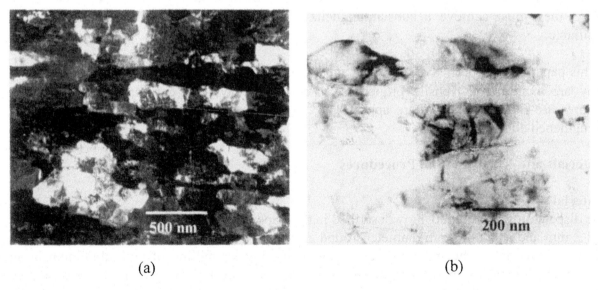

(a)                                                    (b)

Figure 1. a) Dark field TEM image showing grain structure aligned with the rolling direction. Typical grain size is 300nm long by 150nm high. b) Bright field image showing alumina particles (arrows) along grain boundaries.

**Tensile Test Results** Several representative stress-strain curves for the DS sheet are shown in figure 2. The curves illustrate the expected declining strength with increasing temperature. Particular features to note include the very low uniform elongation exhibited at room temperature. The RT curve shows total engineering strain of about 8%, but the tensile strength is achieved at a strain of close to 1%. In all cases, the uniform strain is relatively low and a minimum in elongation to fracture may be present at intermediate temperature (125°C). The yield strength, tensile strength and elongation to fracture are summarized in figure 3 where it can be seen that the reductions in the two strength measures are linear with decreasing temperature. Also in figure 3, the possible ductility dip in the temperature range of 100°C-150°C can be seen. The DS-Al extrusion was tested only at room temperature with the tensile axis corresponding to the radial direction of the extrusion. The average properties of the extrusion were yield strength=254 MPa, tensile strength=273 MPa, and elongation to fracture=1%: substantially worse than the sheet in terms of ductility.

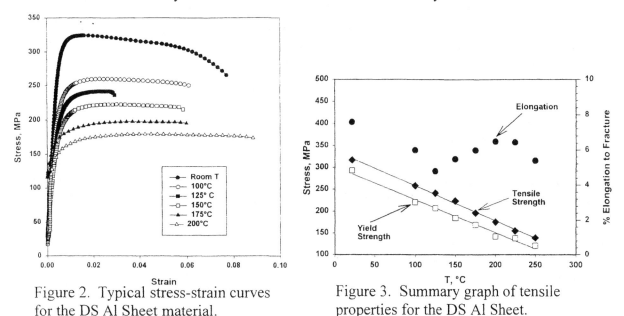

Figure 2. Typical stress-strain curves for the DS Al Sheet material.

Figure 3. Summary graph of tensile properties for the DS Al Sheet.

**Fracture Properties** The DS-Al extrusion exhibits low toughness. The average stress intensity at the onset of unstable fracture (2 tests) was 7.4 MPa√m and no stable tearing was observed. On the other hand, room temperature R-curve testing of the DS-Al sheet resulted in substantial stable tearing and rising R-curve behavior. In figure 4 representative R-curves for L-T and T-L orientations are shown. Also shown in figure 4 is a 2024-T3 R-curve (identical CT specimen with W=50.8 mm but B=3.2 mm). The DS-Al sheet exhibits isotropic fracture behavior: L-T and T-L specimens behave identically. The level of the DS-Al R-curves is below that of the 2024-T3, but the DS sheet shows substantial increase in tearing resistance with increasing crack

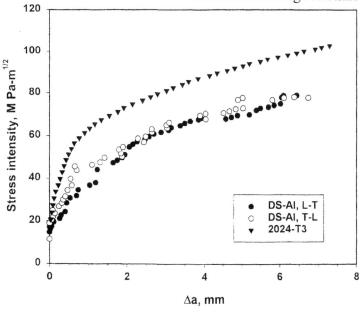

Figure 4. R-curves for L-T and T-L oriented DS-Al and 2024-T3 , L-T.

length. Unstable behavior was not observed in the sheet and larger specimens will be required to fully explore the fracture properties.

**Fatigue Crack Growth Results**   K-increasing and K-decreasing da/dN-$\Delta$K curves at stress ratios of 0.05 and 0.8 are shown for test temperatures of 22°C and 150°C in figures 5 and 6 respectively.   The fatigue behavior of the DS-Al sheet is typical of that reported for many other very fine grain, dispersion strengthened aluminum alloys.   The presumed closure free, high R, threshold value for the DS-Al sheet is approximately 1.1 MPa$\sqrt{m}$ at both room temperature and 150°C. This is similar to closure free threshold values reported for many IM and PM aluminum alloys.   The low R (0.05) threshold for the DS-Al is about 2 MPa$\sqrt{m}$ at room temperature and about 1.5 MPa$\sqrt{m}$ at 150°C. These values are substantially lower than typical IM alloy values, but consistent with low R thresholds for most PM and DS aluminum alloys.   This behavior in DS aluminum alloys is generally ascribed to low crack roughness that reduces closure effects relative to IM alloys.   The difference in crack surface roughness is probably linked to the submicron grain size of the DS alloys.[4,5]

At room temperature and R=0.8, the transition from Paris law to overload dominated fatigue crack growth rate occurs at a $\Delta$K of about 6 MPa$\sqrt{m}$.   This $\Delta$K corresponds to a $K_{max}$ of 30 MPa$\sqrt{m}$ which is well above the apparent initiation of stable tearing observed in the R-curve testing (see figure 4).   At R=0.05 and room temperature, the Paris law slope appears to increase continuously between $\Delta$K=6 and $\Delta$K=20 MPa$\sqrt{m}$, possibly indicating a mixed fatigue and overload fracture mode in this range.

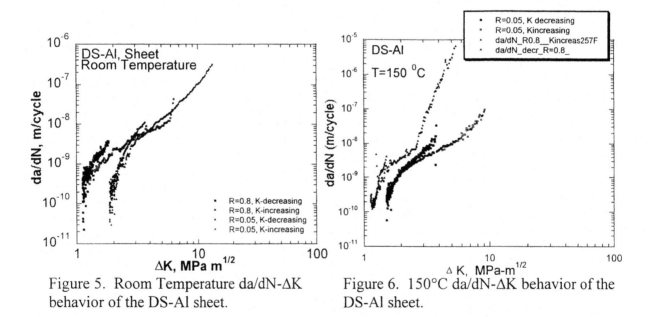

Figure 5.  Room Temperature da/dN-$\Delta$K behavior of the DS-Al sheet.

Figure 6.  150°C da/dN-$\Delta$K behavior of the DS-Al sheet.

Transitions from power law (Paris regime) crack growth to overload failure occur at substantially lower $\Delta$K (hence $K_{max}$) at a test temperature of 150°C than at room temperature. At R=0.8 and 150°C, this transition occurs at a $\Delta$K of about 3 MPa$\sqrt{m}$, or a $K_{max}$ of 15 MPa$\sqrt{m}$. This indicates a reduction in the resistance to stable tearing at elevated temperature relative to room T.  For R=0.05 at 150°C, substantial deviations from linear Paris regime behavior occur at $K_{max}$=7.4 MPa$\sqrt{m}$ ($\Delta$K=7 MPa$\sqrt{m}$).   The difference between the $K_{max}$ values at which deviations from Paris law type behavior are observed may be due to a synergistic effect between DK and $K_{max}$, this type of behavior has been previously seen in other DS alloys.[6]

**Fractography** Scanning electron microscopy was used to examine the fracture surfaces of sheet tensile specimens produced at room temperature and at 125°C. Near threshold fatigue fracture surfaces from high and low R testing at room temperature were also examined. Figure 7 a and b are images of tensile fracture surfaces produced at room temperature and at 125°C respectively. The photomicrographs show that the dimple structure is more pronounced at room T than at the elevated temperature. Dimple depths are greater and the margins of the dimples are better defined. This is consistent with the greater ductility exhibited at room temperature as compared to 125°C.

(a)

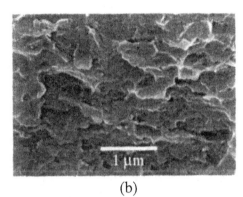
(b)

Figure 7. Scanning electron micrographs of (a) a room temperature tensile specimen fracture and (b) a 125°C tensile specimen fracture.

Figure 8 is a relatively low magnification image of a near-threshold, room temperature, fatigue fracture surface. The salient feature of the surface is its low roughness. This is consistent with the measured low threshold values at low R due to lack of roughness induced closure. In figure 9a and 9b, the near threshold fatigue surfaces produced at room temperature and stress ratios of 0.05 and 0.8 are compared. The features observed at high magnification on the R=0.05 surface are consistent with the impression of a low-roughness surface obtained from low magnification viewing. The R=0.8 surface, on the other hand, exhibits many features of an overload type

Figure 8. Room temperature, near-threshold, fatigue surface, R=0.05.

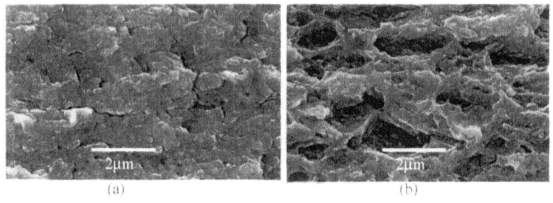
(a)                                    (b)

Figure 9. High magnification images of (a) an R=0.05 near threshold fatigue surface and (b) an R=0.8 near threshold fatigue surface.

fracture, even though $K_{max}$ is substantially below the value required to initiate stable tearing in an R-curve test.

## Discussion

The key microstructural features of the DS-Al alloy examined in this paper may be summed up as follows. The alloy has a grain size of less than 0.5μm. In the alloy, there is a small but significant volume fraction of very fine particles that are primarily associated with the grain boundaries. There is no significant solute in the aluminum matrix. These features are to a greater or lesser extent shared by essentially all dispersion strengthened aluminum alloys. Perhaps the most universal trait is that of having a grain size of approximately 0.5μm or less.

One of the commonly observed behaviors of this class of alloys is that they tend to exhibit declining ductility with increasing temperature. This is in direct contrast with typical ingot metallurgy aluminum alloys (and most alloys of other classes e.g. ferrous, etc.) that tend to show increased ductility with increasing temperature.[7,8,9] For essentially all materials, of course, increasing temperature is accompanied by declining strength. The rate of decline in strength with increasing temperature will depend on the mechanisms of strengthening in a particular alloy. It has been shown previously that for at least some DS alloys, 8009 in particular, that the proportional limit is quite low and that all of the subsequent work hardening is kinematic rather than isotropic in nature. That is, the bulk of the strengthening is due to the Bauschinger effect.[10]

Substantial Bauschinger effect strengthening in turn requires a material with non-homogeneous plastic properties (e.g. a soft matrix with reinforced with non-deformable particles). It has been proposed that isotropic mechanisms of work hardening do not occur in alloys with grain size less than 0.5μm. Isotropic mechanisms of work hardening require that as plastic deformation increases, the number of barriers to dislocation motion increases. These barriers normally take the form of dislocation intersections that trap additional dislocations, finally leading to the formation of walls and dislocation cells. For normal deformation rates (i.e. not shock loading) in a variety of alloys types, including aluminum and ferrous based, the minimum dislocation cell size that is observed is between 0.3 and 0.5 μm. This minimum cell size implies that no dislocation intersections are created when the mean free path of dislocations is on the order of 0.5 μm. The logical conclusion is that in materials that possess a grain size of 0.5mm or less, work-hardening by creation of additional dislocation barriers is highly unlikely.[11]

In figure 10, the room temperature stress strain behavior of the DS-Al alloy is compared to a similar, cryo-milled aluminum/alumina alloy (CM-Al), and alloy 8009. Each of these alloys possesses a grain size of 0.5μm or less. Alloy 8009 contains approximately 25% by volume of Aluminum-iron-vanadium-silicide particles of 50-80 nm diameter. The two other alloys have similar volume fractions of alumina particles as reinforcement. The two alumina reinforced materials have similar strengths: the strength of 8009 is substantially higher. The higher strength of 8009 is to be expected based on the much greater volume fraction of reinforcing phase present (25% vs. 2-4%). None of the alloys exhibits large uniform elongations. The DS-Al alloy and 8009 in particular reach UTS at similarly low levels of strain (1%-1.5%) and then exhibit high levels of post uniform elongation. The low uniform strain levels are consistent with a lack of isotropic hardening. As soon as the capacity for kinematic hardening is exceeded in one of these alloys, work hardening ceases (work softening has been reported in some DS alloys) and the criterion for necking is met.

Figure 11 is a comparison between the temperature dependent yield strengths of the two alumina reinforced DS alloys. While the DS-Al alloy exhibits higher strength at room temperature than the CM-Al alloy, as the temperature is increased, the DS-Al alloy exhibits a more rapid decline in strength.[7] This is somewhat unexpected as one would suppose that for a kinematically strengthened material, the strength should scale as the modulus. To illustrate this point, in figure 12, the data shown in figure 11 have been re-plotted after normalization by the temperature dependent modulus (taken as the modulus of pure aluminum).[12] As can be seen in figure 12, the CM-Al alloy exhibits an essentially constant ratio of yield strength to elastic modulus as the temperature increases. This has been observed previously for 8009 as well. The DS-Al material, on the other hand, exhibits a declining ratio of yield strength to elastic modulus with increasing temperature. This is a somewhat puzzling behavior particularly in that the decline is linear, indicating that no critical temperature, above which some change in microstructure or deformation mode occurs, has been exceeded. In order to fully explore the elevated temperature behavior of the DS alloy, it would be desirable to characterize the strain rate sensitivity as a function of temperature; however, because of the low uniform elongations

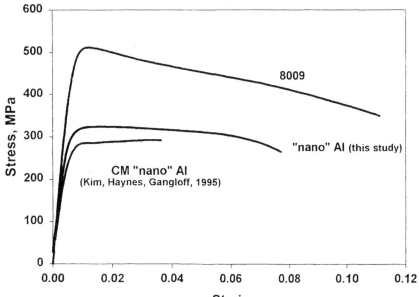

Figure 10. Comparison of the room temperature tensile properties of three dispersion strengthened aluminum alloys.

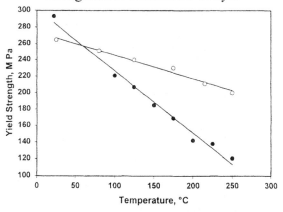

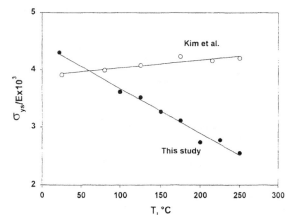

Figure 11. Temperature dependent yield strength of the currently studied alloy (solid symbols) and the Exxon alloy (open symbols).

Figure 12. Temperature dependent yield strength normalized by the corresponding elastic modulus.

exhibited by the material, this would best be done by compression testing and compression testing of the 1.2 mm thick sheet is problematic.

95

Another example of similarity in temperature dependent behavior between the DS-Al alloy and other dispersion strengthened alloys may be gleaned from the fatigue data. Although R-curves were produced only for room temperature, the fatigue data indicate that the critical stress intensity for onset of overload type fracture declines with increasing temperature. This is evidenced by the different values of $K_{max}$ at which substantial deviation from Paris regime behavior is observed in the R=0.8 testing at room temperature (30 MPa$\sqrt{m}$) and at 150°C(14 Mpa$\sqrt{m}$).

## Summary and Conclusion

1) The DS-Al alloy possesses a microstructure that is similar to that of most other dispersion strengthened aluminum alloys.

    a) a grain size of less than 0.5 μm.
    b) negligible solute.
    c) insoluble, nano-sized particles.

2) The fine grain size of the alloy contributes to high fatigue crack growth rates at low R, presumably due to reduced roughness induced closure compared to IM alloys with typical grains sizes greater than 20 μm.

3) Mechanical working (rolling of sheet from extrusion) greatly improves tensile ductility and fracture toughness. The mechanical work has a lesser effect on strength. After rolling, the fracture properties of the sheet are isotropic.

4) The DS-Al alloy exhibits temperature dependent tensile ductility and fracture toughness that are similar to those exhibited by other very fine-grain, dispersion strengthened aluminum alloys.

5) The temperature dependence of the strength is inconsistent with the temperature dependence of the elastic modulus, possibly indicating the operation of an novel deformation mechanism.

## References

1. P. S. Gilman and S. K. Das, *Proc. Int. Conf. on PM Aerospace Materials*, 1987, pp. 27.1-27.11.

2. R. L. Bye, N. J. Kim, D. J. Skinner, D. Raybould, and A. M. Brown, *Proc. Symp. Enhanced Properties in Structural Materials via Rapid Solidification*, F. H. Froes and S. J. Savage, eds., ASM, Metals Park, OH, pp. 283-289.

3. D. J. Skinner and K. Okazaki, *Scripta Metall*, 1984, vol. 18, pp. 905-909.

4. A. P. Reynolds, *Fatigue and Fracture of Engineering Materials and Structures*, vol. 15, no. 6, pp. 551-562, 1992.

5. G. H. Bray, A. P. Reynolds, and E. A. Starke, Jr. *Metallurgical Trans A*, vol. 23A, pp. 3055-3066, 1992.

6. W. C. Porr, Jr., A. Reynolds, Y. Leng, and R. P. Gangloff, *Scripta Met et Mat*, vol. 25, pp. 2627-2632, 1991.

7.  S. Kim, M. J. Haynes, and R. P. Gangloff, *Materials Science and Engineering,* A203 (1995) 256-271.

8.  E. Bouchaud, L. Kubin, and H. Octor, *Met Trans. A,* 1991, vol. 22A, pp. 1021-1028.

9.  Y. Leng, W. C. Porr, Jr., and R. P. Gangloff, *Scripta Met et Mat,* vol. 24, pp. 2163-2168.

10.  A. P. Reynolds and J. S. Lyons, *Met Trans A,* vol. 28A, May 1997, pp. 1205-1211.

11.  B. Bay, N. Hansen, D. A. Hughes, and D. Kuhlmann-Wilsdorf, *Acta Met Mater,* 1992, vol. 40, pp. 205-219.

12.  M. E. Fine, *The Review of Scientific Instruments,* vol. 28, number 8, 1957, pp. 643-645.

# LIGHTWEIGHT ALLOYS FOR AEROSPACE APPLICATION

*Edited by:*
*Dr. Kumar Jata, Dr. Eui Whee Lee,*
*Dr. William Frazier and Dr. Nack J. Kim*

## ALUMINUM ALLOYS

## The Effect of Retrogression and Reaging on 7249 Aluminum Alloy

*P. Fleck, K. Koziar, G. Davila, H. Pech, E. Fromer,*
*M. Leal, J. Foyos, E.W. Lee and O.S. Es-Said*

Pgs. 99-108

184 Thorn Hill Road
Warrendale, PA 15086-7514
(724) 776-9000

# THE EFFECT OF RETROGRESSION AND REAGING
## ON 7249 ALUMINUM ALLOY

P. Fleck, K. Koziar, G. Davila, H. Pech, E. Fromer, M. Leal, J. Foyos,
E.W. Lee*, and O.S. Es-Said

National Science Foundation
Research Experience for Undergraduates Program
Loyola Marymount University
Los Angeles, CA 90045-8145

*Naval Air Systems Command
Naval Air Warfare Center Code 4342
MS Bldg 2188
Patuxent River, MD 20670-1908

## Abstract

A study of retrogression and reaging (RRA) of aluminum alloy 7249-T76 was performed. The primary goal of this experiment was to achieve a T6-RRA optimization treatment for 7249, which included T6 and T76 optimizations. The retrogression heat treatments consisted of a number of times between 20 and 120 minutes and a number of temperatures between 170°C and 210°C. Reaging treatments were done in accordance with the optimized T6 condition (121° C for 24 hours).

Lightweight Alloys for Aerospace Applications
Edited by Kumar Jata, Eui Whee Lee,
William Frazier and Nack J. Kim
TMS (The Minerals, Metals & Materials Society), 2001

# Introduction

Retrogression and Reaging (RRA) is a heat treatment, developed and patented by B. Cina in the early 1970's, which increases the stress corrosion cracking (SCC) resistance of the 7xxx aluminum alloy series to that of the T7 temper, while maintaining the strength of the T6 temper, [1]. It is a multi-step process which involves taking the alloy in the T6 temper, retrogressing it by heating it to a high temperature for a short time, quenching, then reaging the alloy at a low temperature for a long, similar to the T6 aging time, [2]. This results in the stress corrosion cracking resistance of the T7 temper and the strength of the T6 temper.

Park [3] and Park and Ardel [4,5] concluded that the resistance to stress corrosion cracking without loss of strength, which is obtained through the RRA process, results from the coarsening of second-phase particles at the grain boundaries and the increase in volume fraction of second-phase particles at the grain interior.

Talianker and Cina [6] performed research in 1989 to confirm Cina's original hypothesis. They performed RRA treatments on three different alloys of aluminum. In their research, they showed that for all samples, the grain boundary dislocations decreased in density after RRA treatment. Also, RRA treatment was accompanied by significant improvement in the resistance to stress corrosion and exfoliation corrosion. Since dislocations are believed to be the cause of susceptibility to stress corrosion cracking, they concluded that the decrease in the density of the grain boundary dislocations is the cause of the increased resistance to stress corrosion cracking.

The purpose of this research is to find the optimum T6 and T6 RRA condition of the new aluminum alloy 7249. Al-7249 is an Al-Zn-Mg-Cu-Cr alloy developed as a derivative from 7149. It was developed as a replacement material for 7075-T6 forgings, which are susceptible to SCC; it also has higher strength at the higher thickness ranges and higher ductility than 7075-T6, [7]. Since it is relatively new, T6 optimization treatments will be established, as well as the RRA treatments.

# Experimental Procedure

## Machining

7249 was received as both wide and narrow extrusions. The wide extrusions were received as plates with fins. The dimensions are as shown in Figure 1; the horizontal thickness of the fins was 20.3 mm (0.80in.) for the outside fins, and 17.0 mm (0.67in.) for the inside fins. The horizontal distance between the fins was 33 mm (1.30in.). A cross-sectional area of the narrow extrusions is shown in Figure 2, along with dimensions; the length of the extrusions varied from 711.2 mm (28.00in.) to 1219.2 mm (48.00 in.). There were several heat treatment optimizations, both for the wide and narrow extrusion. For both, the samples were machined parallel to the extrusion direction. Each condition was evaluated by at least three tensile samples.

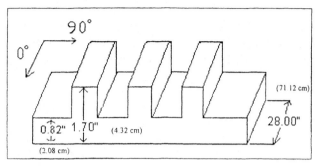

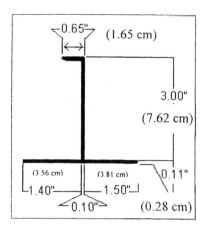

| Figure 1 | Figure 2 |

## Optimization of T76 Temper

T76 optimization consisted of solution treating, naturally aging, then artificially aging the samples at various temperatures and times. Tables I (a) and I (b) show the solution treatments and aging treatments times and temperatures. All samples were solution treated at their respective time and temperature, then water quenched, Table I (a). They were then naturally aged for 48 hours, after which they were kept in a freezer until they were artificially aged, Table I (b), air-cooled, and then tensile tested.

The reason for the multi-step solution treatments is that in previous research, Ikei et al [8], showed that multiple step solution treatments on aluminum alloy 6061 improved the mechanical tensile properties of the material.

| Table I (a) | Table I (b) |
|---|---|
| Solution Treatment | Aging |
| $^1/_2$ hr @ 463°C | 8 hrs @ 121°C + |
| $^1/_2$ hr @ 474°C | 8 hrs @ 163°C |
| 1 hr @ 363°C + 1 hr @ 413°C + 2 hrs @ 463°C | 8 hrs @ 121°C + 16 hrs @ 163°C |
| 1 hr @ 374°C + 1 hr @ 424°C + 2 hrs @ 474°C | 16 hrs @ 121°C + 32 hrs @ 163°C |

## Optimization of T6 Temper for the Wide Extrusions

T6 optimization for the wide extrusions consisted of solution treating, naturally aging, then artificially aging 60 samples at various temperatures and times. All samples were solution treated at their respective time and temperature, Table II(a), then water quenched. They were then naturally aged for 48 hour, after which they were kept in the freezer until they were artificially aged at 121°C for their respective times, Table II(b). After artificial aging, all samples were air-cooled, and then tensile tested.

| Table II (a) |
| --- |
| Solution Treatment |
| 1 hr @ 463°C |
| 1 hr @ 474°C |
| 1 hr @ 363°C + 1 hr @ 413°C + 2 hrs @ 469°C |
| 1 hr @ 374°C + 1 hr @ 424°C + 2 hrs @ 474°C |

| Table II (b) |
| --- |
| Aging @ 121°C |
| 24 hrs |
| 36 hrs |
| 48 hrs |
| 60 hrs |

Again, multi-step solution treatments were performed to observe if there were any improvements in the mechanical tensile properties of the alloy from the different solution treatments, [8].

## Optimization of T6 Temper for the Narrow Extrusions

T6 optimization for the narrow extrusions consisted of solution treating, naturally aging, then artificially aging the samples at various temperatures and times. The solution treatments were at 463°C for 1 hour and 474°C for 1 hour. The artificial aging treatments were at 121°C for 24 and 36 hours. The solution treatments which were chosen for the narrow extrusions were chosen based on the results from the T6 optimization treatments of the wide extrusions.

## Retrogression and Reaging

After the optimum T6 condition was reached, 60 wide extrusion samples, which were machined in the 0° direction, and 60 narrow extrusion samples, were treated to the T6 temper (i.e. they were solution treated for 1 hour at 474°C, water quenched, naturally aged for 48 hours, then artificially aged at 121°C for 24 hours). Optimization of retrogression and reaging then consisted of various retrogression times and temperatures. The reaging time and temperature was the same as that used for the T6 condition, (i.e. 121°C for 24 hours). The samples were retrogressed at 170°C, 180°C, 190°C, and 210°C. The retrogression times were 20, 40, 60, 90, and 120 minutes. There were three samples per combination. All samples were retrogressed at their respective time and temperature then water quenched, after which they were kept in a freezer until they were reaged. After reaging, all samples were air-cooled, then tensile tested.

## Exfoliation Tests

Exfoliation tests, performed according to ASTM G34 (EXCO Test), were used to evaluate the exfoliation corrosion resistance of various conditions by exposing specimens to an acid saline solution for 48 hours. The EXCO solution consists of NaCl (4.0M), $KNO_3$ (0.5M), and $HNO_3$ (0.1M). At least two coupons were prepared for each material/condition tested. The EXCO Test classifications in order of decreasing corrosion resistance are N (no appreciable attack), P (pitting), and EXCO (exfoliation corrosion) A, B, C, and D.

## Results and Discussion

The "As Received" wide extrusions had a 0.2% yield and ultimate strength of 531.2 and 591.0 MPa (77.1 and 85.8 ksi), respectively, while the "As Received" narrow extrusions had a 0.2% yield and ultimate strength of 506.4 and 551.2 MPa (73.5 and 80.0 ksi), respectively.

For the optimization of the T76 temper, the best 0.2% yield and ultimate strength was 571.2 and 602.9 MPa (82.9 and 87.5 ksi), respectively. The solution treatment and aging which gave this result was 1 hour at 363°C plus 1 hour at 413°C plus one hour at 463°C for the solution treatment, and 8 hours at 121°C plus 8 hours at 163°C for aging. As shown in the 0.2% yield and ultimate strength graphs for the T76 optimization, Figure 3, none of the solution treatments had much effect on the strength properties of the material; this is evident by the fact that the lines for each aging treatment is nearly a straight line. However, the aging treatments did have considerable effect on the strength values.

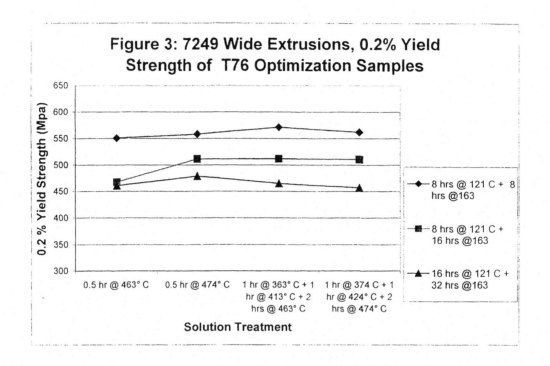

Figure 3: 7249 Wide Extrusions, 0.2% Yield Strength of T76 Optimization Samples

For the optimization of the T6 temper for the wide extrusions, the best 0.2% yield and ultimate strength was 616.7 and 659.4 MPa (89.5 and 95.7 ksi), respectively. The solution treatment and aging which gave this result was 1 hour at 474° C for the solution treatment, and 60 hours at 121° C for aging. Even though these were the best results for the T6 optimization, the solution treatment and hours of aging which were chosen to be used for the T6 optimization was actually 1 hour at 474° C and 24 hours @ 121°C, respectively. The reason for this is the 0.2% yield and ultimate strength for this combination was 591.9 and 655.2 MPa (85.9 and 95.1 ksi), respectively. This difference in strength is not significant, and thus it was decided that the savings in time of the aging (36 hours) would be worth the slight loss in strength. Similar to the T76 optimization, the multi-step solution treatments did not have much effect on the strength properties of the material, as shown in the 0.2% yield and ultimate strength graphs for the T6 optimization, Figure 4. For this reason, the multi-step treatments were not performed for the T6 optimization of the narrow extrusions, Figure 5. The solution treatment of 1 hour at 474°C and aging treatment of 24 hours at 121°C was chosen as the optimum T6 temper for both wide and narrow extrusions.

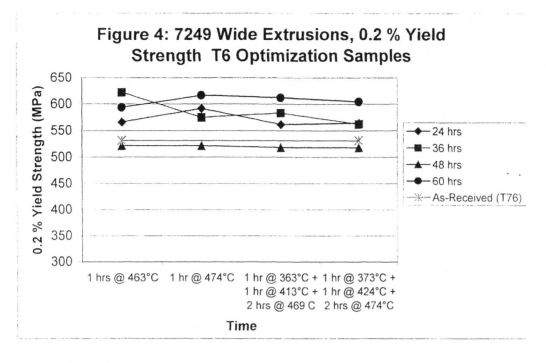

Figure 4: 7249 Wide Extrusions, 0.2 % Yield Strength T6 Optimization Samples

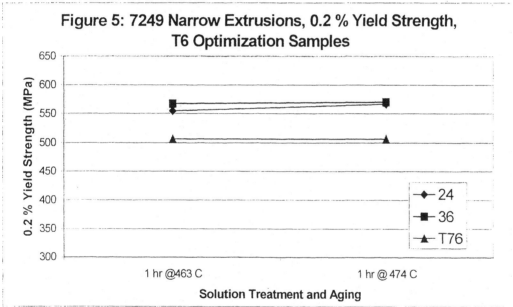

Figure 5: 7249 Narrow Extrusions, 0.2 % Yield Strength, T6 Optimization Samples

For the optimization of the T-6RRA temper for the wide extrusions, the best 0.2% yield and ultimate strength was 640.1 and 662.1 MPa (92.9 and 96.1 ksi), respectively, Figure 6. The retrogression time and temperature were 60 minutes at 170°C. The percent elongation for this time and temperature was 7.60%, Figure 7.

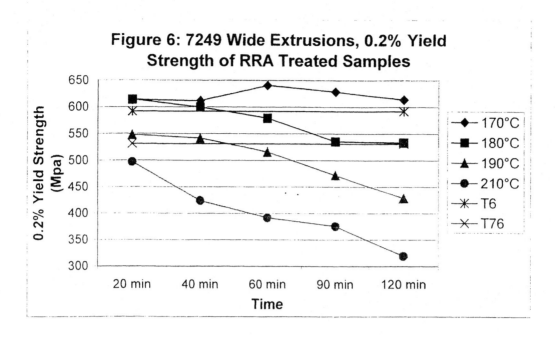

Figure 6: 7249 Wide Extrusions, 0.2% Yield Strength of RRA Treated Samples

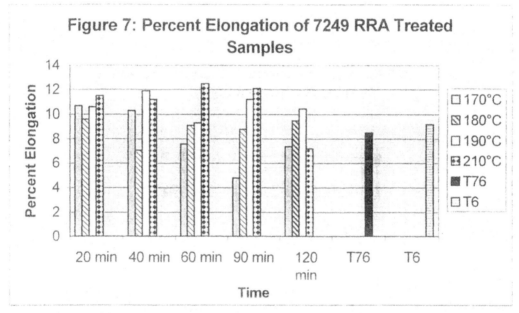

Figure 7: Percent Elongation of 7249 RRA Treated Samples

For the optimization of the T-6RRA temper for the narrow extrusions, the best 0.2% yield and ultimate strength was 623.6 and 632.5 MPa (90.5 and 91.8 ksi), respectively. The retrogression time and temperature were 20 minutes at 180°C for the 0.2% yield strength, Figure 8, and 20 minutes at 190°C for the ultimate strength. The percent elongation for the retrogression time and temperature of 20 minutes at 180°C was 10.7%.

EXCO tests were performed on the wide extrusion samples retrogressed at 180°C for 60 minutes and on the narrow extrusion samples retrogressed at 180°C for 30 minutes. The results indicated that the RRA samples had better exfoliation corrosion properties than the T6 and T76 tempers, Figure 9.

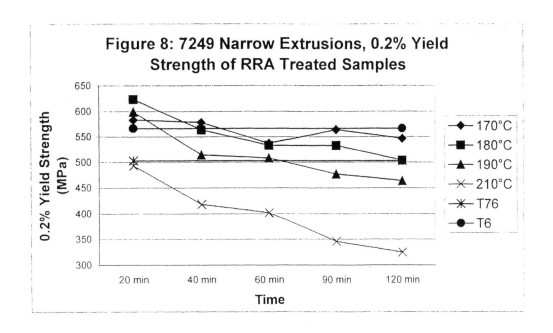

Figure 8: 7249 Narrow Extrusions, 0.2% Yield Strength of RRA Treated Samples

## Figure 9: EXCO Corrosion Test

|  |  | Wide Extrusions | Narrow Extrusions |
|---|---|---|---|
| 5 hours | T6 | Pitting and copper re-depositing on the surface | Pitting and copper re-depositing on the surface |
|  | T76 | Light pitting and copper re-depositing on the surface | Light pitting, no copper deposits |
|  | RRA | Very light pitting with copper in the scratch pattern | Very light pitting, no copper deposits |
| 24 hours | T6 | EB EB EB | EB EB EB |
|  | T76 | EB EB EB | EB EB EB |
|  | RRA | EB EB EB | EA EA EA |
|  |  | All the EB surfaces with copper layers dissolved off or lifted and floated away from the surface | |
| 48 hours | T6 | EC EC EC | EC EC EC |
|  | T76 | EC EC EC | EC EC EC |
|  | RRA | EC EC EC | EB EB EB |
|  |  |  | EB - Surface layer with copper re-deposited still in place |

**Conclusions and Recommendations**

1. The multi-step solution treatments in both the T76 and T6 optimization treatments of the wide extrusions gave no significant advantages in strength.

2. The recommended T6 heat treatment for both the wide and narrow extrusions is solution treating at 474°C for one hour, water quenching, naturally aging for 48 hours, then artificially aging for 24 hours at 121°C.

3. For strength, the best retrogression time and temperature for the wide extrusions is 170°C for 1 hour, while it is 180°C for 20 minutes for the narrow extrusions.

4. The purpose of retrogression and reaging is to find the best strength and resistance to stress corrosion cracking; because of this, the best retrogression and reaging treatment is not always the one which exhibits the greatest strength. Therefore, the recommended retrogression treatment that might provide good strength and stress corrosion cracking resistance is 180°C for 30 minutes for the narrow extrusions, and 180°C for 60 minutes for the wide extrusions, then reage each for 24 hours at 121°C.

5. The EXCO results indicated that the RRA treatments had better exfoliation corrosion properties as compared to the T6 and T76 tempers.

## Acknowledgements

This work was funded by the National Science Foundation (NSF) Grant No. EEC-9732046, Research Experience for Undergraduates (REU) site.

## References

1. B. Cina, "Reducing Stress Corrosion Cracking in Aluminum Alloys." *U.S. Patent 3856584*, December 24, 1974.

2. R.S. Kaneko, "RRA: Solution for Stress Corrosion Problems with T6 Temper Aluminum." *Metal Progress*, 1980. Volume 117. pp 41-43.

3. J.K. Park, "Influence of Retrogression and Reaging Treatments on the Strength and Stress Corrosion Cracking Resistance of Aluminum Alloy 7075-T6." *Materials Science and Engineering*, 1988. Volume A103. pp 223-231.

4. J.K. Park and A.J. Ardell, "Microstructures on the Commercial 7075 Al Alloy in the T651 and T7 Tempers." *Metallurgical Transactions A*, 1983. Volume 14A. pp 1957-1965.

5. J.K. Park and A.J. Ardell, "Effect of Retrogression and Reaging Treatments on the Microstructure of Al-7075-T651." *Metallurgical Transactions A*, 1984. Volume 15A. pp 1531-1543.

6. M. Talianker and B. Cina, "Retrogression and Reaging and the Role of Dislocations in the Stress Corrosion of 7000-Type Aluminum Alloys." *Metallurgical Transactions A*, 1989. Volume 20A. pp 2087-2092.

7. MIL-HDBK-5G, "3.7.7 7249 Alloy", *Metals Handbook*, 1998. pp 3-422a

8. C. Ikei et al, "The Effect of Processing Parameters on the Mechanical Properties and Distortion Behavior of 6061 and 7075 Aluminum Alloy Extrusions." *Aluminum Alloys, Their Physical and Mechanical Properties – Proceedings ICAA7*, 2000. Part 1.

# LIGHTWEIGHT ALLOYS FOR AEROSPACE APPLICATION

*Edited by:*
*Dr. Kumar Jata, Dr. Eui Whee Lee,*
*Dr. William Frazier and Dr. Nack J. Kim*

## ALUMINUM-LITHIUM ALLOYS

Grain Boundary Corrosion and Stress
Corrosion Cracking Studies of Al-Li-Cu
Alloy AF/C458

*R.G. Buchhei, D. Mathur and P.I. Gouma*

Pgs. 109-118

184 Thorn Hill Road
Warrendale, PA 15086-7514
(724) 776-9000

# GRAIN BOUNDARY CORROSION AND STRESS CORROSION CRACKING STUDIES OF AL-LI-CU ALLOY AF/C458

R. G. Buchheit, D. Mathur

Department of Materials Science and Engineering,

Ohio State University, Columbus Ohio 43210.

P. I. Gouma

Department of Materials Sciences and Engineering,

Stony Brook University, Stony Brook, New York 11794

## Abstract

The effect of artificial aging on intergranular attack (IGA) and stress corrosion cracking (SCC) of an experimental alloy AF/C458 (Al-2.8Cu-1.8Li-0.6Zn-0.3Mg) were evaluated. Artificial aging (AA) treatments consisted of isothermal treatment at either 150° and 190°C for times ranging from 2.4 to 36 hours. Artificial aging was conducted on samples that were either in the solution heat treated (SHT) and quenched, or naturally aged (T3) condition initially. Overall, results showed high SCC resistance for AF/C458 with indications of greater susceptibility near the peak aged condition. Transmission electron microscopy showed $T_B$ ($Al_{15}Cu_8Li_2$) present on boundaries when the alloy was IGA and SCC resistant, and $T_1$ ($Al_2CuLi$) when the alloy was susceptible. This observation suggests that increased precipitation of electrochemically active $T_1$ on boundaries makes the alloy more susceptible to intergranular and intersubgranular attack.

## Introduction

A new Al-Li-Cu alloy, designated AF/C458 was recently fabricated by Alcoa/Davenport under US Air Force contract. This alloy was designed to overcome some of the main shortcomings of the previous Al-Li alloys; namely low short transverse (ST) fracture toughness, and high anisotropy. Additionally, this alloy offers 10 to 40% weight savings by virtue of its mechanical properties and density [1]. Research conducted so far indicates that AF/C458 has a low anisotropy and excellent fracture toughness in ST orientation [2]. Mahoney suggests that AF/C458 plate exhibits good friction-stir weldability and high SCC resistance, though work in this regard has been limited [2].

The objective of this study is to define corrosion and stress corrosion behavior of AF/C458 in standard tests, and compare its to that of earlier Al-Li alloys. Along these lines, studies were conducted to evaluate the effect of artificial aging on corrosion behavior of the alloy, originally in naturally aged (T3) and solution-heat treated (SHT) tempers. Aging was conducted at 150°C and 190°C for times ranging from an underaged condition of 2.4 h to overaged condition of 36 h. Results reported here include transmission electron microscopy characterization, alternate immersion exposure testing and constant extension strain rate testing. Overall, results indicate that this alloy possesses good SCC resistance. Results also shed light on the effect of different Al-Cu-Li precipitate phases in determining IGA and SCC susceptibility in these alloys.

Lightweight Alloys for Aerospace Applications
Edited by Kumar Jata, Eui Whee Lee,
William Frazier and Nack J. Kim
TMS (The Minerals, Metals & Materials Society), 2001

## Experimental Procedures

### Materials, Preparation and Characterization

The unrecrystallized 1.8" thick plate of AF/C458 (Al-2.8Cu-1.8Li-0.6Zn-0.3Mg-0.3Mn-0.07Zr) used in this study, was supplied by Wright Patterson Air Force Base, Dayton, OH. The plate was solution heat treated and subjected to 6% stretch and natural aging to a T3 condition prior to shipping. Some of the samples were cut from the plate, solution heat treated by heating to 549°C ± 3°C (SHT), then immediately quenched in cold-water. Further artificial aging (AA) of the SHT and T3 samples was done at 150°C and 190°C for 2.4, 10.4, 18.4, 24 and 36 h representing underaged to overaged conditions. Samples will be referred to as "T3 + AA", and "SHT + AA" to reflect condition of the alloy prior to artificial aging.

Transmission electron microscopy (TEM) was conducted on thin foil samples cut from the center of T3 + AA and SHT + AA samples. Microstructures were imaged using darkfield and brightfield imaging conditions. Precipitate phase identification was carried out by selective area diffraction. To prepare foils, samples were polished and ground to a 150 μm thickness. Discs, 3 mm in diameter, were then punched and dimpled to less than 30 μm, and subsequently perforated in an ion mill (6 kV, 1mA, 12°).

### Intergranular Attack (IGA) and Stress Corrosion Cracking (SCC) Evaluations

IGA evaluation was conducted using the $NaCl+H_2O_2$ solution specified in ASTM G110 [3]. Following a two-hour immersion, the specimens were removed and rinsed in distilled water. Samples having greater accumulation of corrosion product on the surface were placed in an ultra-sonic cleaner and polished very lightly with 1μm alumina slurry in order to facilitate inspection. Finally, the tested samples were examined with an optical microscope and attack morphology was assessed.

SCC was evaluated using alternate immersion (AI) and constant extension rate testing (CERT). AI tests closely followed the ASTM Standard G44 [4], which prescribes cyclic 10-minute immersion in 3.5% NaCl followed by 50 minutes drying in air. Exposure was conducted for 60 days. Round tensile bars with a 1-inch gauge length and 0.125 inch diameter were pre-stressed to 65% of yield strength in self-stressing frames. CER testing was carried out on samples of the same dimensions as those used for AI test. Tests were performed in 3.5% NaCl solution at an extension rate of $10^{-6}$ in/s. The degree of embrittlement was characterized by a ductility ratio, which was taken as the strain to failure in solution divided by the strain to failure in air.

## Results and Discussion

### TEM Characterization

TEM characterization of boundary and matrix precipitation was carried out on SHT+AA and T3+AA samples aged for 24 h at both 150° and 190°C. SHT + AA samples aged at 150°C, showed presence of δ' (Al₃Li) and β' (Al₃Zr) with some θ' (Al₂Cu) inside the grains (Fig. 1). Streaking was also observed along (111) suggesting presence of small amounts of T₁ (Al₂CuLi). Grain and subgrain boundaries were populated with plate-like precipitates of T₁ (Fig. 2). Grains in SHT+AA samples aged at 190°C, had δ' and an increased density of T₁ plates. The grain and subgrain boundaries again showed presence of $T_1$, and a distinct $\delta'$ PFZ along grain boundaries was detected (Fig 3). Several other phases were also detected on grain boundaries in this case, including R ($Al_5CuLi_3$) and $T_2$ ($Al_6CuLi_3$). T3 + AA samples aged at 150°C showed presence of $\delta'$, $\beta'$, and some $T_2$ inside the grains. The subgrain boundaries showed presence of $T_B$ phase (Fig. 4), while grain boundaries exhibited the presence of $T_B$ and $T_2$ phase. In T3 + AA samples aged at 190°C, the $T_1$ phase was again detected at subgrain boundaries, while grain boundaries exhibited a mixture of $T_2$ and R phases (Fig. 5). The matrix contained $T_1$, $T_2$, $\delta'$, and $\beta'$. Table I summarizes the findings from TEM characterization performed of the different samples.

IGA

Intergranular attack (IGA) or intersubgranular attack (ISGA) was induced during exposure to the ASTM G110 solutions depending on the sample temper. In some cases, both IGA and ISGA were detected simultaneously. In these evaluations, none of the samples showed coarse intergranular attack of the kind that has been observed during interrupted quenching of this alloy from SHT temperatures [5]. Representative micrographs showing the extent of intersubgranular attack for artificial aging times of 24 are shown in Figs. 6-9. ISGA was most severe in SHT + AA samples aged at 190°C. It was least severe in T3 + AA samples aged at 150°C. There was an increased incidence of pitting with increased by artificial aging time. Also SHT + AA samples pitted more than T3 + AA samples.

**Table I.** Phases in AF/C458 identified with TEM.

| TEMPER | SHT | | T3 | |
|---|---|---|---|---|
| Aging Temp. | 150°C | 190°C | 150°C | 190°C |
| SGB | $T_1$ | $T_1$ | $T_B$ | $T_1$ |
| GB | $T_1$ | $T_2$, R | $T_B$, $T_2$ | R, $T_2$ |
| Matrix | $\delta'$, $\beta'$, $\theta'$ | $T_1$, $\delta'$, $\beta'$ | $T_2$, $\delta'$, $\beta'$ | $T_1$, $T_2$, $\delta'$, $\beta'$ |
| ISGC | High | High | Low | High |
| IGC | High | Low | Low | Low |

SGB: Sub-grain boundary
GB : Grain boundary
IGC : Intergranular corrosion
ISGC: Inter subgranular corrosion
$\delta' = Al_3Li$     $\beta' = Al_3Zr$     $T_1 = Al_2CuLi$     $\theta' = Al_2Cu$
$T_2 = Al_6CuLi_3$   $T_B = Al_{15}Cu_8Li_2$   $R = Al_5CuLi_3$

Alternate Immersion and Constant Extension Rate Testing

In alternate immersion testing no SHT + AA samples failed, and no macroscopic cracks were evident on the surfaces for exposures up to 60 days. For comparison, tests results from similar

**Figure 1.** Brightfield micrograph shown the fine precipitation of δ', β' and θ' within grains and subgrains for SHT+AA samples aged for 24 hours at 150°C.

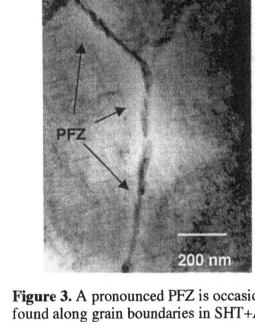

**Figure 3.** A pronounced PFZ is occasionally found along grain boundaries in SHT+AA samples aged for 24 hours at 190°C.

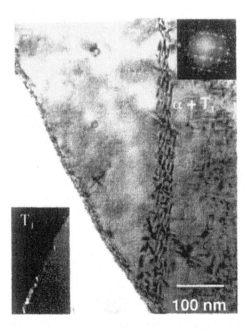

**Figure 2.** Brightfield micrograph and a darkfield inset illustrating the occurrence of $T_1$ ($Al_2CuLi$) precipitation on subgrain boundaries in SHT+AA samples aged for 24 hours at 150°C.

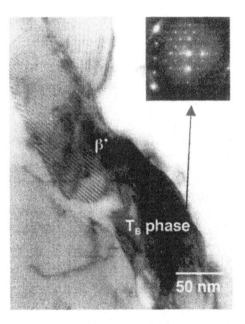

**Figure 4.** Brightfield micrograph illustrating the occurrence of $T_B$ ($Al_{15}Cu_8Li_2$) precipitation on subgrain boundaries in T3+AA samples aged for 24 hours at 150°C.

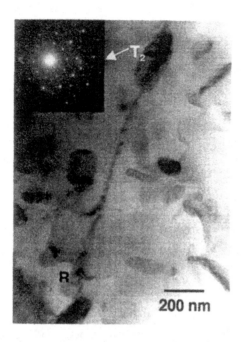

**Figure 5.** T₂ (Al₆CuLi₃) and R (Al₅CuLi₃) phase precipitation on a grain boundary in T3+AA sample aged for 24 hours at 190°C.

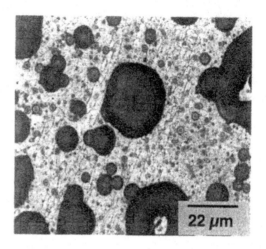

**Figure 7.** Optical micrograph of a SHT+AA aged for 24 h at 190°C after exposure to ASTM G110 solution for 2h. ISGA is visible in between corrosion pits.

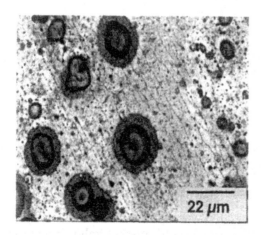

**Figure 6.** Optical micrograph of a SHT+AA aged for 24 h at 150°C after exposure to ASTM G110 solution for 2h. ISGA is visible in between corrosion pits.

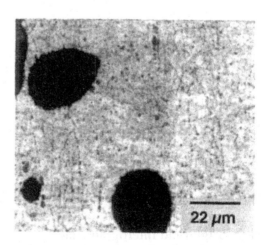

**Figure 8.** Optical micrograph of a T3+AA aged for 24 h at 150°C after exposure to ASTM G110 solution for 2h. Slight, discontinuous ISGA is visible in between corrosion pits.

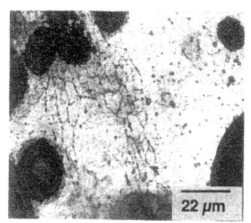

**Figure 9.** Optical micrograph of a T3+AA aged for 24 h at 190°C after exposure to ASTM G110 solution for 2h. ISGA is visible in between corrosion pits

studies on 2219, 2090-T81 and 2090B are shown in Table II. Most of these alloys failed under similar conditions and exposure times. CER testing of SHT + AA samples aged at 150° and 190°C showed that AF/C458 was not severely embrittled in 3.5 % NaCl environment under free corrosion conditions. In fact, the ductility ratios observed in these experiments were always greater than

0.50. Rather, the CER response was noticeably affected by artificial aging-induced changes in the mechanical behavior. Environmental effects were superimposed on these trends (Fig. 10). For samples artificially aged at both temperatures, the ductility versus aging time measured in NaCl solution closely mirrored the trends observed with the samples were tested in laboratory air. The lowest ductility ratio was observed for samples aged for 18.4 hrs at 150°C (0.68). By comparison, the sample aged for 24 h at 150°C had a ductility ratio of 0.87. It is interesting to note that the aging condition that showed the lowest ratio (18.4 hrs) was also the one that showed the most intersubgranular attack in IGA tests. The ductility ratio was close to 1.0 for samples aged at 190°C, but absolute ductilities were very low for aging times greater than about 10 h. Table III compares the CERT results for AF/C458 to some other related alloys. These data show that AF/C458 compares very favorably with antecedent alloys confirming earlier indications of high SCC resistance.

**Table II.** Alternate immersion tests data for AF/C458 and related alloys (3.5wt.% NaCl).

| Alloy | Stress (% of YS) | Duration (days) | Failures |
|---|---|---|---|
| 2219[1] | 75 | 40 | Failed |
| 049[2] | 75 | 40 | Failed |
| 049[3] | 75 | 40 | No failures |
| 2090[4] | 40 | 30 | 1-10 days |
| 2090[4] | 55 | 30 | 2-9 days |
| 2090[5] | 70 | 60 | 2 days |
| AF/C458[6] | 65 | 60 | No failures |

1 – Al-6.3Cu-1.35Li-0.41Ag, aged 160°C for 24 hrs, LT [6].
2 – Al-5.4Cu-1.35Li-0.41Ag, aged 160°C for 24 hrs, LT [6].
3 – Al-5.0Cu-1.35Li-0.41Ag, aged 160°C for 24 hrs, LT [6].
4 – 2090 T81 plate, 1.5" thick, ST [7].
5 – 2090B, Al-2.08Li-2.35Cu, 6% stretch, aged 162°C for 10 hrs, ST [8].
6 – Al-1.9Li-2.7Cu, 150°C and 190°C, ST-LT, 2.4 to 36 hrs.

# Summary

Overall, the results of this evaluation indicate that AF/C458 possesses SCC resistance that is as good or better than its immediate Al-Cu-Li alloy antecedents. Specific findings in this investigation include:

- The presence of SGB or GB precipitation is not a sufficient cause for IGC or SCC susceptibility in AF/C458.
- The degree of supersaturation affects GB and SGB precipitation and IGC tendencies.
- IGC test and TEM studies associated the presence of $T_1$ at subgrain and grain boundaries with IGC and SCC susceptibility.
- The ability of AF/C458 to resist SCC derives from the low rates of grain boundary dissolution under many conditions, and damage tolerance conferred by robust isotropic properties

**Table III.** Data comparing CER tests results for various alloys.

| Alloy | Strain rate ( sec$^{-1}$) | Potential (mV$_{SCE}$) | Environment | Ductility loss $\varepsilon_{env}/\varepsilon_{ref}$ |
|---|---|---|---|---|
| 2090C – UA[1] | 1.25x10$^{-6}$ | -700 | Aerated 3.5% NaCl | 0.57 |
| A[2] | 10$^{-6}$ | -750 | Aerated 0.5M NaCl | 0.77 |
| A[3] | 7.2x10$^{-6}$ | -2000 | Pre-charging in 0.1N HCl, pH1 at –3000 mV$_{SCE}$ for 10 hrs, followed by cathodic charging in 0.1N HCl. | 0.64 |
| 2090[4] | 5.6x10$^{-5}$ | -- | Cathodic charging in 0.04N HCl + As$_2$O$_3$ for 12 hrs | 0.82 |
| 2219-T87[5] | 7.6x10$^{-6}$ | -720 (OCP) | 3.5% NaCl | ~0.9* |
| 2195-T3[6] | 7.6x10$^{-6}$ | -615 (OCP) | 3.5% NaCl | ~0.9* |
| 2195-T8[7] | 7.6x10$^{-6}$ | -747 (OCP) | 3.5% NaCl | ~0.95* |
| AF/C458[8] | 10$^{-6}$ | -710 (OCP) | 3.5% NaCl | 0.96 |
| AF/C458[9] | 10$^{-6}$ | -690 (OCP) | 3.5% NaCl | 0.87 |
| AF/C458[10] | 10$^{-6}$ | -730 (OCP) | 3.5% NaCl | 0.99 |

\* approximate
1) Al-2.12Li-2.60Cu, solution heat-treated, aged @ 162°C for 4 hrs [8].
2) Al-2.05Li-2.15Cu, peak aged for 24 hrs at 190°C [9].
3) Al-2.2Li-2.9Cu, peak aged at 190°C [10].
4) Al-2.08Li-2.35Cu, T8E41 (overaged), 12.7mm thick [11].
5) Al-6.3Cu-0.3Mn-0.18Zr, 1.6 mm thick, overaged [12].
6) Al-6.3Cu-0.3Mn-0.18Zr [12].
7) Al-3.94Cu-0.96Li-0.39Ag, overaged 24 hrs at 190°C [12].
8) Al-1.9Li-2.7Cu, under aged 2.4 hrs at 150°C.
9) Al-1.9Li-2.7Cu, peak aged 24 hrs at 150°C.
10) Al-1.9Li-2.7Cu, over aged 36 hrs at 150°C.

## Acknowledgements

This work was supported by the Air force Research Laboratory, Wright Patterson AFB under contract no. TMC-96-5835-0026-06, K.V. Jata Program Manager. This paper was originally published in the Proceedings of the Symposium on Corrosion and Corrosion Prevention of Light Metals and Alloys, held at the 198th meeting of the electrochemical Society, Phoenix AZ, October 2000 (in press). This paper is published with the permission of the Electrochemical Society, Pennington, NJ.

## References

1. E. L. Colvin, G. E. Stoner, G. L. Cahen, Jr., Corrosion, 42, 416 (1986).
2. "Workshop on characterization of Al-Li alloy AF/C458 plate"; Summary, K. Jata, Aluminum-Lithium Workshop, Eds: Kumar Jata, Aug 18, 1998, collection of slides.
3. ASTM G110-92, " Standard Practice for Evaluating Intergranular Corrosion Resistance of Heat-Treatable Al alloys in NaCl plus $H_2O_2$ Solution," American Standards for Testing and Materials, Philadelphia, PA, 1992.
4. ASTM G-44, "Standard Practice for Evaluating Stress Corrosion Cracking Resistance of Metals and Alloys by Alternate Immersion in 3.5 % Sodium Chloride Solution," American Standards for Testing and Materials, Philadelphia, PA, 1988.
5. J. E. Kertz, P. I. Gouma, R. G. Buchheit, "Localized Corrosion Susceptibility of Al-Li-Cu-Mg-Zn Alloy AF/C458 Due to Interrupted Quenching from Solutionizing Temperatures," submitted to Met. Trans. A, November, 2000.
6. K. Moore, T. J. Langan, F. H. Heubaum, J. R. Pickens, Aluminum-Lithium V, p.1261, E. A. Starke, Jr., T. H. Sanders, Jr., Eds., MCEP Oxford (1989).
7. R. J. Bucci, R. C. Malcolm, E. L. Colvin, S. J. Murtha, R. S. James, Eds., ALCOA Technical Report, p.143, NSWC TR 98-106, Aluminum Company of America, Alcoa Laboratories, Aloca Center , PA (1989).
8. J. P. Moran, "Mechanism of Localized Corrosion and Stress Corrosion Cracking of an Al-Li-Cu Alloy 2090", Ph.D. Dissertation, Univ. of Virginia (1990).
9. R. Balasubramaniam, D. J. Duquette, K. Rajan, Acta Mater., 39, 2597 (1991).
10) W. Hu, E. I. Meletis, Materials Science Forum, 331-337, 1683 (2000).
11. S. S. Kim, E. W. Lee, K. S. Shin, Scripta Metall., 22, 1831 (1988).
12. A. Frefer, G. E. Bobeck, T. A. Place, J. F. Bailey, "Advances and Applications in the Metallography and Characterization of Materials and Microelectronic Components, " p. 49, Proc. of the 28th Annual Technical Meeting of the International Metallographic Society, D. W. Stevens, E. A. Clark, D. C. Zipperian, E. D. Albrecht, Eds., Materials Park OH (1996).

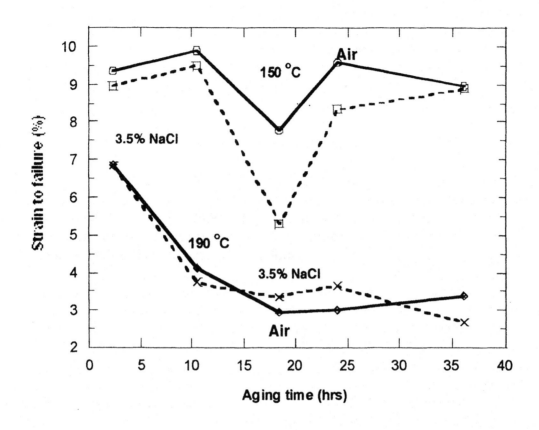

**Figure 10.** Strain to failure in 3.5% NaCl and in air versus artificial aging time for SHT+AA AF/C458 aged at 150° and 190°C.

# LIGHTWEIGHT ALLOYS FOR AEROSPACE APPLICATION

*Edited by:*
*Dr. Kumar Jata, Dr. Eui Whee Lee,*
*Dr. William Frazier and Dr. Nack J. Kim*

## ALUMINUM-LITHIUM ALLOYS

Localized Corrosion Mechanisms
of Al-Li-Cu Alloy AF/C458 After
Interrupted Quenching from
Solutionizing Temperatures

*P.I. Gouma, J.E. Kertz and R.G. Buchheit*

Pgs. 119-127

184 Thorn Hill Road
Warrendale, PA 15086-7514
(724) 776-9000

# LOCALIZED CORROSION MECHANISMS OF AL-LI-CU ALLOY AF/C458 AFTER INTERRUPTED QUENCHING FROM SOLUTIONIZING TEMPERATURES

P. I. Gouma
Department of Materials Science and Engineering,
SUNY at Stony Brook, Stony Brook, NY 11794

J. E. Kertz, and R. G. Buchheit
Department of Materials Science and Engineering,
The Ohio State University, Columbus, OH 43210

## Abstract

The effect of slow or delayed quenching on localized corrosion mechanisms for the Al-2.05Li-2.70Cu-0.6Mg-0.3Zn-0.08Zr alloy AF/C458 was assessed. Alloy samples were subject to a series of systematic interrupted quenching experiments conducted at temperatures ranging from 480°C-230°C for times ranging from 10 to 1000 seconds. Samples were then exposed to an oxidizing aqueous chloride environment to induce localized attack. The alloy exhibited pitting, intersubgranular attack (ISGA), or intergranular attack (IGA), depending on the time at temperature. Detailed microstructural characterization by means of transmission electron microscopy was employed, which clearly showed that ISGA and IGA were related to the precipitation of a Zn-modified $T_1(Al_2(Cu,Zn)Li)$ at high and low angle boundaries respectively. In situations where the $T_B(Al_7Cu_4Li)$ phase was present on boundaries instead of $T_1$, IGA and ISGA susceptibility was comparatively diminished.

## Introduction

The alloy examined in this work is a new Al-Li-Cu alloy developed by the Alcoa under a US Air Force contract (designated as: AF/C458) which is targeted for aerospace applications. One of the most important features of this alloy is that it exhibits mechanical properties with significantly reduced anisotropy, compared to the earlier generations of Al-Li alloys [1-2].

Plate processing is typically used for Al-Li-type alloys. The quench sensitivity is likely to affect the corrosion behavior of these materials, including inter-granular attack (IGA) and intersubgranular attack (ISGA) and stress corrosion cracking (SCC). In this study, the pitting, IGA, and ISGA behavior of the alloy AF/C458 were studied for various time - temperature combinations that are relevant to quenching of thick plates from high temperature thermal processing. The localized corrosion behavior of this alloy has been mapped using interrupted quenching experiments for temperatures ranging from 480°C-230°C and times ranging from 10-1000s.

The results showed that the maps for IGA and ISGA are almost coincident which suggests that a similar corrosion mechanism is in effect in both cases. Detailed investigations using transmission electron microscopy showed that the precipitation of $T_1$ phase at subgrain and grain boundaries is associated with the observed localized corrosion behavior of this alloy.

Lightweight Alloys for Aerospace Applications
Edited by Kumar Jata, Eui Whee Lee,
William Frazier and Nack J. Kim
TMS (The Minerals, Metals & Materials Society), 2001

## Experimental Procedures

### Materials, Processing, and Characterization

The samples used in this study were prepared from unrecrystallized 0.5 in. plate of AF/C458 (Al-2.8Cu-1.8Li-0.6Zn-0.3Mg-0.3Mn-0.07Zr) prepared from direct chill cast billets (scalped to 0.375 in.). This material was produced by Alcoa and the Air Force Research Laboratories. The plate was subsequently solution treated, quenched, stretched to 6%, and allowed to naturally age, yielding a material in a T36 temper. The grain size and morphology of the plate resembled those of a rolled product.

Interrupted quenching procedures used in this work were based on the method followed in reference [3]. Thirty samples of the alloy AF/C458 were cut into 1.4 x1.4 x1.0cm blocks. Type K thermocouples werre attached to the center of these blocks. The thermocouples were then connected to National Instruments Virtual Bench-Logger software for data acquisition. All specimens were initially solution heat-treated (SHT) at 549°C for one hour, in a salt pot. Following the SHT, each specimen was quenched in Wood's metal to a second desired temperature ( 482°, 425°, 349°C, 282°, and 234°C), then held for 10, 50, 100, or 1000 seconds, and then quenched in cold water to room temperature to terminate the experiment. The quench rate between steps was on the order of 50°-100°C/s.

Following heat-treatment, IGA evaluation was conducted using the $NaCl+H_2O_2$ solution as specified in ASTM G110 [4].Following a six hour immersion, the specimens were removed and rinsed in distilled water. Samples were then polished on a nylon cloth with 1$\mu$m alumina slurry for about ten seconds for excess NaCl and surface product removal to make inspec-

tion easier. Finally, the tested samples were examined by optical microscopy.

Transmission electron microscopy (TEM) was employed to assess the microstructural development of the specimens. A Philips CM200 TEM and a Philips CM300 FEG STEM equipped with EDAX detectors and the EmiSpec software were used for these studies. Disks of 3mm diameter were prepared from samples treated at 3 conditions, as shown in Table 1. The discs were punched from thinned foils, dimpled (using the VCR Group 500i Dimpler) and ion-milled to perforation in Ar- 6kV, at 12°(using a Gatan Dual Ion Mill) Microstructures were imaged using bright field and dark field imaging conditions. Electron diffraction techniques were used to study the structural details of the various precipitates present in these samples. Elemental maps were produced using Energy Dispersive X-ray Spectroscopy.

**Table 1:** Specimens Studies in TEM

| Specimen | Treatment Conditions |
|----------|----------------------|
| 1 | 350°C / 84 sec |
| 2 | 476°C / 6sec |
| 3 | 288°C / 7sec |

### Results and Discussion

### Localized Corrosion Mapping

Depending on the time at temperature during the interrupted quenching experiments, one or more modes of localized corrosion were observed after exposure to the $NaCl/H_2O_2$ solution. These modes included pitting at constituent particles, *intra*granular pitting probably associated with coarse *intra*granular $T_1$ precipitates, IGA, and ISGA (see figure 1). IGA and ISGA are readily distinguishable. Figure 1a illustrates the typical characteristics of ISGA in these tests, which is manifested as fine, continuous etching that outlines subgrains that range from 5 to 40 μm in diameter. There is also evidence of constituent particle pitting that is not directly associated with subgrain boundary dissolution. TEM analysis of subgrains in samples susceptible to SGB attack shows that these boundaries are populated with a high density of fine $T_1$ precipitates. Figure 5b is representative of the IGA mode of attack. Attack at high angle grain boundaries (IGA) is considerably more coarse than ISGA, and thus clearly distinguishable from it. Figure 1c is representative of samples in which pitting, but no IGA or ISGA was detected. In these samples, large equiaxed pits were randomly dispersed throughout the alloy. These pits were associated with large constituent particles that would have been unaffected by the solution heat treatment and quenching procedures. This mode of localized corrosion is expected with solution heat treated and quenched alloys, or alloys in which the natural aging response is limited. Since there is no appreciable precipitation at grain boundaries, there is little or no localization of attack there.

The results from the interrupted quenching and exposure tests were used to map out the localized corrosion behavior of AF/C458 according to the various attack modes described above. In this way time-temperature-localized corrosion mechanism plots were constructed. Plots illustrating the occurrence of pitting, IGA and ISGA are shown in Figure 2.

The curves drawn on each plot approximately separate domains of either IGA or ISGA from regions where only pitting was detected. The

domains for IGA and ISGA are very nearly coincident suggesting that the origins of these two types of attack are related to the same precipitation process.

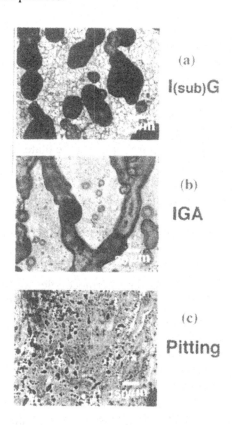

(a)

I(sub)G

(b)

IGA

(c)

Pitting

**Figure 1:** Optical micrographs showing representative modes of localized corrosion observed in this work.

Microstructural Characterization

Three different samples processed under the conditions shown in Table I, were chosen for microstructural analysis by TEM. Particular emphasis was given to characterizing the various phases present, both within the grains as well as at the grain and subgrain boundaries.

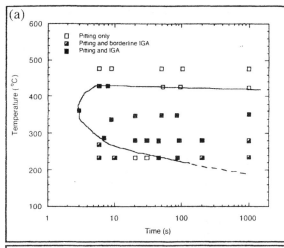

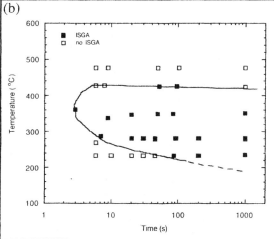

**Figure 2:** Time-temperature-property curves for localized corrosion: (a) indicates domains of IGA susceptibility; (b) indicates domains of ISGA susceptibility.

TEM characterization of AF/C458 isothermally treated at 350°C for 84 seconds: In exposure testing this sample exhibited pitting at constituent particles, IGA and ISGA. Imaging and electron diffraction analysis showed the distribution of fine, plate-like $T_1$ precipitates within the matrix grains, a fine dispersion of $\delta'$ ($Al_3Li$) and a lesser concentration of spherical $\beta'$ ($Al_3Zr$) particles.

These two precipitates are isostructural and form a simple superlattice of the fcc $\alpha$-Al matrix [5]. This sample contained an extensive network of low angle boundaries, which were decorated by small $T_1$ precipitates (see figure 3a). It was often observed that $T_1$ plates shared a common orientation (inset in figure 3a is a selected area diffraction pattern showing the relative orientation of the two subgrains and the precipitate plates). This may enable a high concentration of $T_1$ to develop along subgrain boundaries making them especially susceptible to preferential dissolution as the active $T_1$ phase forms a nearly continuous pathway. No Cu depletion was detected in EDX profiles collected across the boundaries suggesting that ISGA susceptibility is due to selective $T_1$ dissolution in this case. $T_1$ was also observed to occur on high angle grain boundaries as shown in Figure 3b. The $T_1$ precipitates along this particular boundary appear to have an orientation relationship with at least one of the two grains that form the boundary (Figure 3c). Furthermore, the boundary has a step-wise appearance, as if it is pushed locally by precipitate growth. Sometimes, the sequence of $T_1$ phase precipitates along a grain boundary is seen to be interrupted by the presence of $T_2$ ($Al_6CuLi_3$) phase particles. The $T_2$ particles were found to have an icosahedral structure and the lattice spacings measured for this phase were consistent with the JCPDS files [6]. As with the subgrain boundaries, EDX profiles did not indicate Cu depletion.

Elemental mapping of the grain boundary precipitates revealed that $T_1$ precipitates contain Zn in this alloy. Bright and dark field scanning transmission electron micrographs were obtained and the areas of interest for compositional analysis were determined. Figure 4 shows a scanning transmission micrograph and companion Al, Cu, and Zn maps for a grain boundary precipitate that electron diffraction showed was $T_1$. Pending fur-

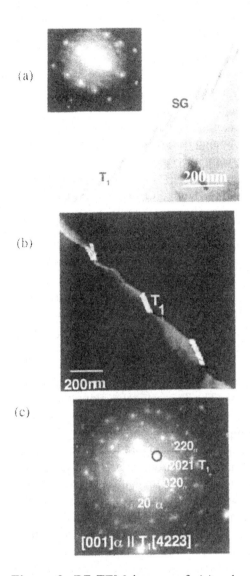

(a)

SG

$T_1$

200nm

(b)

$T_1$

200nm

(c)

220

2021 $T_1$

020

20 α

$[001]α \parallel T_1[4223]$

**Figure 3:** BF TEM images of: (a) subgrain boundary decorated by T1 platelets; (b) a high-angle boundary. (c) SAD of $T_1$ and matrix.

ther characterization, the notation $Al_2(Cu,Zn)Li$ is used to describe Zn-modified $T_1$. How Zn affects the $T_1$ electrochemistry and alloy localized corrosion has not yet been determined. Zn has not been detected in smaller $T_1$ precipitates on subgrain boundaries and in the matrix phase,

however, this may be due to sensitivity limitations imposed by the small precipitate volumes compared to the beam-specimen interaction volume, Zn dissolved in the matrix phase, and instrumentation.

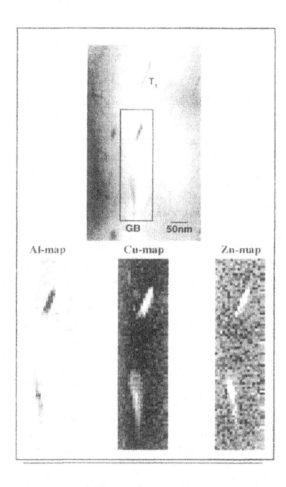

$T_1$

GB    50nm

Al-map    Cu-map    Zn-map

**Figure 4:** The picture in the top is a BF STEM image of a grain boundary containing $T_1$ precipitates; the pictures at the bottom represent elemental maps obtained using X-ray spectroscopy.

TEM characterization of AF/C458 isothermally treated at 476ºC for 6 seconds. Samples in this condition exhibited pitting at constituent particles, but no IGA or ISGA. In the matrix phase fine spherical β′ phase precipitates and overlapping distributions larger, elongated $T_1$ and θ′ phase particles were detected The θ′ was found to have a distinct orientation relationship with the matrix: ([112]α // [001]θ′). Although this sample exhibited no signs of IGA of ISGA, a significant amount of precipitation was detected on both grain and subgrain boundaries. Often, precipitates formed continuous or nearly continuous networks, as shown in Figure 5. Electron diffraction analysis showed that these networks were comprised of overlapping $T_B$ and θ'. There is very little information about the $T_B$ phase in the literature. But, this phase is considered to be unstable in lower temperatures (<350°C) [7]. It has also been reported (in overaged commercial 2020 alloys) that a transition takes place from the θ′ phase to $T_B$ [8]. This transition has been assumed to associated with Li incorporation into θ′ and subsequent development of $T_B$. The electrochemistry of $T_B$ has not been rigorously studied, but in their early work, Hardy and Silcock noted that $T_B$ remained unattacked after etching in 1%HF solution for short times, whereas the $T_1$ phase was clearly etched by the same treatment [8].

TEM characterization of AF/C458 isothermally treated at 288ºC for 7 seconds: This sample exhibited pitting associated with constituent particles, and IGA. ISGA was not detected. Analysis of the sample treated under this condition revealed that the material has an almost identical microstructural evolution to that of the sample 1 in table 1. There was a fine dispersion of δ′ particles within the matrix, as well as few β′ particles scattered in the material. Although the subgrain boundaries were found to

be depleted of θ′ precipitates, occasionally β′ particles were seen decorating these boundaries along with numerous fine plates of the $T_1$-phase. The grain boundaries were decorated by large $T_1$ plates and some $T_2$ phase particles.

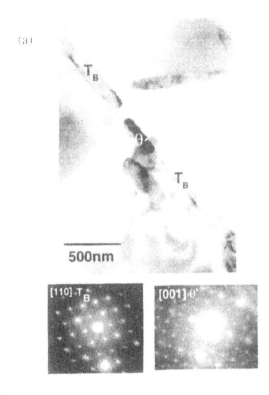

**Figure 6:** (a) BF TEM image showing $T_B$ and θ′ phase precipitation on SGBs; (b) and (c) are SADs from zone axes of $T_B$ and θ′ phases.

Implications of the findings: AF/C458 is a wrought, precipitation hardenable alloy that derives its strengthening from the development of complex microstructure. Elevated temperature exposure induces microstructural changes, some of which alter any subsequent accumulation of localized corrosion damage. The only mode of localized corrosion that does not appear to be significantly affected by elevated tem-

125

perature exposure is pitting associated with coarse constituent particles. The common observation is for pits to form in the matrix phase at the periphery of the particle. Dissolution of the particle may or may not be apparent. This pitting can sometimes lead to intergranular or intersubgranular attack when grain and subgrain precipitation has occurred.

Of all the phases identified in these samples, the occurrence of IGA and ISGA correlate most strongly with the presence of $T_1$ on grain and subgrain boundaries. This effect has been noted for other Al-Li-Cu alloys where $T_1$ precipitation on boundaries is clearly the predominant microstructural influence on intergranular and stress corrosion cracking susceptibility [9]. The AF/C458 samples used in this study were taken from material that was solution-treated, stretched, and naturally aged. However, individual samples were again solution treated presumably eliminating the influence of prior stretching and natural aging. Nonetheless, $T_1$ was observed to precipitate within grains. There also was clearly a tendency for coarse $T_1$ precipitation on high angle boundaries, and high-density $T_1$ precipitation on subgrain boundaries. However, EDX line profiles and elemental mapping did not reveal the presence of either continuous or discontinuous Cu depleted regions in IGA- and ISGA-susceptible samples. These observations indicate that selective $T_1$ dissolution dominates boundary attack, but are still somewhat unsatisfying because they do not account for why continuous IGA is observed in instances where TEM shows that coarse $T_1$ precipitation on boundaries is relatively discrete.

It is worth noting that results from this study suggest that Zn is incorporated in to coarse grain boundary $T_1$. It is possible that Zn is also incorporated into subgrain boundary and ma-

trix $T_1$ as well. Although the effect has not yet been measured, Zn incorporation would be expected to have a direct effect on the electrochemistry of the $T_1$ phase.

## Summary

- Isothermal TTT diagrams for IGA and ISGA were determined for the experimental alloy AF/C458. The domains of IGA and ISGA susceptibility were coincident and occurred when samples were exposed in the 280° top 430°C temperature range for times greater than about 10 seconds.

- On the basis of selected TEM observations, IGA and ISGA appeared to occur when $T_1$ precipitates were detected on boundaries. No evidence for Cu depletion along grain or subgrain boundaries could be detected in any of the samples examined. The presence of $T_1$ on boundaries appears to be a necessary condition for boundary attack.

- Zn was detected in $T_1$ precipitates on high angle grain boundaries in AF/C458. Whether Zn was present in $T_1$ in the matrix phase or on subgrain boundaries remains to be determined. Additionally, the effect of Zn on $T_1$ electrochemistry, and subsequent effects on IGA or ISGA remain to be determined.

- $T_B$ and $\theta'$ were detected on grain and subgrain boundaries under conditions where the alloy was resistant to IGA and ISGA. Presuming the boundary attack depends on selective precipitate dissolution, this result suggests that these phases are electrochemically more benign than the $T_1$ phase.

## Acknowledgments

This work was supported by the Air Force Research Laboratories under contract no. TMC96-5835-71-04 through Technical Management Concepts, Inc. The support and technical interactions with K.V. Jata, AFRL are gratefully acknowledged.

## References

1. A. K. Hopkins, K. V. Jata, R. J. Rioja, Materials Science Forum, 5, p. (1996).
2. V. K. Jain, K. V. Jata, R.J. Rioja, J. T. Morgan, A. K. Hopkins, J. Materials Proc. Tech., 73, 108 (1998).
3. W. L. Fink and L. A. Willey, Trans. AIME, 175, 414 (1948).
4. ASTM G110-92, *Standard Practice for Evaluating Intergranular Corrosion Resistance of Heat-treatable Al alloys in NaCl plus $H_2O_2$ solution*, ASTM, Philadelphia, PA, (1992).
5. J. M. Silcock, J. Inst. Metals, 88, 357, (1959-60).
6. JCPDS file 40-1158.
7. R. G. Buchheit, J. P. Moran, G.E. Stoner, Corrosion, 46, 610 (1990).
8. H. K. Hardy, and J. M. Silcock, J. Inst. Metals, 84, 423, (1955-56).
9. R. G. Buchheit, J. Electrochem. Soc., 142, 3394 (1995).

# LIGHTWEIGHT ALLOYS FOR AEROSPACE APPLICATION

*Edited by:*
*Dr. Kumar Jata, Dr. Eui Whee Lee,*
*Dr. William Frazier and Dr. Nack J. Kim*

## ALUMINUM-LITHIUM ALLOYS

Effect of Single and Duplex Aging on
Microstructure and Fatigue Crack
Growth in Al-Li-Cu Alloy AF/C-458

*J. Fragomeni, R. Wheeler, K.V. Jata and S. Geoffrey*

Pgs. 129-139

184 Thorn Hill Road
Warrendale, PA 15086-7514
(724) 776-9000

# EFFECT OF SINGLE AND DUPLEX AGING ON MICROSTRUCTURE AND FATIGUE CRACK GROWTH IN Al-Li-Cu ALLOY AF/C-458

J. Fragomeni, †R. Wheeler, *K.V. Jata and ‡S.Geoffrey

*Air Force Research Laboratory, Materials and Manufacturing Directorate, AFRL/MLLMD,
2230 Tenth Street, Wright Patterson Air Force Base, Ohio
†UES, Inc., 4401 Dayton-Xenia Rd., Dayton, Ohio
‡Ohio University, Department of Mechanical Engineering, Athens, Ohio
The University of Detroit, Department of Mechanical Engineering , Detroit, Michigan

## Abstract

Microstructure and fatigue crack growth (FCG) rates in Al-Li-Cu alloy AF/C-458 were studied following single and duplex aging treatments on specimens that were given a 6 percent stretch after solution heat treatment. Aging response was determined using hardness and compression yield strength measurements. Quantitative transmission electron microscopy methods were used to characterize average size, volume fraction, number density, and interparticle spacing of strengthening precipitates, $\delta$'($Al_3Li$) and $T_1$ ($Al_2CuLi$). Strength and fatigue crack growth rates for select heat treatments were obtained and were related to the precipitate microstructure and compression strength data. The relationship between precipitate characteristics and mechanical properties are discussed.

## Introduction

Aluminum-lithium alloys are being commercially used in military aircraft and space in several critical applications. There is still, however, an interest in developing next generation aluminum-lithium alloys with improved specific strength and damage tolerance and reduced mechanical property anisotropy. Recently, the U.S. Air Force and Alcoa developed two aluminum-lithium alloys designated AF/C-489 and AF/C-458 (1 -5 ) with 2.1 and 1.8 weight percent lithium, respectively. The AF/C-458 alloy has been shown to possess superior strength, damage tolerance properties (6 ) and stress corrosion cracking resistance (7 , 8 ). Studies by Csontos and Starke recently (4, 5) compared precipitation and slip behavior between the AF/C-489 and AF/C-458 alloys by employing single and duplex aging. They concluded that the lower ductility in AF/C-489 compared to AF/C-458 is due to a large grain size, a higher volume fraction of the $\delta$' precipitates and enhanced planar slip and stress concentration at grain boundaries.

Lightweight Alloys for Aerospace Applications
Edited by Kumar Jata, Eui Whee Lee,
William Frazier and Nack J. Kim
TMS (The Minerals, Metals & Materials Society), 2001

The objective of this work was to understand the effect of single and duplex artificial aging on the precipitate microstructure in the advanced aerospace Al-1.8wt%Li-2.7Cu-0.3Mg-0.5Zr-0.3Mn-0.8Zn alloy AF/C-458 and to study the effects of aging on FCG.

## Background

Microstructures of Al-Li alloy AF/C-458 were recently characterized by Csontos and Starke (4, 5). Results of their studies show that precipitation in AF/C-458 is typical to that of other Al-Li-Cu alloys. Various precipitates encountered in this alloy are briefly described below.

### The $T_1$ ($Al_2LiCu$) phase

The crystal structure of the $T_1$ ($Al_2LiCu$) phase is known to have an hexagonal crystalline structure. The $T_1$ phase was initially discovered by Hardy and Silcock (9) who determined its crystal structure as hexagonal with a = 0.4965 nm and c = 0.9345 nm. Noble and Thompson (10) observed the formation of plate shaped precipitates on {111} in ternary Al-Li-Cu alloys comparable to 2090 and C458. Noble and Thompson (10) identified these plate shaped precipitates on {111} as the $T_1$ phase and also verified the orientation relationship originally proposed by Hardy and Silcock (9) as {0001}$T_1$ //{111}. The thin plates lie on {111} matrix planes, which means that three matrix slip planes intersect the plate at angles of 70.5°. The $T_1$ particles provide a substantial strengthening contribution to Al-Li-Cu alloys.

### The $\delta'$ ($Al_3Li$) phase

The precipitates are ordered, spherical, and coherent with the aluminum lattice and they also nucleate and coarsen/grow homogeneously in the matrix. The $\delta'$ ($Al_3Li$) phase has the ordered face-centered-cubic (fcc) or $L1_2$ ($Cu_3Au$) superlattice crystal structure and has a lattice parameter of a = 0.404±0.003 nm. The cubic lattices of the d' particles and the d' aluminum matrix are geometrically similar, having a small misfit strain on the order of −0.08±0.02% (2, 6, 7). The $\delta'$ ($Al_3Li$) phase orientation is the same as the fcc matrix that is {100}$_{\delta'}$ // {100}$_{Al}$.

### The $Al_3Li/Al_3Zr$ Composite Precipitates

Zirconium additions to aluminum-lithium alloys form the dispersoid $\beta'$ ($Al_3Zr$) which substantially decreases the recrystallization of grains during ingot breakdown and hot rolling. The composite precipitates contain an inner core of $Al_3Zr$ surrounded by an outer shell of the $Al_3Li$ phase and resemble a ring shaped or doughnut shaped spherical type particles (11).

## Experimental

AF/C-458 plate was obtained from Alcoa in T3 temper. The plate had a 6 percent stretch after solution heat treatment. Samples were machined from the plate and later artificially aged for both single and duplex aging practices. Heat treatments were carried out on one inch square coupons in ambient atmosphere under static air conditions. A number of test coupons were aged at 150°C for various times and Rockwell B hardness measurements recorded. Hardness readings were made on the rolling plane with parallel surfaces ground to a 600 grit finish. After determining the peak age condition for 150°C, a number of coupons were given duplex heat treatments consisting of 24 hours to achieve the peak age hardness at 150°C followed by various aging times at 190°C. Artificial aging was followed by hardness and compression yield strength measurements.

The average size, distribution, number density, volume fraction, and spacing of the intermetallic strengthening precipitates $\delta'$ and $T_1$ were directly measured from TEM dark field images. The volume fraction was also calculated based on the TEM foil thickness and average precipitates size.

TEM analysis was carried out on a Philips CM200 instrument equipped with a field emission source. Thin foil specimens were prepared by electropolishing 3 millimeter discs in a solution of 5% perchloric acid in methanol at a temperature of –40°C. Bright field and dark field images were recorded from each sample with the grain of interest oriented near <110> crystallographic directions. Foil thicknesses used for the quantification of particle volume fractions and densities were measured using fringe images of inclined $T_1$ particles belonging to other crystallographic variants.

Particle size measurements were performed for both $T_1$ and $Al_3Li$ precipitates directly from TEM images. Projected images of the actual particle sizes were measured. Based on standard quantitative microscopy theory and methods, the measurements of the projected images were converted to actual real particle sizes. Two projected particle size diameter images were measured directly from TEM photmicrographs for each particle. Thus an average size and aspect ratio was determined for each individual particle of the entire distribution of particle sizes. For the nonspherical $T_1$ particle size distribution histograms were determined from the size measurements. The average particle size was also determined based on the measurements of the projected particle images. The alloy was stretched 6% deformation after solution heat treating but prior to artificial aging.

## Results

**Optical Microstructure, Hardness, Strength:** The optical microstructure of the alloy AF/C-458, shown in Figure 1. This figure illustrates the conventional pancake shaped grains observed in aluminum-lithium alloy plate products. The aging response of the

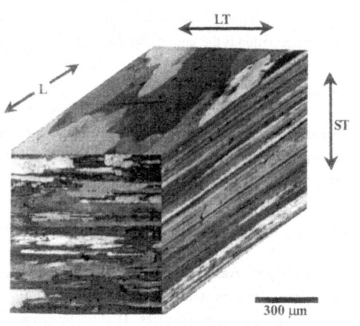

*Figure 1. Optical microstructure of AF/C-458.*

AF/C-458 plate material is shown in Figure 2. The open circles indicate the Rockwell B hardness of samples isothermally aged at 150°C as a function of time. The closed circles represent samples given an initial 24 hour heat treatment at 150°C followed by aging for additional times at 190°C. Here, times are given as a total of the 24 hour-150°C plus the time at 190°C. Data in Figure 2 shows that even with duplex aging hardness loss in AF/C-458 is minimal. Compressive buckling strengths obtained as a function of aging time are shown in this figure as well.

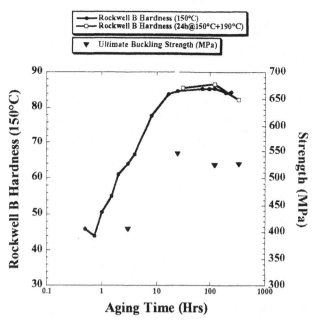

Figure 2. Hardness of AF/C-458 as a function of single and duplex aging.

The precipitate phases that were observed included δ' ($Al_3Li$), $T_1$ ($Al_2LiCu$), β'($Al_3Zr$), and θ' ($Al_2Cu$). For all of the aging times studied here the composite $Al_3Zr$/ $Al_3Li$ particles were found to be larger in size than the δ' but much fewer in number density than the δ'precipitates. Here, only observations on the $T_1$ and δ' precipitates are discussed.

**Microstructure of sample aged at 150°C for 3 hours:**

Figure 3(a) shows a bright field image of the uniform distribution of dislocations across a 2 micron square area of the microstructure. The SAD pattern in the insert of Figure 3(b) does not contain the

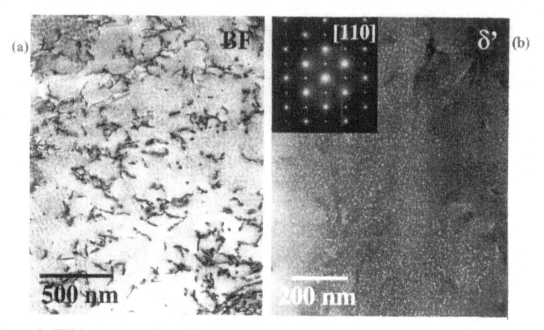

Figure 3. TEM micrographs of the AF/C-458 sample aged for 3 hours at 150°C showing (a) bright field image of dislocation segments and (b) dark field image of δ' precipitates. Inset is a [110] diffraction pattern with {100} superlattice intensity used to image the δ' particles.

streaking associated with the $T_1$ phase. Some degree of diffusivity is present, however, suggesting the existence of very early stage nucleation events of precipitates related to the line defects as reported in the literature. The $\delta'$ particles are very fine in Figure 3(b) with a high number density. Strong superlattice intensity consistent with the $L1_2$ ordering of this phase is present in the diffraction pattern. Finally, under weak beam imaging conditions, not shown, a high density of dislocations are evident but $T_1$ has not begun to nucleate. Higher magnification imaging would be required to resolve such events.

**Microstructure of samples aged at 150°C for 24 hours:**

The microstructure for this aging condition (peak aging) is shown in Figure 4. Coarsening of the $\delta'$ phase, in (a), and nucleation and growth of $T_1$ particles, in (b), has occurred as seen by comparing the TEM micrographs in Figure 3 for the 3 hour aged sample. The [110] diffraction pattern now indicates well- developed streaking along g=111 vectors related to the thin $T_1$ particles. These were absent in the pattern from the 3 hour aged sample. It should be noted that each $\delta'$ precipitate now coexists with a very thin plate of the $\theta'$ phase within each particle. This thin sheet does not persist at the longer aging times.

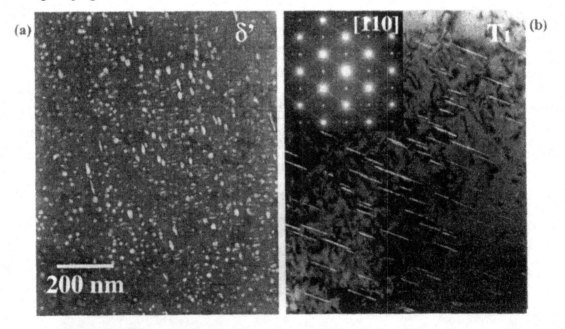

*Figure 4. Dark field TEM micrographs and the associated [110] diffraction pattern for AF/C-458 alloy aged at 150 °C for 24 hours; (a) $\delta'$ precipiates and (b) $T_1$ precipitates.*

**Microstructure of duplex aged samples:**

Long aging times at 150°C did not decrease the hardness, as illustrated in Figure 2. Hence a duplex aging treatment was implemented that consisted of initially aging the sample at 150°C sample for 24 hours followed by an additional treatment for various times at 190°C. Interestingly, from Figure 2, long aging times at 190°C caused little change in hardness once the peak age microstructure was established. The microstructures were still expected to coarsen. Figure 5 shows the precipitate structure of the duplex aged sample given a secondary age for 96 hours at 190°C. Here, both the $\delta'$

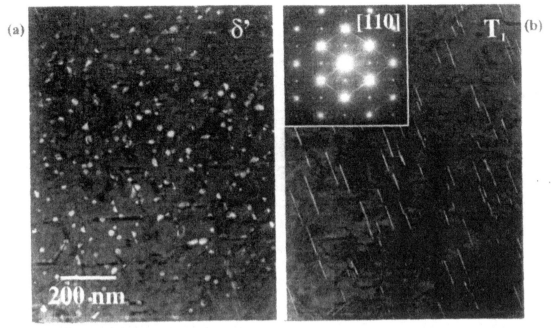

*Figure 5. TEM dark field images and the associated diffraction pattern showing coarsening of the δ' particles in (a) and of $T_1$ precipitates in (b) after aging at 150°C for 24 hours followed by 190°C for 96 hours.*

and $T_1$ particles have coarsened. Particle separation has also increased as is particularly noticeable for the δ' precipitates. The degree of such changes are the subject of this work where quantitative measurements of the strengthening particles has been carried out.

**Quantitative Microscopy Measurements:**

Particle volume fraction, volume number density and particle size for $T_1$ and δ' measured from TEM micrographs are shown in Figures 6, 7 and 8, respectively. The experimentally measured planar particle sizes were converted into true particles sizes determined from quantitative theory based on the TEM foil thickness and the area and number densities of the precipitates.

The harmonic diameters used in the calculating the measured values were determined from the planar particle size measurements since the harmonic diameters are used to correlate with particle size surface area. The harmonic diameter is inversely equal to the sum of the reciprocals of the measured diameters divided by the total num-

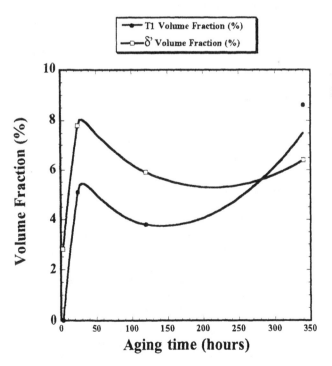

*Figure 6. Volume fraction of $T_1$ and δ' precipitates as a function of aging for both single and duplex aging treatments of the alloy AF/C-458.*

135

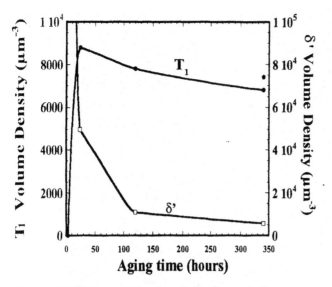

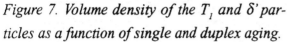

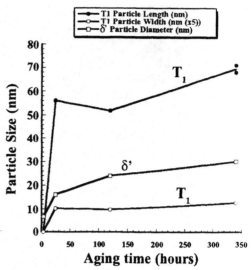

Figure 7. Volume density of the $T_1$ and $\delta'$ particles as a function of single and duplex aging.

Figure 8. Particle size are shown for both $T_1$ and $\delta'$ particles as a function of aging single and duplex aging time.

ber of measured diameters. The harmonic diameters were thus calculated using the expression given by Underwood (12),

$$D_h = \{[\Sigma(N_i/D_i)][1/\Sigma N_i]\}^{-1} \qquad (1)$$

where $D_h$ is the harmonic average particle size diameter, and $N_i$ is the number of particles for a give particle size diameter. The reciprocal relationship can be seen more readily in the form (12),

$$1/D_h = (N_1/D_1 + N_2/D_2 + \ldots + N_N/D_N)/(N_1 + N_2 + N_3 + \ldots + N_N) \qquad (2)$$

The volume fraction of the precipitates in the microstructure were also calculated from the experimental data. For the $Al_3Li$ precipitates both overlapping and truncation effects were incorporated into the calculations. However for $T_1$, $Al_2LiCu$, only the truncation effect was included in the volume fraction determinations since the overlapping was not evident. From Underwood (12), for spherical particles of diameter D, both overlapping and truncation effects must be taken into account using the equation,

$$f_{\delta'} = [-2\ln(1-A)][D/(D+3t)] \qquad (3)$$

where $f_{\delta'}$ is the volume fraction of $\delta'$ precipitates, D is the average particle size diameter, A is the projected area fraction measured from the dark-field TEM micrographs, and t is the TEM foil thickness. For the situation where there are very large spherical particles with very overaged alloys, the effect of particle overlapping is small, but the effect of truncation at the foil surface needs to be considered. Thus for foil truncation effects alone, the above equation becomes (12),

$$f_{\delta'} = 2DA/[(2D+3t)] \qquad (4)$$

For the $T_1$ precipitates the particle overlapping is not important so only the truncation effects need to be considered. The $T_1$ $Al_2LiCu$ intermetallic precipitates are not spherical so thus considering the measured planar length of the T1 precipitates, the expression for the volume fraction can be written as

$$f_T = (4A_{pl} N_{total})[2l_{pl}/(2l_{pl}+3t)]/A_{total} \qquad (5)$$

Where $l_{pl}$ is the experimentally measured planar length of the $T_1$ precipitates, $N_{total}$ is the total number of precipitates on the measured plane of TEM image area, t is the TEM foil thickness, $A_{total}$ is the total area of the microstructure field of view from which the particles are being measured, and $A_{pl}$ is the experimentally measured planar particle area. For the $T_1$ intermetallic precipitates, $A_{pl}$ can be expressed as

$$A_{pl} = l_{pl} \, w_{pl} \tag{6}$$

Where wpl is the experimentally measured planar particle width of the $T_1$ precipitates and $l_{pl}$ is the experimentally measured planar length of the $T_1$ precipitates.

The edge effect were also taken into account when doing the particle size measurements for the planar particle size areas with respect to particles extending beyond the field of view. The edge effects were incorporated using the expression (12),

$$N_{total} = \{N'' + (N'/2)\}/A_{total} \tag{7}$$

Where N' is the number of particles intersected by the edges of the field of the microstructure, N'' is the number of particles lying in field of view microstructural area, $A_{total}$ is the total area of the microstructure field of view from which the particles are being measure, and $N_{total}$ is the total number of particles corrected for edge effect within the microstructural area image field of view.

Based on the harmonic diameters and the foil thickness, the number of particles per unit volume i.e., the volume number density, of the precipitates can be determined. Hilliard (13) showed that the volume number density can be determined from the formula,

$$N_{volume} = N_{total} \{t + (\pi D_h/2)\}^{-1} \tag{8}$$

Where $N_{volume}$ is the total number of intermetallic precipitate particle per unit volume i.e the volume number density, and t is the thin foil thickness. Thus, based on the volume number density of precipitates, the true particle size diameter can be predicted from the equation given by Hilliard (13),

$$D_{true} = \{N_{total}/N_{volume}\} - t \tag{9}$$

Where $D_{true}$ is the actual true average particle size diameter for the distribution of intermetallic precipitate particles within the microstructure. For very small thin foil the true particle size diameter can be estimated by the equation (13),

$$D_{true} = \{N_{total}/N_{volume}\} = \{\pi/2\} D_h \tag{10}$$

where $D_h$ is the harmonic average particle size diameter.

The harmonic diameter, is used to convert the experimentally measured planar particle sizes to the actual true particle sizes. Thus, the harmonic diameter is very important to determining the true average particle size of a distribution of particle sizes.

**Fatigue Crack Growth Rates**: FCG rates for the single and duplex aging treatments were obtained for R=0.1 at 30 Hz in laboratory air. FCG results show that the underaged alloy has the best FCG resistance and the most overaged condition showing the least FCG resistance.

## Discussion

Fatigue thresholds, $\Delta K_{th}$, observed in this work are shown as a function of aging time in Figure 9, as is interparticle spacing for the $\delta$' and $T_1$ particles. Precipitate diameter is then related to $\Delta K_{th}$ in Figure 10. These graphs show that the strongest relationship of FCG thresholds is observed with $\delta$' particle size. Neither the $T_1$ particle size nor the spacing appears to significantly affect the fatigue threshold for the aging conditions studied here. However, as $\delta$' particle size decreases a higher fatigue threshold is observed. This is consistent with observations on the effects of particle shearing and bypassing on fatigue thresholds in Al-Li-Cu alloys (14). It is shown in the literature that when precipitates are coherent and precipitate diameter is less than a critical-size, precipitates are sheared by dislocations resulting in slip pla-

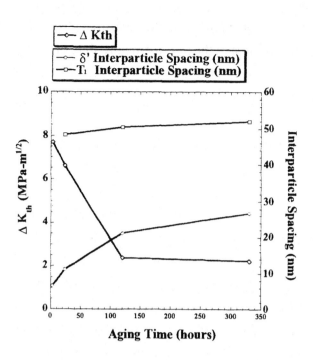

*Figure 9. Fatigue threshold of AF/C- 458 as a function of aging time, (b) Fatigue thresholds as a function of interparticle spacing and (c) fatigue threshold as a function of precipitate diameter.*

narity and when the precipitate diameter exceeds the critical (precipitate) diameter dislocation looping mechanism takes over. Planar slip has been shown to increase reversible slip at the crack tip and crack deflection or crack path is more tortuous (15). Both these factors contribute to the slower fatigue crack growth rates in the less aged conditions. On the other hand when the mechanism changes over to a more homogeneous slip more irreversible slip (compared to the planar slip condition) is accumulated at the crack tip resulting in faster crack growth rates. A much flatter fracture path (or lower crack path deflections) also accompanies this.

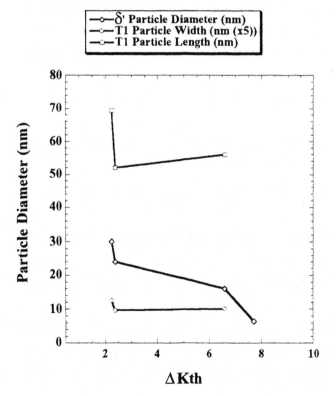

*Figure 10. The relationship of particle diameter and fatigue threshold for the four aged samples studied here.*

## Conclusions

Four aging treatments were used to quantify microstructure and precipitate characteristics to the fatigue crack growth thresholds for the alloy AF/C-458. As aging progressed, $\delta'$ interparticle spacing and precipitate diameter increased more strongly than did the $T_1$ spacing and size. FCG results on AF/C-458 are consistent with the observation of the fatigue threshold being higher for conditions of planar slip. The present aging conditions used here suggest that $\delta'$ particles may have a stronger effect than the $T_1$ precipitates. However, work is being carried out to validate these results to a much broader aging condition window.

## Acknowledgements

This work was supported in part under USAF contract F33615-97-C-5274.

## References

1. Isotropic Wrought Aluminum-Lithium Plate Development Technology, A. K. Hopkins, K.V. Jata, and R.J. Rioja in Fifth International Conference on Aluminum Alloys, Grenoble, France, Materials Science Forum, pp. 5, 1996.

2. The Anisotropy and Texture of Al-Li Alloys, K.V. Jata, A. K. Hopkins and R.J. Rioja in Fifth International Conference on Aluminum Alloys, Grenoble, France, Materials Science Forum , pp. 647-652, 1996.

3. Processing of an Experimental Aluminum-Lithium Alloy for Controlled Microstructure, V.K. Jain, K.V. Jata, R.J. Rioja, J.T. Morgan and A. K. Hopkins, Journal of Materials Processing Technology, vol. 73, pp.108-118, 1998.

4. A.A. Csontos and E.A. Starke, Jr., Metallurgical transactions, Vol. 31A, August 2000, pp. 1965-1976.

5. A.A. Csontos, Ph.D. Thesis, "Microstructural effects on the slip and fracture behavior of isotropic Al-Li-Cu-X alloys," University of Virginia , February, 2001.

6. Aluminum-Lithium Workshop Report, August 18, 1998, Bass Lake Lodge, Wright-Patterson Air Force Base, Ohio.

7. D. Mathur, M.S. Thesis, "Localized corrosion and stress corrosion cracking studies of Al-Li-Cu alloy AF/C-458, Ohio State University, 2000.

8. P.I. Gouma, J.E. Kertz and R.G. Buchheit, Localized corrosion mechnaisms of Al-Li-Cu alloy AF/C458 after interrupted quenching from solutionizing temperatures, These proceedings.

9. H.K. Hardy and J.M. Silcock, Journal Institute of Metals, Vol. 84, p. 423, 1955-1956.

10. B. Noble and G.E. Thompson., Metal Science Journal, Vol. 5, pp. 357-364, 1971.

11. F.W. Gayle and J.B. Vander Sande, Scripta Metallurgica, Vol. 18, pp. 473-478, 1984.

12. Underwood, E.E., Quantitative Stereology, Addison-Wesley Publishing Company, Reading, Massachusetts, 1970.

13. Hilliard, J.E., Transactions of the Metallurgical Society of AIME, Vol. 224 , pp. 906-917, 1962.

14. K.V. Jata and E.A. Starke, Metallurgical Transactions, vol.17A, p.1011, 1986.

15. A.K. Vasudevan, K. Sadananda, Metall. & Matls.Trans., vol.26,p.1221, 1995.

# LIGHTWEIGHT ALLOYS FOR AEROSPACE APPLICATION

*Edited by:*
*Dr. Kumar Jata, Dr. Eui Whee Lee,*
*Dr. William Frazier and Dr. Nack J. Kim*

## ALUMINUM-LITHIUM ALLOYS

## Effect of Friction Stir Welding on the Superplastic Behavior of Weldalite Alloys

*H. Salem, A. Reynolds and J. Lyons*

Pgs. 141-150

184 Thorn Hill Road
Warrendale, PA 15086-7514
(724) 776-9000

# EFFECT OF FRICTION STIR WELDING ON THE SUPERPLASTIC BEHAVIOR OF WELDALITE ALLOYS

By

H. Salem[1], A. Reynolds[2], and J. Lyons[2]

[1] American University in Cairo, Department of Mechanical Engineering
[2] University of South Carolina, Department of Mechanical Engineering

## ABSTRACT

Al-Cu-Li alloys offer attractive property combinations of low density, high specific strength and modulus and exceptional cryogenic properties. This makes them excellent candidates for a variety of aerospace applications. Weldability of the aluminum alloys becomes of great concern when pressurized fuel tanks are manufactured for space launch systems. Friction Stir Welding (FSW) of sheet aluminum alloys has proven advantages over fusion welding processes in certain applications. In the current research, the effect of friction stir welding (FSW) on the superplastic behavior of Weldalite 2095 dynamically recrystallized, superplastic, rolled, sheets is investigated. Uniaxial superplastic behavior of the alloys is characterized before and after FSW. Microstructural evolution is assessed through light optical microscopy and transmission electron microscopy. It is shown that superplasticity can be retained in friction stir welded 2095 sheet, but that the choice of welding parameters will affect the behavior.

Lightweight Alloys for Aerospace Applications
Edited by Kumar Jata, Eui Whee Lee,
William Frazier and Nack J. Kim
TMS (The Minerals, Metals & Materials Society), 2001

# INTRODUCTION

Because mechanical fastening of parts has been the primary joining method in the aircraft industry, weldability hasn't been a concern in aluminum alloy development for aeronautical applications. On the other hand, welding is the primary joining method of parts in aerospace applications, so one must be concerned not only for mechanical properties, but also for hot tearing, and other weldability parameters necessary to obtain a sound joint. Welding is a necessity, especially when pressure service conditions are involved. For example, propellant tankage, which represents the bulk dry weight of space launch systems, in which $O_2$, and $H_2$ fuel are contained under high pressures at cryogenic temperatures of $-196^{\circ}$C [1]. Fusion welding of aluminum alloys is always associated with heat affected zone cracking, solidification cracking, and porosity.

In order to address the weldability and mechanical property requirements for the Space Shuttle External Tank, the Weldalite series of Al-Li alloys was developed in the late 80's. Weldalite alloys are very complex due to the variety of second phase particles that precipitate within the matrix. Higher Cu:Li ratio promotes the precipitation of $\theta^{/}$ ($Al_2Cu$), $T_1$ ($Al_2CuLi$), $T_2$($Al_6CuLi$), and $T_B$ ($Al_{15}Cu_8Li$), depending on the Cu-content [2-4]. In addition $\delta^{/}$ ($Al_3Li$), $S^{/}$ ($Al_2CuMg$), and $\beta^{/}$ ($Al_3Zr$) phases precipitate within the Al-matrix due to the addition of Li, Mg and Zr elements, respectively. Silver is added promote precipitation of $T_1$ phase, providing very high strength in the peak-aged condition [5]. $\beta^{/}$ ($Al_3Zr$) fine-coherent precipitates are the key-phase for prevention of grain growth during the elevated temperature forming associated with superplastic forming (SPF) [6].

Due to the recent development of Friction Stir Welding (invented at TWI, 1991), there has been increased interest in implementation of welding for production of aerospace structure. A major driver for use of welded structure is reduction in cost and part count. Further reductions in part count and improvements in manufacturing flexibility might be realized through the combination of welding and superplastic forming. Superplasticity (large homogeneous deformation at very low flow stress) enables net-shape fabrication of complex components. Cost savings associated with this technique are based on high materials utilization, reduced part count and resulting low assembly costs. However, one of the principal requirements for successful superplastic forming is fine grain structure, $< 10\mu$m. This is generally achieved through complex thermomechanical processing. If a conventional fusion welding technique were used to join superplastic sheets, then the desired microstructure would be destroyed at the weld and the superplastic flow behavior would be lost. On the other hand, one of the salient features of all friction stir welds in aluminum alloys is a highly refined grain structure in the weld region. Therefore, FSW could be a good candidate for welding Al-Cu-Li alloys intended for subsequent SPF processing. The development of a method to join superplastic alloy blanks prior to forming would provide practically unlimited component design flexibility. For example, one limitation on shape complexity is the tendency for failure by local thinning or fracture in regions that experience high deformation as the flat blank is stretched [7]. Using FSW to join sheets in a rough pre-form shape prior to SPF processing would result in more uniform deformation during processing. With FSW, it is also possible to join sheets of different thickness and/or different compositions, thereby enabling the customization of mechanical properties through out the formed product. The objective of the current research is to assess the superplastic flow behavior and tensile properties of the friction stir welded superplastic Al-Cu-Li base alloys. In addition, microstructural evolution will be studied in order to understand the effect of FSW on the superplastic behavior of the investigated alloy.

## MATERIALS AND EXPERIMENTAL PROCEDURE

Weldalite 2095 alloy, provided by Reynolds Metal Company, was selected for the current research. The alloy was received in the form of dynamically recrystallized, superplastic sheets. Table 1 shows the chemical composition and thickness of the sheets.

### Table 1. Chemical Composition of Weldalite 2095 SP Sheets

| ALLOY | THCKNESS (IN) | COMPOSITION (WT%) | LI | CU | MG | ZR | AG | TI | ZN |
|-------|---------------|-------------------|------|------|------|------|------|-------|-------|
| 2095  | 0.063         | Nominal           | 1.14 | 4.6  | 0.38 | 0.17 | 0.33 | 0.03  | 0.02  |
|       |               | Measured          | 1.08 | 4.52 | 0.31 | 0.15 | 0.34 | 0.028 | 0.065 |

Some minor additions: Mn, Fe, & Si less than 0.06 wt% each

Strips about 50 mm wide were cut from the Weldalite sheets and prepared for butt welding using the FSW technique. Welding was conducted at 1000 rpm and 2.1, 3.2, and 4.2 mm/s welding rates. Samples were cut from the welded sections and prepared for superplastic tensile testing and for optical and transmission electron microscopic examination. The hardness as a function of distance from the friction stir weld centerline was measured using a Vickers hardness tester with a 100 gm load.

### Characterization of Superplastic Behavior

The tensile stress/strain behavior of the as received superplastic sheet and of the friction stir welded sheet was examined. The welded sheets were tested with the welds transverse to the loading direction. The tensile tests were conducted at 495°C and a constant strain rate of $2 \times 10^{-4} s^{-1}$ without back pressure. The specimens from the friction stir welded sheet had a gage length of 13 mm and a width of 4.6 mm with the welded region centered within the gage length. A gage length of 7.5 mm was used for the as-received superplastic sheets. Temperature was controlled by a SATEC TCS 3203 furnace equipped with a three-zone temperature controller and thermocouples located around the sample. Reported elongations for the SP and the FSW sheets are based on the average of at least two tests.

### Microstructural Analysis

Microstructure of the as-received superplastic sheets was investigated using optical microscopy to reveal the rolled structure and the second phase particle size, shape, and distribution at low magnification before and after FSW within the weld nugget, heat affected zone (HAZ), and the base metal. Transverse sections through the welds were cut and prepared using standard metallographic procedures.

Transmission electron microscopy (TEM) was performed to reveal the substructure size and shape of the as-received SP sheets prior to FSW and the microstructural evolution that took place after FSW within the center of the weld nugget for the three different welding rates in the as-welded condition. TEM samples discs were prepared by grinding followed by jet-electropolishing in a nitric acid-methanol solution.

## RESULTS

### As-Received Superplastic 2095

Examination of the as-received dynamically recrystallized SP 2095 sheets using the light optical microscope revealed the presence of coarse, second phase particles scattered within the aluminum matrix and along the grain boundaries (Figure 1a). Transmission electron microscopy showed details of the grain structure, where fine and coarse sub-grain structures of about 1 and 2.5 μm were observed, as shown in Figure 1b. Dislocation activity was observed within the grains around bright and dark-spherical precipitates corresponding to $\delta'$ (Al$_3$Li) and $\beta'$ (Al$_3$Zr) precipitates, respectively.

144

Dislocations were also observed along grain boundaries (marked A), other sides of the grain boundaries (marked B) revealed the formation of extinction contours.

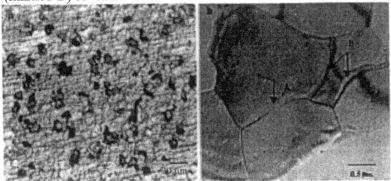

**Fig. 1.** 2095 SP sheet microstructure. (a) Optical micrograph, (b) TEM photomicrograph showing sub-grain structure and dislocation activity.

Microstructure of FSW-SP Sheets

Optical microscopic examination of the welded sheets at low magnification revealed sound and un-defected welds for the 3.2 mm/s and 4.2 mm/s welding rates. Figure 2 shows an optical micrograph for FSW 2095 SP sheet at 1000 RPM and 4.2 mm/s welding rate.

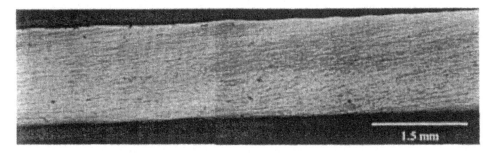

**Fig. 2.** Optical photomicrograph of FSW 2095 SP sheets at 1000 RPM & WR of 4.2 mm/s.

Conversely, sheets FSW at 2.1 mm/s showed evidence for incipient melting and segregation within the weld nugget starting at the top surface of the sheet and extending about 0.37 mm through the thickness. Particle coarsening and agglomeration was clearly observed within that region, while finer particles were revealed outside the region (below and to the sides of the weld crown).

TEM examination revealed different response to FSW depending on the welding rate. The sheets welded at 2.1 mm/s showed an obvious coarsening in the sub-grain structure where an average sub-grain size of 2.93 $\mu$m was measured (Figure 3a). In addition, dislocation free, fine recrystallized grains ranging in size between 1.2 $\mu$m and 2.95 $\mu$m were observed. A non-uniform sub-grain structure was revealed for the sheets welded at 3.2 mm/s. The average size is about 2 $\mu$m as shown in Figure 3b. Examination of the sheets welded at 4.2 mm/s revealed the formation of almost equiaxed structure about 1.58 $\mu$m in average size with a relatively high dislocation density within the sub-grains and along their boundaries as shown in Figure 3c. Selected area diffraction patterns (SADP) were captured from areas of 5 $\mu$m for the three welding rates as shown in Figure 3 a-c. No evidence for the presence of any second phase particles except for the superlattice reflection spots corresponding to the coherent $\delta'(Al_3Li)$ and $\beta'(Al_3Zr)$ phases was observed.

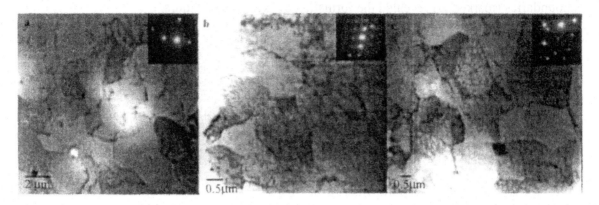

**Fig. 3.** TEM photomicrographs for 2095 SP sheets FSW at 1000 RPM and (a) 2.1 mm/s WR with a SADP corresponding to $[110]_{Al}$ zone axis, (b) 3.2 mm/s WR with a SADP corresponding to $[112]_{Al}$ zone axis, and (c) 4.2 mm/s WR with a SADP corresponding to $[110]_{Al}$ zone axis.

Figure 3a shows minor dislocation activity associated with the un-dissolved-fine-coherent $\beta'$ ($Al_3Zr$) phase in addition to $\delta'$(Al3Li) fine-coherent phase which could have precipitated during cooling of the welded sheets by a reversion process. Moreover, the presence of extinction contours at grain boundaries indicates an advanced stage of dynamically recovered structure with well defined-unrelaxed boundaries and grains about 1.5 μm in average size within the original dynamically recrystallized sub-grain (DRSG) structure of the as-received SP sheets. Figure 4b shows the SP sheets structure welded at 3.2 mm/s where a non-uniform degree of dynamic recovery took place, since high density of dislocations was observed along the boundaries of some of the structure formed (dislocation cells), while other regions revealed the formation of sub-grains (about 0.5 μm in average size) with extinction contours along their boundaries. On the other hand, sheets welded at 4.2 mm/s experienced an early stage of dynamic recovery. This was manifested (Figure 4c) by the formation of dislocation tangles in the form of fine cells about 0.18 μm in average size which consumed almost all the interior of the SP sheet's original DRSG structure.

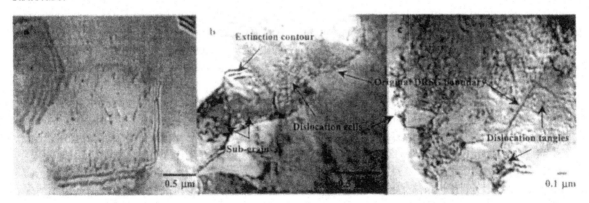

**Fig. 4.** TEM photomicrographs at high magnification showing the effect of increasing the welding rate on the dislocation density and activity within the developed structure of 2095 SP sheets FSW 1000 RPM and WR of (a) 2.1 mm/s, (b) 3.2 mm/s, and (c) 4.2 mm/s.

Hardness of the FSW SP Sheets

Vickers hardness testing results for the FSW sheets are plotted in the graph shown in Figure 5. The hardness data are plotted as a function of distance from the weld centerline (zero distance). The FSW process results in an increase in hardness of the weld region relative to the base metal. The base metal hardness-value (VHN 59) is recovered at a distance of approximately 6.4 mm from the weld centerline. Although the differences due to various

146

welding rates are small, the hardness in the weld region shows a consistent increase with increasing welding rate.

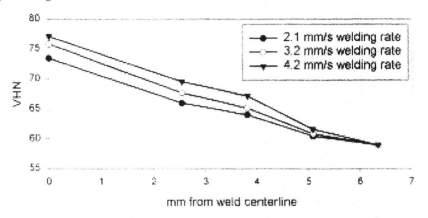

**Fig. 5.** VHN vs. distance from weld centerline for SPF sheet welded at various rates.

Tensile Behavior: Superplastic Elongation

Figure 6 shows the superplastic behavior of the as-received 2095 SP sheets versus that of the FSW sheets tested in transverse tension. Note that the 10 mm wide welded section was centered within the 13 mm gage length of the tension sample that was cut transversely to the welding direction. The results shown in Figure 6 represent the superplastic behavior displayed over the total gage length 13 mm. The SP behavior of the as received sheet is different from that displayed by the FSW sheets especially those welded at 3.2 and 4.2 mm/s. SP sheets displayed a uniform flow stress of 4.5 MPa between true strain-values of 0.2 and 1.6, corresponding to % elongation–to-fracture (% EF) of 550%. On the other hand, the sheets FSW at 3.2 and 4.2 mm/s exhibited strain hardening behavior with maximum true stress-values of 4.5 MPa at a true strain of 1.2 and 4.0 MPa at a true strain of 1.39, respectively. Sheets FSW at 2.1 mm/s displayed a relatively constant and low flow stress of 1.0-1.2 MPa between true strains of 0.25 and 1.3. Note that the higher the welding rate, the higher the % EF, where 250, 337, and 353.7 % EF were measured for the sheets FSW at 2.1, 3.2, and 4.2 mm/s, respectively.

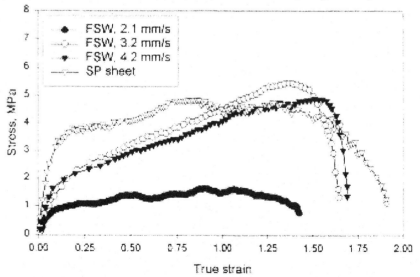

**Fig. 6.** SP behavior for 2095 SP sheet and FSW sheets.

FSW samples fractured within a neck formed on one side of the weld nugget (marked F, Figure 7, specimens c-e). Localized deformation was observed on the other side of the weld nugget at a distance from the centerline equal to that from the fracture region. Conversely, fracture took place within the tensile specimen gage center of the as-received SP sheets (marked F, Figure 7 a).

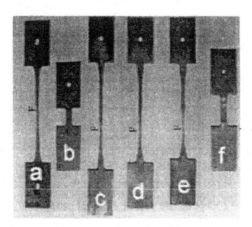

**Fig. 7.** 2095 SP specimens deformed at 490°C, and 2x10$^{-4}$s$^{-1}$. (a) as-received SP sheet, 550 % elongation, (b) Un-deformed as-received specimen with 6.4mm total gage length, (c-e) SP deformed FSW specimens at 4.2, 3.2, and 2.1 mm/s WRs, respectively, (f) un-deformed FSW sample with 13 mm total gage length and10 mm weld nugget. (F = fracture zone).

## DISCUSSION

2095 dynamically recrystallized SP sheets have been successfully friction stir welded at 1000 RPM and welding rates of 3.2 and 4.2 mm/s. A lower welding rate of 2.1 mm/s, was found to cause incipient melting, however, superplastic behavior was retained in this material as well. In general, friction stir welding at higher ratios of tool revolution rate to welding rate results in longer times at higher temperatures. Welding at the higher rates results in a structure with a higher dislocation density because of the limited time available for recovery processes. This agrees with the findings of Strangwood *et al.*, who worked with 6082-T6 and 7108-T79 alloys [8]. In addition, coarse second phase particle dissolution within the weld nugget due to the heat input generated during the welding process may have contributed to solid solution hardening, providing an additional explanation for the elevated hardness observed in the weld nuggets relative to the base sheet. The combined effect of solid solution strengthening and high dislocation density generated at the higher welding rates explains the slightly greater hardness observed in the high welding rate specimens and the higher flow stress observed in these specimens during the superplastic deformation . Results obtained by Backlund *et al.*, [9] and Svensson *et al.*, who investigated the mechanical behavior of FSW AA 6082 alloy [10], corroborate the findings of this work. It is worth noting that precipitation of $\delta^{/}$ (Al$_3$Li) fine-coherent phase after FSW (natural aging) is another possible source of strengthening in the weld nuggets (Figure 4).

Size and shape of the developed sub-structure varies with the intensity of deformation and temperature. The faster the welding rate the higher the dislocation density generated which produces dislocation tangled structure and cells (e.g. 4.2 mm/s welding rate), indicating an early stage of dynamic recovery as explained by Humphries [11]. Sub-grains and dislocation cells formed by dislocation climb similar to that observed for the sheets welded at 3.2mm/s represent an intermediate stage of dynamic recovery, which takes place as a result of longer exposure time to elevated temperature. The true stress-true strain behavior of the SP sheets indicates pseduo-brittle fracture by cavitation, which is a typical behavior of aluminum SP alloys when no backpressure is provided. Conversely, the SP behavior of the FSW sheets indicates a combination of psedou-brittle fracture by cavitation and unstable plastic flow (localized deformation) within the fracture region. The superplastically deformed specimens welded at 3.2 and 4.2 mm/s experience a uniform deformation in regions outside the HAZ, while a non-uniform deformation by diffused necking takes place within the fractured region. The development of localized deformation can occur during SPF especially where a high stress concentration is present such as coarse particles or grains. This applies to the HAZ region where relative coarsening of the sub-grain and second phase particles occur. This is evident by the significant decrease in the hardness-values measured between 0.15 and 0.2 inches from the weld nugget centerline, which correspond to the fracture region. According to Pilling, and Radily [12], a successful grain boundary sliding has to be accommodated with either grain boundary and volume diffusion, glide and climb of grain boundary dislocations, or glide and climb of lattice dislocations both across and/or around the grains. This explains the low flow stress and % EF displayed by the FSW sheets welded at 2.1 mm/s. The relatively severe sub-grain coarsening and dislocation annihilation that takes place during the welding process results into an unaccomodated grain boundary sliding and hence severe cavitation leading to premature fracture.

Mukherjee *et al.* state that a material can be classified as superplastic when it is capable of exhibiting tensile elongations greater than or equal to 200% prior to fracture [13]. Based on this definition, it can be stated that superplastic behavior has been retained in the 2095 sheet after friction stir welding.

## CONCLUSION

1.  Friction Stir welding process has proven its capability of successfully joining dynamically recrystallized superplastic Weldalite alloys.

2.  The higher the welding rates at 1000 RPM the higher % EF. Friction stir welding rate is the controlling parameter for softening by dynamic recovery.

3.  A welding rate 2.1 mm/s at 1000 RPM caused incipient melting, and grain and sub-grain coarsening that results in reduced superplastic capability.

4.  High welding rates increase the density of dislocations and develop microstructures consisting of fine dislocation cells and sub-grains with small misorientations.

5.  The SP behavior of the as-received SP sheets indicates uniform plastic deformation followed by pseduo-brittle fracture by cavitation, which is a typical behavior of aluminum SP alloys when backpressure is not provided.

6.  Conversely, FSW sheets SP behavior reveals a combination of uniform and localized plastic flow resulting in psedou-brittle fracture by cavitation and unstable plastic flow within the fracture region (HAZ), respectively.

7.  The SP sheets FSW at 3.2 and 4.2 mm/s display uniform deformation in regions outside the HAZ, while a non-uniform deformation by diffused necking takes place within the HAZ followed by premature fracture.

## REFERENCES

1.    A. Cho, M. Greene, P. Fielding and H. Skillingberg: Report Reynolds Metal Co., Richmond, VA, 1995, pp. 3-8.
2.    M. Zaidi and J. Wert: *Treatise on Materials Science and Technology*, eds., G. Mahon and R. Ricks, PA, 1989, pp. 139.
3.    S. G. Mazzini: *Scripta Mat.*, 1991, Vol. 25, pp.1987.
4.    A. J. Shakesheff and D. S. Darmaid: *Mat. Sci. and Technol.*, Vol.7, 1991, pp. 276-288.
5.    J. Pickens, F. Heubaum, T. Langdon and L. Kramer: in *Aluminum-Lithium V*, Proc., eds., T. H. Sanders and E. A. Starke. Jr., Materials and Component Engineering Publication Ltd., Birmingham UK, 1989, pp.1397-1414.
6.    T. S. Sivatsan, K. Yamaguchi, and E. A. Starke. Jr.: *Mater. Sci. Eng.*, Vol. 83, 1986, pp. 87.

7.    R. Verma, and R. Freedman, *"Superplastic Forming Characteristics of Fine-Grained 5083 Aluminum"*, J. of Mat. Eng. And Performance, Vol. 4, pp. 543-50, 1995.

8.    M. Strangwood, J. E. Berry, D. P. Cleugh, A. J. Leonard, and P. L. Threadgill, *"Characterization of the Thermo-Mechanical Effects on Microstructural Development in Friction Stir Welded Age Hardening Aluminum-Based Alloys"*1st International Symposium on Friction Stir Welding, 14-16 June 1999.

9.    J. Backlund, A. Norlin, and A. Andersson, *"Friction Stir Welding-Weld Properties and Manufacturing Techniques"* Proc. INALCO' 98, 7th International Conference Joints in Aluminum, Cambridge, UK 16th April 1998.

10.    L. E. Svensson, and L. Karlsson, *" Microstructure, Hardness and Fracture in Friction Stir Welded AA 6082"*, 1st International Symposium on Friction Stir Welding, 14-16 June 1999.

11.    F. Humphreys and M. Hatherly: *Recrystallization and Related Annealing Phenomena,* British Library Cataloguing Publication Data, Pergamon, 1995, pp. 167.

12.    J. Pilling and N. Ridley: *Superplasticity in Crystalline Solids,* Pub. By The Institute of Metals, 1989.

13.    A. K. Mukherjee, R. S. Mishra, and T. R. Bieler: *Materials Science Forum*, 1997, Vol. 233-234, pp. 217-234.

# LIGHTWEIGHT ALLOYS FOR AEROSPACE APPLICATION

*Edited by:*
*Dr. Kumar Jata, Dr. Eui Whee Lee,*
*Dr. William Frazier and Dr. Nack J. Kim*

## ALUMINUM-LITHIUM ALLOYS

On the Effect of Thermal Exposure on
the Mechanical Properties of 2297 Plates

*E. Acosta, O. Garcia, A. Dakessian, K. Aung Ra, J. Torroledo,*
*A. Tsang, M. Hahn, J. Foyos, J. Ogren, A. Zadmehr,*
*H. Garmestani and O.S. Es-Said*

Pgs. 151-160

184 Thorn Hill Road
Warrendale, PA 15086-7514
(724) 776-9000

# ON THE EFFECT OF THERMAL EXPOSURE ON THE MECHANICAL PROPERTIES OF 2297 PLATES

E. Acosta*, O. Garcia*, A. Dakessian*, K. Aung Ra*, J. Torroledo*, A. Tsang*,
M. Hahn**, J. Foyos*, J. Ogren*, A. Zadmehr*, H. Garmestani*** and O.S. Es-Said*

\* National Science Foundation
Research Experiences for Undergraduate Program
Loyola Marymount University
Los Angeles, California 90045-8145

\*\* Northrop Grumman
Materials and Processes Technology
Department 9L26, Zone W5
El Segundo, California 90245-2804

\*\*\* FAMU- FSU
College of Engineering
Department of Mechanical Engineering
Tallahassee, Florida 32310-6046

## Abstract

The objective of this study is to assess the effect of thermal exposure on the tensile properties of the 2297 plates of 5.40 cm (2.125") thickness. The matrix studied included exposure times from 1 hour up to 200 hours and temperatures from 177°C (350°F) through 260°C (500°F). Samples were machined in the quarter (t/4) and half (t/2) thickness. The strength and crystallographic texture components were higher in the half thickness (t/2) samples. All samples experienced a slight age hardening (or plateau) in the initial thermal exposure and as time went on, the material began to experience overaging.

Lightweight Alloys for Aerospace Applications
Edited by Kumar Jata, Eui Whee Lee,
William Frazier and Nack J. Kim
TMS (The Minerals, Metals & Materials Society), 2001

# Introduction

Aluminum-lithium 2297 was developed as a moderate strength, damage tolerant plate material for aircraft bulkheads. In comparison with the 7050 alloy series, 2297 offers clear advantages in thermal stability, spectrum fatigue performance and offers a 7% density reduction [1]. Similarly 2297 also surpasses the 2124 alloy series with a stiffness of around 7% greater, 3-5 times improvements in resistance to fatigue crack growth, 32% higher S-L fracture toughness, 25% higher stress corrosion cracking resistance and a 4-5% density savings [2-5].

The objective of this thermal aging study was to determine the effects of high temperatures over extended periods of time on the mechanical properties of the 2297 alloy. Aerodynamic or engine-related heating that the alloy could experience in service is of large concern, therefore, it is important to evaluate the mechanical properties under these conditions.

# Experimental Procedure

The as-received plate was in the T8 temper. The thermal aging study involved machining tensile samples into 127 mm x 25.4 mm x 6.35 mm (5" x 1" x 0.25") from the plate either from the t/4 (quarter thickness) or t/2 (half thickness) locations, Figure 1(a) and Figure 1(b). The length of the tensile samples was parallel to the rolling direction. The thickness of the samples was further reduced from t/2 (27 mm or 1.1") and t/4 (13.5 mm or 0.53") to 6.35 mm or 0.25" in thickness. This study involved trying to explore a range of potential thermal exposures for this alloy. Accordingly, after the parts were machined they were thermally exposed at various times and temperatures. Table I shows the times and temperatures at which the alloy was exposed in the thermal exposure study. An Instron model 4505 universal testing machine was used to evaluate the tensile properties. Three tensile samples at quarter (t/4) and one half (t/2) thicknesses were evaluated for each condition in Table I. Specimens for texture analysis were cut from (t/4) and (t/2) locations, mechanically polished and etched. The original rolling direction was used as the pole figure reference direction. The crystallographic texture evaluation technique is described elsewhere, [6].

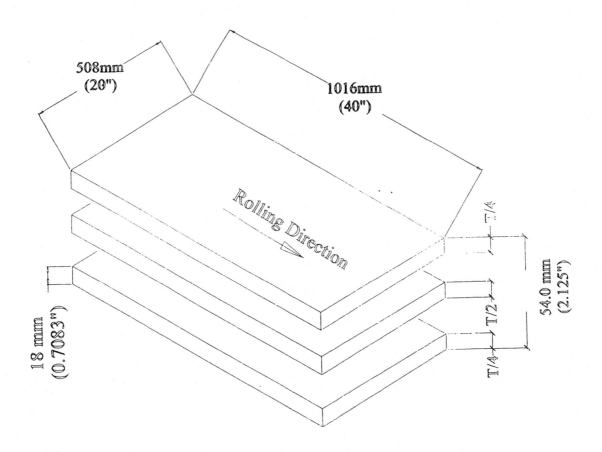

**Figure 1(a): As Received Plates 1016 mm x 508 mm x 54 mm dimensions (20"x 40"x 21.125") divided along t/4, t/2, t/4 thickness sections.**

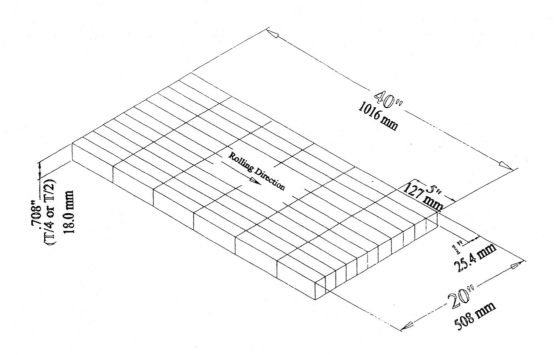

**Figure 1(b): Tensile Samples with 0°orientation of 127 mm x 25.4 mm x 6.35mm dimensions (5"x 1" x .708").**

154

| Table I: Time and Temperature Matrix for Thermal Exposure Study | | | | | | | | |
|---|---|---|---|---|---|---|---|---|
| | Time, h | | | | | | | |
| Temperature, °C (°F) | 1 | 2 | 5 | 10 | 20 | 50 | 100 | 200 |
| 177 (350) | X | X | X | X | X | X | X | X |
| 191 (375) | X | X | X | X | X | X | X | |
| 204 (400) | X | X | X | X | X | X | X | |
| 218 (425) | X | X | X | X | X | | | |
| 232 (450) | X | X | X | X | X | X | | |
| 260 (500) | X | X | X | X | X | X | | |

## Experimental Results

The tensile properties of the as-received 2297-T8 plate in the quarter (t/4) and (t/2) half thicknesses are shown in Table II. The deformation and shear components and the recrystallization components are listed in Table III. The half thickness (t/2) tensile strength is 5-6% higher than the quarter thickness (t/4) strength. The Goss {110} <100>, Brass {110}<112>and x {110}<111> textures are significantly sharper in the half thickness (t/2) grain orientations.

Figures 2-4 after one hour, five hours, and ten hours indicate an initial age hardening at lower temperatures (<191°C or 375°F) prior to overaging at higher temperatures. The half thickness (t/2) sample appeared to be consistently higher in yield and ultimate strengths at lower temperature exposure and longer times as compared to the quarter thickness (t/4) samples, Figure 5. This trend diminished at higher temperatures, Figure 6.

| Table II: As-Received Tensile Properties For t/4 and t/2 Thickness for 2297 Plates | | | |
|---|---|---|---|
| Thickness | Yield Strength MPa (ksi) | Ultimate Strength, MPa (ksi) | % Elongation |
| t/4 | 399.6 (58.0) | 439.6 (63.8) | 9.7 |
| t/2 | 426.5 (61.9) | 462.3 (67.1) | 9.1 |

**Table III: As-received crystallographic texture components for t/4 and t/2 thicknesses for 2297 Plates.**

| Deformation and Shear Components | | | | | |
|---|---|---|---|---|---|
| 2297 | Brass {110} <112> | Copper {112} <111> | Shear {111} <112> | X {110} <111> | Rotated Brass {110} <223> |
| As-Received t/4 | 0.5 | 0 | 1.5 | 2.5 | 0 |
| As-Received t/2 | 12 | 0 | 0 | 16 | 4 |

| Recrystalization Components | | | | | | | |
|---|---|---|---|---|---|---|---|
| 2297 | RCN {001} <013> | RCT {013} <013> | P {110} <112> | Cube {001} <100> | Goss {110} <001> | R {124} <211> | S {123} <624> |
| As-Received t/4 | 1.5 | 0.5 | 4.5 | 4 | 0.5 | 0.5 | 0.5 |
| As-Received t/2 | 1.5 | 3.5 | 4 | 4 | 17 | 0 | 0.5 |

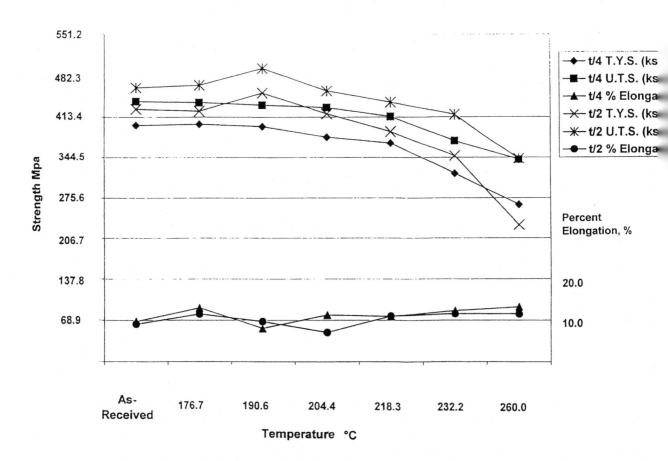

Figure 2: Mechanical Properties of 2297 Alloy for 1hour of Thermal Exposure.

156

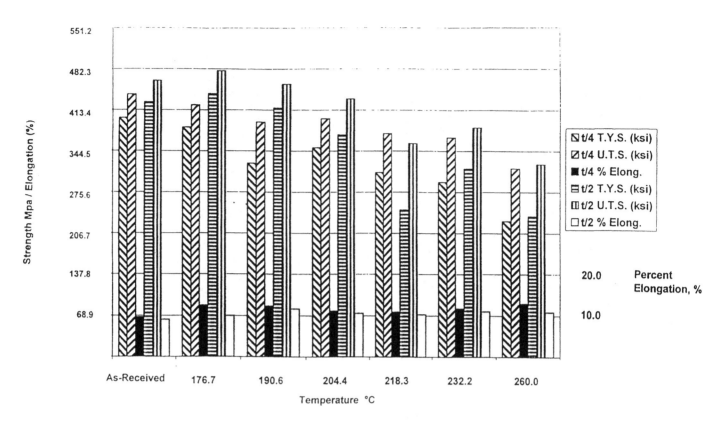

Figure 3: Mechanical Properties of 2297 Alloy for 5 hours Thermal Exposure.

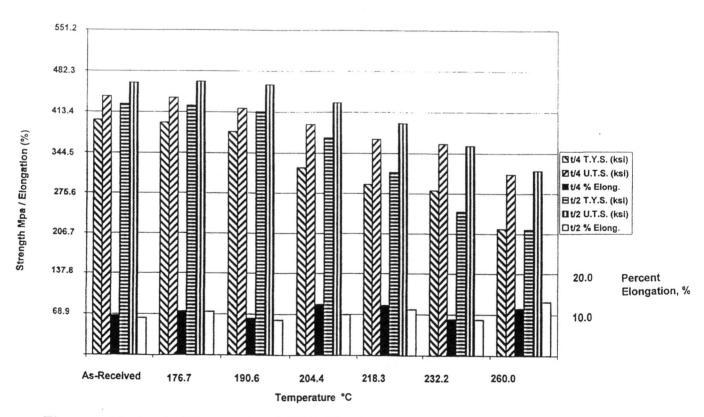

Figure 4: Mechanical Properties of 2297 Alloy for 10 hours Thermal Exposure.

157

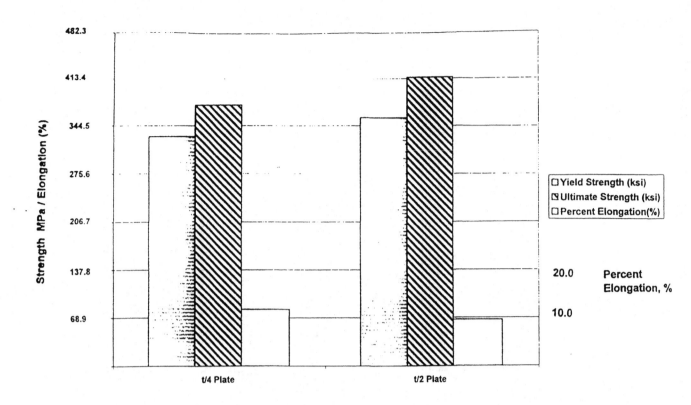

Figure 5: t/4 vs. t/2 Properties at 176.7°C (350°F) for 200 hours of Thermal Exposure.

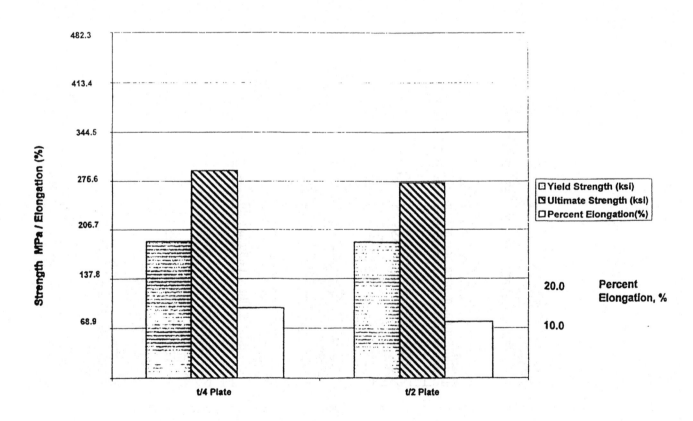

Figure 6: t/4 vs. t/2 Properties at 260°C (500°F) for 50 hours of Thermal Exposure.

158

The Thermal Exposure study clearly showed that the material appeared to experience either a slight age hardening during the initial thermal exposure or a plateau and would then slowly begin to reduce in strength as time and temperature increased. Tables IV and V represent percentage values as compared to the as-received yield strength data.

**Table IV: Yield Strength at t/4 (Temperature vs. Time)**

| | Time, h | | | | | | | | |
|---|---|---|---|---|---|---|---|---|---|
| Temperature, °F | 1 | 2 | 5 | 10 | 20 | 50 | 100 | 200 | 500 |
| 350 | 100.0 | 97.0 | 96.0 | 99.0 | 96.0 | 94.0 | 89.0 | 82.0 | |
| 375 | 99.5 | 97.0 | 81.0 | 95.0 | 90.0 | 83.5 | 76.0 | | |
| 400 | 95.0 | 102.0 | 87.5 | 80.0 | 77.0 | 55.5 | 58.0 | | |
| 425 | 93.0 | 80.0 | 77.0 | 73.0 | 66.0 | | | | |
| 450 | 79.0 | 68.0 | 73.0 | 70.0 | 67.0 | 57.0 | | | |
| 500 | 66.0 | 69.0 | 57.0 | 54.0 | 54.0 | 47.0 | | | |

Values are percentages as compared to as-received data at room temperature

| | 100%-110% | As-Received | | |
|---|---|---|---|---|
| | 90%-100% | Y.S. (ksi) | U.S. (ksi) | % Elongation |
| | Below - 90% | 58.0 | 63.8 | 9.7 |

Table 4 and 5 clearly show that as time and temperature increases the yield strength decreases, in a very consistent manner.

**Table V: Yield Strength at t/2 (Temperature vs. Time)**

| | Time, h | | | | | | | | |
|---|---|---|---|---|---|---|---|---|---|
| Temperature, °F | 1 | 2 | 5 | 10 | 20 | 50 | 100 | 200 | 500 |
| 350 | 99.0 | 104.0 | 103.0 | 99.0 | 98.0 | 95.0 | 88.0 | 84.0 | |
| 375 | 106.0 | 99.0 | 98.0 | 97.0 | 89.0 | 81.0 | 75.0 | | |
| 400 | 98.0 | 99.0 | 87.0 | 87.0 | 78.0 | 61.0 | 69.0 | | |
| 425 | 91.0 | 66.0 | 58.0 | 73.0 | 73.0 | | | | |
| 450 | 81.0 | 79.0 | 74.0 | 57.0 | 68.0 | 65.0 | | | |
| 500 | 54.0 | 59.0 | 55.0 | 50.0 | 51.0 | 44.0 | | | |

Values are percentages as compared to as-received data at room temperature

| | 100%-110% | As-Received | | |
|---|---|---|---|---|
| | 90%-100% | Y.S. (ksi) | U.S. (ksi) | % Elongation |
| | Below - 90% | 61.9 | 67.1 | 9.1 |

## Conclusions

1. The thermal exposure study showed that samples experienced a slight age hardening or a plateau in the initial thermal exposure and as time went on the material began to experience overaging.

2. The half thickness (t/2) samples appeared to be consistently higher in yield and ultimate strengths at lower temperature exposures as compared to the quarter thickness (t/4) samples. This correlated with the crystallographic texture of the grains.

3. The half thickness (t/2) and quarter thickness (t/4) samples had similar mechanical properties at longer times and higher temperatures.

## Acknowledgement

This work was funded by the National Science Foundation (NSF) Grant No. EEC-9732046, Research Experience for Undergraduates (REU) Site.

## Reference

1. Rioja, R.J., and Venema, G.B. "Heavy Gauge 2097 Plate Product for use in Fighter Bulkhead Application." Aeromat 98, Virginia, June 15-18.

2. Fielding, P.S., and Wolf, G.J. "Aluminum-lithium for aerospace." Journal of Advanced Materials and Processes. October 1996.

3. Witters, J. K. "Status of Al-li Alloys" Alcoa, Inc. (Aluminum Company of America), February 1996.

4. Webster, D., "Aluminum-Lithium: The Next Generation." Journal of Advanced Materials and Processes. May 1994, 18-24.

5. Rioja, R. "Aluminum-lithium plate alloy resists fatigue, lasts longer." Journal of Advanced Materials and Processes. September 1998, 9

6. Lee, E.W. et al. "The Effects of Off-axis Thermomechanical Processing on the Mechanical Behavior of Textured 2095 Al-Li Alloy." Journal Material Science and Engineering Vol. 265, 1999, 100-109.

# LIGHTWEIGHT ALLOYS FOR AEROSPACE APPLICATION

*Edited by:*
*Dr. Kumar Jata, Dr. Eui Whee Lee,*
*Dr. William Frazier and Dr. Nack J. Kim*

## TITANIUM ALLOYS

## The Effect of Crystal Orientation and Boundary Misorientation on Tensile Cavitation During Hot Tension Deformation of Ti-6Al-4V

*Thomas R. Bieler and S.L. Semiatin*

Pgs. 161-170

184 Thorn Hill Road
Warrendale, PA 15086-7514
(724) 776-9000

# THE EFFECT OF CRYSTAL ORIENTATION AND BOUNDARY MISORIENTATION ON TENSILE CAVITATION DURING HOT TENSION DEFORMATION OF Ti-6Al-4V

Thomas R. Bieler* and S.L. Semiatin

Air Force Research Laboratory, Materials and Manufacturing Directorate
AFRL/MLLM, Wright Patterson AFB, OH 45433-7817

*Department of Materials Science and Mechanics,
Michigan State University, East Lansing, MI 48824-1226, bieler@egr.msu.edu

*Abstract*

Cavitation during hot tensile deformation in Ti alloys is commonly observed. Prior work has indicated that cavities are often located at grain boundaries perpendicular to the tension axis where the lamellae appear nearly parallel to the tension axis. Since lamellar colony interfaces tend to include the c-axis, it had been hypothesized that cavities develop preferentially at boundaries that have the hard c-axis orientation parallel to the tension axis. A tension specimen deformed at 815°C and 0.1/s strain rate was analyzed to evaluate this hypothesis using orientation imaging microscopy, by measuring the crystal orientations surrounding nucleated cavities. Small cavities observed far from the fracture surface were in agreement with the hypothesis. In regions with more strain, more cavity growth was apparent when a small fraction of soft orientations with a non-c-axis orientation was also at or near the cavity. The constraint of the hard orientations and the preferential deformation of the softer orientations around relatively undeformable c-axis orientations was concluded to stimulate tensile conditions that favor cavity growth.

Lightweight Alloys for Aerospace Applications
Edited by Kumar Jata, Eui Whee Lee,
William Frazier and Nack J. Kim
TMS (The Minerals, Metals & Materials Society), 2001

# Introduction

Primary hot working to convert the large-grain cast microstructure to a fine, two-phase microstructure in alpha/beta Ti alloys (known as globularization) is an important technological process that leads to a highly formable microstructure suitable for secondary hot working via forging or superplastic forming. In some areas of a workpiece, tensile stress conditions are generated, such as at the outer edge of an upset forging or at the edges of rolled plate, where cracking can occur. Cavity formation in these areas has been examined in the context of a research program on the primary hot working of Ti-6Al-4V conducted at the Air Force Research Laboratory [1-11].

This prior work included studies to quantify the kinetics and mechanisms of cavitation during hot working of Ti-6Al-4V with a colony alpha microstructure [1-3]. The research revealed that many cavities nucleate on grain boundaries perpendicular to the tension axis, at which the lamellar lath orientations in a polished section appear nearly parallel to the tension axis. It was thus hypothesized that cavities are more likely to form at boundaries between grains that have the c-axis nearly parallel to the tension axis, which is a hard orientation. When the c-axis is aligned with the tension axis, <a> slip on basal or prism planes has no resolved shear stress, so only slip systems with much higher CRSS values such as <c+a> slip (or twinning) can operate [12]. The $\alpha$ lamellar laths tend to have {hk.0} interface planes with $\beta$ laths [e.g. 13], and these planes include the c-axis, so optical micrographs are not sufficient to identify the orientation of the c-axis. Laths that are parallel to the tension axis can have the c-axis in any direction in a plane that includes the tension axis.

In this paper, the specific crystallographic conditions that are associated with cavity formation during hot tension testing were investigated using orientation imaging microscopy (OIM) as a primary investigative tool. To this end, OIM was conducted on a tension specimen that had a strong component of texture with the c-axis parallel to the tension axis.

## Experimental Procedures

The tension specimen investigated was described by Semiatin et al., [1]. The specimen was deformed to failure in tension at 815°C at a nominal strain rate of 0.1 s$^{-1}$, and air cooled. It was sectioned and polished mechanically to measure cavity density along the tension axis. These data from Semiatin et al., 1998 are shown in the plot in Figure 1. The specimen was mill annealed (705°C/1 hr) to assist recovery in order to improve the quality of electron backscattered patterns used in OIM. The specimen was eletropolished in a solution of 590 ml methanol, 350 ml 2-butoxy ethanol, 60 ml HClO$_4$ (60%) at 20 VDC with a stainless steel cathode with moderate agitation for about 1 hour to remove about 5 $\mu$m of material. The electropolishing exaggerated cavity size, so the low magnification image of the specimen in Figure 1 has an enhanced apparent cavity volume fraction, and cavities presented subsequently are enlarged.

Orientation imaging was conducted using a Leicascan 360FE scanning electron microscope (SEM) using TexSEM software version 2.6 run on a Unix machine. Details of the imaging conditions are found in ref [10]. Two types of scans were made - short scans around one particular cavity that included about 1000 pixels

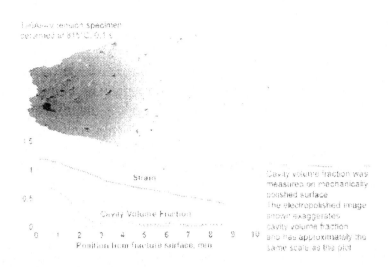

Figure 1: Image of tension specimen, with quantified cavity volume fraction and local strain.

and overnight scans that collected about 40,000 pixels in a larger neighborhood where a number of cavities were present. The step size used was between 0.5 and 2 μm. Data sets were cleaned up by using a nearest neighbor criterion to replace the orientation of pixels in which the confidence in correct indexing was low, to the orientation of the neighbor with the highest confidence index value, thereby altering about 15% of the pixels. Subsequent post processing used PC versions 2.7 and 2.8 of the TexSEM analysis software.

## Results and Discussion

The microstructure of the specimen shoulder region (that had about 5% strain) is shown in Figure 2a. The colony microstructure had about 30-100 μm colonies that were smaller than the 100-200 μm prior β grain size that was generated during a β heat treatment [1]. Within the colonies, 1-μm thick α laths were separated by thin layers of β. Also, ~3-μm thick grain boundary α (GB α) was observed (evident in Figure 2a). The grain boundaries also had a thin layer of beta phase between the grain boundary alpha layer and the colony-alpha side plates. Many colonies were observed within a prior β grain (which is delineated by the GB α), due to the 12 possible orientation variants that can form as the BCC β transforms to HCP α. Some colonies extended across GB α interfaces, as indicated by the arrows in Figure 2b.

Prior to testing, the specimen material had a texture with about 8 x random for orientations with the c-axis parallel to the tension axis and about 3 x random for orientations with the c axis tilted 60° from the tension axis. The dominant c-axis texture is apparent in the overlaid Taylor factor map in Figure 2b that shows a high area fraction of light gray grains, which have the c-axis nearly parallel to the tension axis. The Taylor factor maps were computed using CRSS ratios of 0.7 : 1 : 3 for prism <a>, basal <a>, and <c+a> slip, that were identified as optimal for the simulation of forging anisotropy

[9]. The orientations of many colonies are indicated with unit cell prisms, which show that the lower Taylor factor (darker) regions have the c-axis tilted more than 15° from the tension axis. The low Taylor factors for tilted orientations imply a soft orientation because both prism and basal slip systems have a high resolved shear stress to facilitate their operation. (Neither prism nor basal slip can operate in c-axis oriented grains, where only <c+a> slip is possible.)

Cavities investigated in 12 regions in the tension specimen using OIM scans are summarized in Table I. Some of the scanned areas are presented with overlaid Taylor factor maps and unit cell prisms in Figures 3-8 to illustrate the descriptions in Table I. The local strain increases with successive entries in the table, so that a stochastic sense of cavity nucleation and growth characteristics is gained. Many additional cavities were examined with manual evaluations of EBSP patterns around the circumference.

The majority of cavities were nearly or completely surrounded by c-axis grain orientations. Triple points as well as boundaries perpendicular to the tension axis were common

(b)

Figure 2: (a) Secondary electron image of electropolished region near shoulder of specimen 22 mm from fracture surface (~0.05 local strain), tension axis is vertical. (b) Taylor factor map and sample orientations are overlaid with hard orientations as lightest shade of gray. Arrows indicate examples of GB α with the same crystal orientations on both sides.

locations for cavity nucleation, whether as different variant orientations within a prior β grain or at GB α triple points, even if nearly the same orientation was present in all three grains as in Figure 4. Larger cavities tended to have non-c-axis orientations around the perimeter (e.g. Figure 3). Since cavities that also had non c-axis orientations were more interesting to scan, the list in Table I represents a bias toward cavities that were not completely surrounded by c-axis oriented grains.

## Table I  Summary of Cavitation Observations in Tension Specimen

| *mm from fracture, local strain, % cavitation, number, and size of cavities, μm* |
|---|
| Figure #, description of cavities (GB α orientations are given with respect to vertical tension axis and c-axis orientations have the c axis parallel to the tension axis). |
| *22mm (shoulder), <0.05, 0%, none* |
| Figure 2 illustrates bright (high Taylor factor) areas with c axis near tension axis.  Most bright areas have nearly vertical lamellae orientations. |
| *15.3mm, 0.15, <1%, 12, 2-10 μm* |
| Figure 3 illustrates cavities in Taylor factor map on both sides of horizontal or inclined GB α. 8 cavities have a non c-axis orientation near, 4 others are wholly in c oriented grains. |
| *15.3mm, 0.15, <1%, 4, 4-20 μm* |
| Figure 4 illustrates cavities clustered on horizontal GB α with c-axis oriented grains on either side.  Non c-axis oriented regions are present on opposites sides of the cavity and boundary. |
| *14.4mm, 0.15, <1%, 3, 3-10 μm* |
| Figure 5a shows a cluster at a GB α triple point.  The smallest cavity is at the triple point with 20, 140, -135° GB α orientations from tension axis.  C-axis orientations are on all three sides, but cavities are on the -135° leg.  Following horizontal GB α boundary to next triple point to right with 80, -35, -135° orientations, there are no cavities (Figure 5b), but 3 non-c axis orientations are present around triple point. |
| *13.9mm, 0.15, <1%, 3, 4-5 μm* |
| Cluster at GB α triple point with 70, -45, -175° GB α orientation to tension axis; all three grains having c-axis orientation.  Cluster is on lower side of 70° leg.  Adjacent to farthest cavity from triple point are two non c-axis oriented grains. |
| *10.1mm, 0.3, <2%, 1, 4 μm* |
| Cavity in GB α 20° from tension axis.  Left side is c-axis orientation, right side is soft orientation. |
| *9.9mm, 0.3, <2%, 6, 3-17 μm* |
| Figure 6 shows a GB α triple point with 30, 170, -90° with c-axis orientations on lower and left grains, and upper right grain has orientations with c-axis 70-90° from tension axis, but 20 microns above the -90° GB α, the orientation is c-axis.  Two large 15 μm cavities are on the -90° GB α segment.  The 30° GB α has soft orientation despite hard orientations on either side, and has two cavities in it.. |
| *9.7mm, 0.3, <2%, 6, 4-10 μm* |
| Two of 6 cavities adjacent to c-axis orientations, one in -40° GB α with soft orientation on one side, other at a variant interface.  Two cavities in a soft orientation have a lamellar ligament bent between them.  One small cavity in soft 110° GB α interface area.  A larger cavity at base of highly distorted soft oriented variant that was stretched around hard orientation. |
| *9.6mm, 0.32, <2%, 1, 6 μm* |
| 40° soft GB α with c-axis grains on both sides.  Left of cavity is soft orientation similar to GB α. |
| *8.56mm, 0.38, 2%, 21, 4-21,* |
| Figure 7 shows three large clusters of larger cavities oriented at 45° from TA  Two are surrounded by c orientations, the third is parallel to the edge of a c-axis grain but bounded by a soft orientation.  Shear banding is suggested, perhaps initiated by softer orientation variants initially inclined at about 45°. |
| *6.7mm, 0.55, 4%, 2, 4 μm* |
| Distorted horizontal GB α with c-axis orientations on either side, but near cavity there are many small, soft, and varied orientations. |
| *6.4, 0.56, 6%, 1, 3 μm* |
| Cavity in vertical 0° GB α, surrounded by about 10 μm orientations but none are dominant. |
| *3.8mm, 0.77, 6%, 10, 1-10* |
| Figure 8 show distorted microstructure.  There are many small orientations, but the largest patch of microstructure is c-axis oriented.  Most cavities are close to near-c-axis orientations. |

The process of cavity nucleation and growth that emerged from this investigation confirmed the hypothesis that grain boundaries with hard c-axis orientations (c-axis within 10-15° of the tension axis) were most likely to nucleate cavities. The presence of cavities along GB α that was completely surrounded by hard orientations is somewhat puzzling because such regions could be expected to behave as a single crystal, and hence not generate substantial hydrostatic tensile components of stress. Small differences in α orientation and the presence of a thin, soft layer of beta phase adjacent to the grain boundary alpha may be conjectured to give rise to cavities in these areas due to a constrained plasticity mechanism analogous to the failure of brazed joints (Ghosh et al., 1999).

In contrast, the larger cavities (that nucleated earliest or grew fastest) had a small fraction of soft orientations at or close to the perimeter (Figures 3,4,6). With increasing strain, cavity nucleation also occurred adjacent to non c-axis oriented grains, but these nucleations were not as significant to the failure process as the larger cavities that have a mixture of c-axis and soft orientations surrounding them. The largest observed cavities were in regions within or around large 300-500 μm c-axis oriented grains (or grain

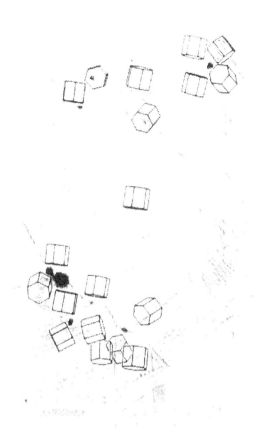

Figure 3: 15.3 mm from fracture surface (local strain of ~0.15). Cavities form preferentially on GB α that is perpendicular to the tensile axis. 8 of 12 cavities have lower Taylor Factor (darker) orientations adjacent to cavity in an area dominated by c-axis texture (light orientations).

clusters) as indicated in Figures 2 and 7. In these larger c-axis clusters, minority variants with soft orientations may trigger instabilities that lead to shear banding, as suggested in Figure 7 (and discussed further in [10]). The soft orientations were typically broken into small grains with large misorientations that flowed even more easily than the parent soft orientation, due to the small grain size [10]. This suggests that cavity nucleation and growth was stimulated by heterogeneous flow of soft oriented grains in the vicinity of hard orientations that constrain material flow so that large local hydrostatic tensile stress states developed.

## Conclusions

Cavity nucleation during hot working at 815° C and 0.1s⁻¹ in Ti-6Al-4V with a colony microstructure is stimulated in regions where the crystals on either side of a tensile grain boundary are in a hard orientation with the c-axis within 10-15° of the tension axis. The larger cavities nucleated in triple points and tensile grain boundary regions that also have a small fraction of soft orientations. The largest cavities are found in regions that have a large group of c-axis oriented grains in a cluster. Shear banding coupled with constraint effects imposed by the hard grains leads to heterogeneous flow that can stimulate high local hydrostatic stress conditions that favor cavity growth.

Figure 4: 15.3 mm from fracture surface (local strain of ~0.15). Largest cavity on GB α perpendicular to the tensile axis has c-axis grains (light) and soft grains (dark) on opposite sides of the boundary.

### Acknowledgements

This work was conducted as part of the in-house research activities of the Processing Science Group of the Air Force Research Laboratory's Material and Manufacturing Directorate. The support and encouragement of the laboratory management and the Air Force Office of Scientific Research (Dr. C.S. Hartley, program manager) are gratefully acknowledged. TRB was supported through Air Force contract F33615-94-C-5804.

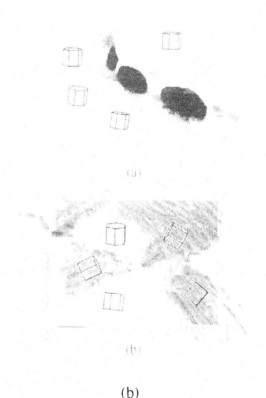

(b)

Figure 5: 14.4 mm from fracture surface (local strain of ~0.15). Cavity cluster at triple point between three c-axis grains (a), no cavity at next triple point to the right, where there is a balanced set of c-axis (light) and soft (dark) grains (b).

### References

1. S.L. Semiatin, V. Seetharaman, A.K. Ghosh, E.B. Shell, M.P. Simon, and P.N. Fagin, "Cavitation During Hot Tension Testing of Ti-6Al-4V," Materials Science and Engineering, 1998, vol. A256, (1998), 92-110.

2.  S.L. Semiatin, R.L. Goetz, E.B. Shell, V. Seetharaman, and A.K. Ghosh, "Cavitation and Failure During Hot Forging of Ti-6Al-4V," Metall. and Mater. Trans. A, vol. 30A, (1999), 1411-24.

3.  A.K. Ghosh, D-H. Bae, and S.L. Semiatin, "Initiation and Early Stages of Cavity Growth During Superplastic and Hot Deformation," Materials Science Forum, 1999, vol. 304/306, p. 609-16.

4.  S.L. Semiatin, V. Seetharaman, and I. Weiss, "Flow Behavior and Globularization Kinetics During Hot Working of Ti-6Al-4V With a Colony Alpha Microstructure," Materials Science and Engineering, vol. A263, (1999), 257-71.

5.  E.B. Shell E.B. and S.L. Semiatin, "Effect of Initial Microstructure on Plastic Flow and Dynamic Globularization During Hot Working of Ti-6Al-4V. Metall. Trans. A. Vol. 30A, 3219-29.

6.  P.D. Nicolaou and S.L. Semiatin, "Modeling of Cavity Coalescence During Tensile Deformation." Acta Mater., vol. 47, (1999), 3679-86.

7.  P.D. Nicolaou, S.L. Semiatin, and A.K. Ghosh. "An Analysis of the Effect of Cavity Nucleation Rate and Cavity Coalescence on the Tensile Behavior of Superplastic Materials," Metall. and Mater. Trans. A. vol. 31A, (2000), 1425-34.

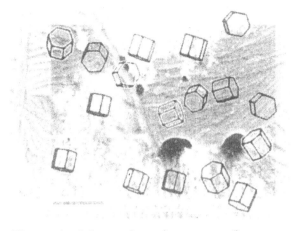

Figure 6: 9.9 mm from fracture surface (local strain of ~0.3). Two large cavities on GB α perpendicular to the tensile axis have hard c-axis grains beneath (light) and soft grains (dark) above the boundary.

Figure 7: 8.56 mm from fracture surface (local strain of ~0.38). Large cavities are located within or between c-axis (lighter) grains, but also in softer (darker) shear regions just below the hard grain zone.

8. P.D. Nicolaou and S.L. Semiatin, "An Analysis of the Effect of Continuous Nucleation and Coalescence on Cavitation During Hot Tension Testing," Acta Mater., vol. 48, (2000), 3441-50.

9. S.L. Semiatin, and T.R. Bieler, "Effect of Texture and Slip Mode on the Anisotropy of Plastic low and Flow Softening During Hot Working of Ti-6Al-4V" submitted to Metall. Mater. Trans. 2001.

10. T.R. Bieler and S.L. Semiatin, "The Origins of Heterogeneous Deformation During Primary Hot Working of Ti-6Al-4V", submitted To International Journal Of Plasticity, 2001.

11. S.L. Semiatin, and T.R. Bieler, 2001, unpublished.

12. N.E. Paton, J.C. Williams, and G.P. Rauscher, "The deformation of alpha-Phase Titanium", in: Titanium Science and Technology, (Proceedings of the second international coference), R.I. Jaffee and H.M. Burte, Eds., Plenum Press, New York, (1973), 1049-69.

13. S. Suri, G.B. Viswanathan, T. Neeraj, D.-H. Hou, and M.J. Mills, "Room Temperature Deformation and Mechanisms of Slip Transmission in Oriented Single-Colony Crystals of an α/β Titanium Alloy", Acta Mater. 47, (1999), 1019-34.

Figure 8: 3.8 mm from fracture surface (local strain of ~0.77), in region between large cavities. Nucleating cavities form near but not in (lighter) near-c-axis grains.

# LIGHTWEIGHT ALLOYS FOR AEROSPACE APPLICATION

*Edited by:*
*Dr. Kumar Jata, Dr. Eui Whee Lee,*
*Dr. William Frazier and Dr. Nack J. Kim*

## TITANIUM ALLOYS

## Processing and Properties of Gamma Titanium Aluminides and Their Potential for Aerospace Applications

*J.D.H. Paul, M. Oehring, F. Appel and H. Clemens*

Pgs. 171-181

184 Thorn Hill Road
Warrendale, PA 15086-7514
(724) 776-9000

# PROCESSING AND PROPERTIES OF GAMMA TITANIUM ALUMINIDES AND THEIR POTENTIAL FOR AEROSPACE APPLICATIONS

J.D.H. Paul, M. Oehring, F. Appel and H. Clemens

GKSS Research Center
Institute for Materials Research
Max-Planck-Str., D-21502, Geesthacht, Germany

## Abstract

The development of high-temperature materials is key to the technological advancements in engineering areas where materials have to withstand extremely demanding conditions. Examples for such areas are the aeroengine and aerospace industries. Intermetallic $\gamma$ (TiAl) based alloys offer many attractive properties for use in various hypersonic space and aerospace vehicles. This paper reviews the present status in alloy development of $\gamma$ (TiAl) based alloys and thermomechanical processing on industrial scale.

## Introduction

Over the last ten years considerable research effort has been expended in developing a basic understanding of the physical metallurgy and related processing of gamma titanium aluminides. The work has been driven by the need for a low density, high strength structural material in the aerospace and aeroengine industries [1-4]. The implementation of such material as a replacement for the heavier nickel-based superalloys currently used would result in significant direct and indirect weight savings combined with better engine efficiency and reduced pollution emissions.

## Alloy development

Titanium aluminides are being considered for long term applications where the temperature ranges from about 650°C to 750°C and for temperatures up to 900°C for shorter periods of time [5]. At such temperatures the thermally activated nature of deformation becomes important, this leads to diffusion assisted climb of dislocations and strain rate dependent behavior [6]. Thus during service, creep and microstructural stability become issues. To reduce such problems alloys must be designed so as to contain stable phases and microstructures together with alloying additions which reduce diffusion rates and which lead to precipitation hardening mechanisms [7].

Current designs are based on alloys that are slightly lean in aluminium and which contain several other alloying elements. These alloys have the general composition (in at.%) of:

$$\text{Ti-(44-49)Al-(0.3-10)X}$$

Lightweight Alloys for Aerospace Applications
Edited by Kumar Jata, Eui Whee Lee,
William Frazier and Nack J. Kim
TMS (The Minerals, Metals & Materials Society), 2001

where X refers to elements such as Cr, V and Mn for enhanced ductility; Ta, W, Si and C for improved creep resistance and B for grain refinement. Such alloys contain $\gamma$ (TiAl) and $\alpha_2$ (Ti$_3$Al) as major phases possibly together with other minor phases depending on composition and thermal-history.

Recently, alloys with the composition Ti-45Al-(5-10)Nb have received much attention [8]. Suitably processed material typically exhibit flow stresses at room temperature in excess of 800 MPa. At present two hypotheses have been proposed to explain the high strength of these alloys. The first considers niobium to act as a solid solution strengthening element. It is argued that the atomic size difference between Ti and Nb gives rise to lattice strains which inhibit the movement of dislocations and, therefore, the deformation process [9]. However, a significant solid solution strengthening cannot be expected because of the very small size misfit of Ti and Nb atoms. Thus, the second hypothesis is the more likely, and considers the increased strength to be mainly a consequence of the reduced aluminium content and thus the increased $\alpha_2$ content which results in a refinement of the microstructure [10]. Indeed, very fine lamellae spacings were found by TEM investigations. Further, investigation of material deformed at room temperature showed significant twinning deformation, possibly indicating that large niobium additions reduce the stacking fault energy. This seems to be beneficial for the low temperature ductility of the material.

Niobium is also thought to play a significant role in the high temperature strength retention of these alloys. The diffusion rate of niobium has been shown to be significantly slower than that of the Ti for which it substitutes exclusively in the TiAl lattice [11]. This view is also supported by activation enthalpy measurements. Conventional alloys with the base-line composition Ti-48Al-2Cr typically exhibit activation enthalpies of $\Delta H = 3$ eV [12], which compares with values of $4 - 4.5$ eV estimated on alloys containing $5 - 10$ at% Nb [10]. This implies that dislocation glide and climb processes will be more difficult in alloys containing large amounts of niobium which should lead to enhanced creep resistance and improved strength. The microstructural stability of the material should also benefit from the reduced diffusion kinetics. In addition, niobium additions increase the oxidation resistance of the material [13].

Concerning carbon additions significant improvements in strength can be achieved. Through a suitable solution and precipitation heat-treatment, perovskite type precipitates of composition Ti$_3$AlC and dimensions 22 nm in length by 3.3 nm x 3.3 nm in cross-section develop. The precipitates are elongated parallel to <001] TiAl. Significant coherency stresses are developed due to lattice mismatch strains. The precipitates act as strong obstacles to the glide of all types and characters of dislocations in TiAl. Work by Appel et al. has shown that a $0.3 - 0.4$ at.% addition of carbon to two phase alloys can lead to significant improvement of strength [14] and creep resistance [15]. Thus, precipitation hardening seems to be a promising technique to improve creep properties and extend application temperatures. However, the stability of the Ti$_3$AlC precipitates during long-term service needs to be proved.

## Ingot manufacture

For titanium aluminide material to be used in safety critical aerospace and aero-engine applications quality issues with regard to ingot metallurgy and component fabrication must be addressed. The production route for TiAl ingots is similar to that used for conventional titanium alloys. Vacuum arc and induction skull melting techniques have both been used to fabricate TiAl ingots. These methods use clean melting technologies that ensure interstitial pick-up, which can lead to greatly reduced ductility, is minimized.

The vacuum arc process is essentially the same as that used for ingot production of conventional titanium alloys. Small briquettes are cold pressed from a mixture of the elemental alloy components. In order to reduce macroscopic chemical inhomogeneity of the resultant ingot, the briquettes can be made from a suitable mixture of elements and masteralloys [16]. The briquettes are welded together so to form an electrode. The welding process is performed under a protective atmosphere to minimize pick-up of atmospheric impurities. The composition of the electrode is chosen so as to compensate for the loss of volatile elements such as aluminium during the melting process. The electrode is arc melted under vacuum into a water cooled copper crucible to form an ingot. This primary ingot is then remelted to improve chemical homogeneity. The macroscopic variation of aluminium within the final ingot can be reduced from $\pm$ 2 at.% for ingots made from elemental briquettes to $\pm$ 0.7 at.% when the briquettes are made from masteralloys.

For the induction skull melting technique elemental alloying elements are used to load a water cooled copper crucible [17]. Under vacuum, induction heating is used to melt the alloying elements. After melting the alloy is poured into ingot moulds. As a consequence of the high power required, the induction field imparts an intense stirring action in the molten metal, which results in excellent macro-homogeneity of the ingots. As is standard practice in the titanium industry, ingots of TiAl are usually given a hot isostatic pressing treatment to ensure that any casting porosity is closed.

In addition to the ingot metallurgy route, powder metallurgical techniques have also been used in the production of titanium aluminide components [18] and wrought semi-finished products [4]. Pre-alloyed powders can be produced in large quantities by inert gas atomization. In this process, elemental or pre-alloyed components are melted, by a plasma torch for example, on a water cooled copper hearth [19]. Once molten, the liquid metal is allowed to fall as a stream through a nozzle where it is 'atomized' by an inert gas stream into small spherical droplets which then solidify. The whole process, including powder collection takes place under inert gas to minimize interstitial pick-up. After production the powder is usually sieved into different size fractions. The powder particles exhibit cellular microstructures with cell sizes of only a few micrometers [20]. In addition, the composition of the particles varies very little. As a consequence, when the powder is canned and subsequently hot isostatically pressed (HIPed), the resultant material shows much better chemical homogeneity than would be obtained through casting techniques. In addition to this, the greater yield and possibility of near-net shape processing may offset the relatively expensive powder production costs. One disadvantage of the powder route concerns the formation of microscopic argon filled pores in HIPed material after long term thermal exposure [21]. Although these pores seem to have no influence on short-term tensile properties, they reduce the maximum superplastic strain attainable at high temperature [22, 23].


## Thermomechanical processing

### Forging

As-cast ingots usually exhibit a lamellar microstructure with the lamellae orientated perpendicular to the direction of dendritic growth. Depending on the exact alloy composition, the material can be chemically very inhomogeneous as a result of the peritectic solidification path. Thus, to ensure reproducible component properties and homogeneous microstructures, the as-cast material is usually broken down by forging, extrusion or a combination of both techniques.

Isothermal forging of TiAl on an industrial scale is usually performed under vacuum or inert gas at temperatures between 1000°C and 1200°C and employs strain rates of $10^{-3}$ s$^{-1}$ to $10^{-2}$ s$^{-1}$ [24]. Using such forging conditions ingots 270mm in diameter and 250mm high have been forged in a single step to pancakes with a diameter of 600mm [12]. The surface condition of such single step forged ingots can be very good, as seen in Figure 2a. However microstructural inhomogeneities such as striped zones of coarse gamma grains or regions of unrecrystallized lamellar microstructure can remain, Figure 2b.

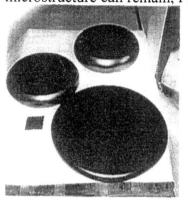 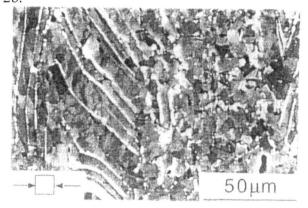

Figure 2: (a) Isothermally forged pancakes of TiAl. (b) Example of striped partially recrystallized microstructure.

A more homogeneous refinement of the cast microstructure can be achieved though two-step isothermal forging with an intermediate static recrystallization heat-treatment. The microstructural homogeneity of forged material may be further enhanced if, prior to forging, the cast ingots are heat-treated for a short time in the $\alpha$-phase field to improve chemical homogeneity [25]. However, this is often not done because of the rapid grain growth at temperatures in excess of 1350°C, which can make subsequent forging difficult.

Extrusion

Extrusion can also be used for ingot break-down and is usually performed at temperatures near the $\alpha$-transus temperature, typically between 1250°C and 1380°C. Due to the severe oxidation of titanium aluminide at such temperatures, the material is usually encapsulated in cans made of conventional titanium alloys or stainless steel. At the extrusion temperature the canning materials have a significantly lower flow stress than the titanium aluminide. This leads to shear stresses developing between the can and ingot material during the extrusion process which may lead to inhomogeneous deformation, cracking or fracture of the titanium aluminide work-piece [26]. To overcome such problems, the flow stress mismatch (which can be as high as 300MPa) can be reduced by the use of thermal insulation between the can and ingot material. The canning material thus cools faster than the ingot material and the flow stresses are more equal. Indeed, sometimes the canning assembly is allowed to cool for a specific dwell time in order to ensure such conditions. It should be emphasized, however, that the temperature of the titanium aluminide must not fall below a critical value where it becomes non-workable. The high hydrostatic pressures involved in the extrusion process facilitate the forming of virtually any composition desired. For example, 80kg ingots of composition Ti-45Al-10Nb were uniformly extruded into a rectangular cross-section with a 10:1 reduction in cross-sectional area, Figure 3. The microstructural and chemical homogeneity of the extruded material can be further improved by using higher reduction ratios or by multi-step processing, Figure 4, the latter of which imposes severe constraints on the geometry of final components.

The room temperature properties are improved by double extrusion. Total extrusion reductions of up to 200:1 lead to highly refined homogeneous microstructures. For example, the refinement and improved microstructural homogeneity of material extruded to 97:1 reduction compared to that extruded to 14:1 is shown in Figure 5.

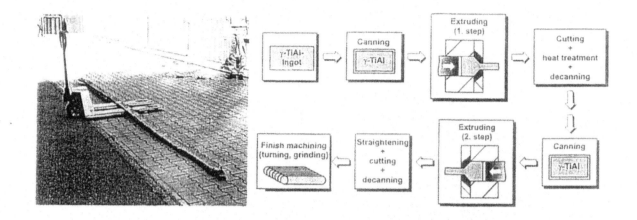

Figure 3: Extruded TiAl workpiece.    Figure 4: Double extrusion technique [27].

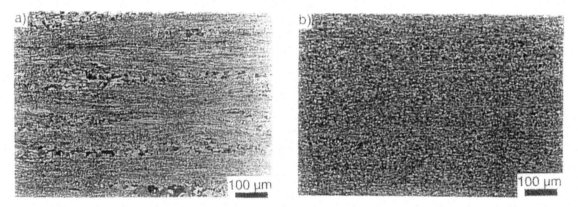

Figure 5: Microstructure of (a) single and (b) double extruded material [21].

Concerning the recently developed high niobium containing alloys discussed earlier, a room temperature tensile test curve of an extruded alloy developed at GKSS showed a plastic elongation of $\varepsilon = 2.5\%$ at a tensile fracture strength of about 1100MPa for a reduction ratio of 7:1 [12]. This is the best combination of room temperature ductility and strength ever reported for titanium aluminide alloys. These types of alloys thus seem to have a promising outlook.

Rolling
The pre-material used for rolling to sheet can be either forged ingot material or HIPed pre-alloyed powder compacts. Forged material is used over as-cast ingot because of the much better chemical homogeneity. Indeed for some alloys, an aluminium variation of $\pm 1$ at.% in the sheet cannot be tolerated because of the large scatter in mechanical properties. In this respect sheet produced from powder is chemically much more homogeneous and subsequently shows much less scatter in mechanical properties.

Concerning the rolling process similar constraints apply as for extrusion, namely the titanium aluminide must be protected from atmospheric oxidation and the flow stress mismatch between the canning material and the TiAl workpiece must be minimized. For the industrial production of large sheet material this is not easy. Special pack designs which allow the canning material to cool while the workpiece remains hot are used. In addition to pack design, the rolling conditions are also extremely important, low rolling speeds of ≤ 10 m/min are used to avoid workpiece fracture [28]. Thus to produce sheet, multiple rolling passes with re-heating of the canning assembly to the rolling temperature between each pass must be performed. Heat losses from the canning assembly can be large, thus the temperature of the material which is last to enter the rolls will be cooler than that which was first to be rolled. This can lead to microstructural differences between the two ends of the sheet [29]. In extreme cases, loss of temperature may lead to fracture of the titanium aluminide sheet during the rolling process. However, Plansee AG has solved these problems and are capable of producing TiAl sheets with uniform microstructure of dimensions up to 1800 x 500 x 1.0 mm. They report that further up-scaling in sheet length and width is technically possible from the rolling standpoint, but that ingot pre-material of sufficient size and homogeneity may be difficult to obtain. In this respect HIPed pre-alloyed powder compacts may be advantageous [4].

The microstructure of the titanium aluminide after rolling and de-canning is not in thermodynamic equilibrium. A primary annealing treatment of 1000°C/2 hrs is therefore performed in order to bring the microstructure into closer thermodynamic equilibrium and simultaneously reduce residual stresses and flatten the sheet. Material in this state is referred to as being in the 'primary annealed' condition. Such sheet shows a 'modified cube' texture with the <001] TiAl lattice direction preferentially orientated in the transverse sheet direction with the <100] axis being preferentially aligned along the rolling direction [23]. The behavior of γ-TAB (composition: Ti-47Al-4(Nb, Mn, Cr, Si, B)) sheet material has been previously investigated [23]. Tensile properties of sheet made from pre-alloyed powder in the primary annealed condition are shown in Figure 6. The tensile properties parallel and transverse to the rolling direction are anisotropic in the temperature range 600°C to 900°C. This has been attributed to the modified cube texture which is present in the primary annealed sheet material. Recent results indicate that the high niobium containing alloys, developed by GKSS, show much less texture after rolling than conventional TiAl alloys.

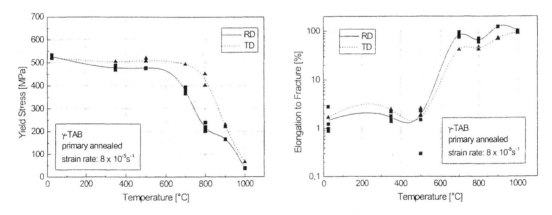

Figure 6: Mechanical properties of rolled TAB sheet material, transverse (TD) and parallel (RD) to the rolling direction [23].

The superplastic forming behavior of primary annealed sheet material was also investigated in the study. It was found that the grain size of superplastically deformed material depends on the deformation conditions. The strain rate sensitivity exponent (m) of γ-TAB sheet fabricated via the powder route is shown for different deformation temperatures as a function of strain rate in Figure 7a. This graph shows that the TAB sheet is potentially superplastically formable at temperatures as low as 950°C (m>0.4). Thus, equipment used for superplastic forming and diffusion bonding of conventional titanium alloys can also be used for titanium aluminide alloys. One significant difference exists in the behavior of sheet material produced from powder and ingot material. The high temperature superplastic elongation to failure of powder derived sheet is significantly lower than that of sheet formed from ingot. The reason for this relates to the evolution of thermally induced microporosity which is thought to arise from argon gas within the pre-alloyed powder. Such pores can lead to grain boundary separation and enhanced microvoid nucleation [21]. Nevertheless, both ingot and powder derived sheet has been successfully superplastically formed, Figure 7b.

## Joining and Machining Techniques

Many different joining techniques including electron beam and laser beam welding [30], rotary friction welding [31], brazing [2, 32] and diffusion bonding [32, 33] have been studied for titanium aluminide materials. For successful crack-free electron and laser beam welding the material to be joined must be pre-heated to temperatures above the brittle to ductile transition temperature so that thermal stresses which lead to crack formation may be relieved by plastic deformation. However, the microstructure in the heat affected zone leads to a reduction of strength and ductility, especially for fine grained and precipitation hardened alloys. Brazing has been successfully carried out on sheets and rods of various compositions. Commercially available Ni and Ti based filler alloys have been used to obtain sound joints [4, 32]. Large TiAl sheet structures have also been successfully assembled using brazing techniques. However, questions remain to be answered with regard to the effect on the mechanical properties of the phases formed at the joints. As with conventional titanium alloys, diffusion bonding techniques have been successfully used to join titanium aluminides. Bonds of excellent quality have been reported for bonding pressures of 5 to 40 MPa at 1000°C for times between 1 and 8 hours [4]. With regard to machining, conventional equipment can be used if appropriate precautions are taken to account for the low ductility [4, 34].

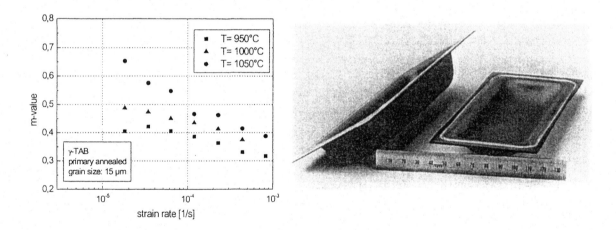

Figure 7: (a) Strain rate exponent of TAB sheet material as a function of strain rate [23]. (b) Parts produced by superplastic forming of sheet from ingot (left) and powder (right) based material. Forming temperature was 1000°C [27].

## Component fabrication

Many types of titanium aluminide aerospace components and structures have been made and tested, in this section two examples will be given to illustrate how the current technology has been implemented. For further information concerning the potential application of titanium aluminides in aerospace the reader is referred to [1-5].

## High pressure compressor blades

More than 200 high-pressure compressor blades have been produced in the framework of a German materials research program [12]. The material used was the GKSS developed γ-TAB alloy. The processing route involved hot-extrusion, closed-die forging and electro-chemical milling to final shape. Prior to the electro-chemical milling process the forged blades were given a super α-transus heat-treatment which resulted in a very homogeneous fully lamellar microstructure with a colony size of 130μm. The milled blades are shown in Figure 8. The same production route was also used to successfully process a limited number of blades from a novel precipitation strengthened high niobium containing alloy [12].

## TiAl sheet structure

It is envisaged that titanium aluminide sheet structures will be exploited as part of the noise abatement strategy in future supersonic civil transport aircraft. BF Goodrich Aerostructures Group has used forming and joining processes to make sub-elements for a proposed engine divergent flap design [2], Figure 9. The corrugations were hot-formed under argon using standard tooling and relatively low temperatures. The parts were joined by brazing under vacuum using a TiCuNi filler [33]. Testing of the structures under mechanical load revealed excellent behavior, thus showing the tremendous potential of such TiAl assemblies in future aircraft designs [2, 4].

Figure 8: High-pressure turbine blades [12].

Figure 9: TiAl sheet sub-elements of the divergent flap design concept with salient features of the full scale flap [2]. Courtesy of NASA/NASA Glenn Research Center.

## Summary

As a result of considerable R&D work over the last decade, the understanding of chemistry-microstructure-properties relationships and of processing techniques has led to successful production of components and structures with attractive engineering properties. Titanium aluminide material thus seems to be on the verge of significant industrial utilization as a structural material for aerospace applications.

## Acknowledgements

The authors would like to thank U. Brossmann, C. Buque, U. Christoph, S. Eggert, V. Küstner, U. Lorenz, J. Müllauer and E. Tretau for technical help and support. The financial support of this work from the HGF-Strategiefonds and BMBF project number 03N3030C0 is gratefully acknowledged.

## References

1. Y-W. Kim, J.Met, 46 (1994), p30 - 39.
2. P.A. Bartolotta and D.L. Krause, in Gamma Titanium Aluminides 1999 (Eds: Y-W. Kim et al., TMS, Warrendale PA, 1999) p3 - 10.
3. R. Leholm et al., in Gamma Titanium Aluminides 1999 (Eds: Y-W. Kim et al., TMS, Warrendale PA, 1999) p25 - 33.
4. H. Clemens and H. Kestler, Advanced Engineering Materials, 2 (9) (2000), p551 – 570.
5. Y-W. Kim and D.M. Dimiduk, in Structural Intermetallics 1997 (Eds: M.V. Nathal et al., TMS, Warrendale PA, 1997) p531 - 543.
6. F. Appel et al., Mater. Sci. Eng. (A), 233 (1997), p1 - 14.
7. F. Appel et al., Thermec 2000, in press.
8. F. Appel et al., Intermetallics, in press.
9. W.J. Zhang et al., Mater. Sci. Eng. (A), 271 (1999), p416 - 423.
10. J.D.H. Paul et al., Acta. Meter, 46 (4) (1998), p1075 – 1085.
11. C. Herzig and T. Przeorski, to be published.
12. F. Appel et al., Advanced Engineering Materials, 2 (11) (2000), p699 - 720.
13. Nickel et al., Mikrochim. Acta, 119 (1995), p23.
14. U. Christoph et al., Mater. Sci. Eng. (A), 239/240 (1997), p39 - 45.
15. F. Appel et al., in Gamma Titanium Aluminides 1999 (Eds: Y-W. Kim et al., TMS, Warrendale PA, 1999), p603 - 618.
16. V. Güther et al., in Gamma Titanium Aluminides 1999 (Eds: Y-W. Kim et al., TMS, Warrendale PA, 1999), p225 - 230.
17. S. Reed, in Gamma Titanium Aluminides, (Eds: Y-W. Kim et al., TMS, Warrendale PA, 1995), p475 - 482.
18. N. Eberhardt et al., Z. Metallkd, 89 (1998), p772 – 778.
19. R. Gerling et al., in Advances in Powder Metallurgy & Particulate Materials, (Eds: J.M. Capus et al., MPIF, Princeton NJ, 1992) p215.
20. U. Habel et al., in Gamma Titanium Aluminides 1999 (Eds: Y-W. Kim et al., TMS, Warrendale PA, 1999), p301 - 308.
21. H. Clemens et al., in Gamma Titanium Aluminides 1999 (Eds: Y-W. Kim et al., TMS, Warrendale PA, 1999), p209 - 224.
22. C.M Lombard et al., in Gamma Titanium Aluminides, (Eds: Y-W. Kim et al., TMS, Warrendale PA, 1995), p579 - 586.
23. H. Kestler et al., in Gamma Titanium Aluminides 1999 (Eds: Y-W. Kim et al., TMS, Warrendale PA, 1999), p423 - 430.
24. S.L. Semiatin et al., Mater. Sci. EngA. 243, (1998), p1 – 24.
25. C. Koeppe et al., Met. Trans A, 24A (1993), p1795 – 1806.
26. H. Clemens et al., Thermec 2000, in press.
27. H. Clemens et al., Z. Metallkd, 90 (1999), p569 – 580.
28. H. Clemens et al., in Structural Intermetallics , (Eds: R. Darolia et al., TMS, Warrendale PA, 1993), p205 - 214.
29. S.L Sematin and V. Seetharaman., Met. Trans A, 25A (1994), p2539 – 2542.
30. M. Klassen et al., in Laser Treatment of Materials (ed. F. Dausinger), Arbeitsgemeinschaft Wärmebehandlung und Werkstofftechnik, Stuttgart, p193.

31. R. Lison, Schweissen, Schneiden 2000, 52, p80.
32. G. Das and H. Clemens., in Gamma Titanium Aluminides 1999 (Eds: Y-W. Kim et al., TMS, Warrendale PA, 1999), p281 - 289.
33. G. Cam et al., Z. Metallkd, 90 (1999), p284 – 288.
34. E. Aust and H. Niemann., Advanced Engineering Materials, 1 (1) (1999), p53 – 57.

# LIGHTWEIGHT ALLOYS FOR AEROSPACE APPLICATION

*Edited by:*
*Dr. Kumar Jata, Dr. Eui Whee Lee,*
*Dr. William Frazier and Dr. Nack J. Kim*

## TITANIUM ALLOYS

Metallurgical and Fabrication Factors
Relevant to Fracture Mechanisms of
Titanium-Aluminide Intermetallic Alloys

*K. Nikbin*

Pgs. 183-191

184 Thorn Hill Road
Warrendale, PA 15086-7514
(724) 776-9000

# METALLURGICAL AND FABRICATION FACTORS RELEVANT TO FRACTURE MECHANISM OF TITANIUM –ALUMINIDE INTERMETALLIC ALLOYS

**K. Nikbin**

*Mechanical Engineering Department, Imperial College, London SW7 2BX, UK*

## Abstract

Titanium aluminides are seen as candidate materials for high temperature applications where weight is the prime concern. A summary of the history of γ-TiAl from its discovery in the 1950's and the effects of alloying additions and their effects on its properties and behaviour are first presented. The paper reviews the recent developments relating to this intermetallic alloy and relates the high temperature mechanical creep and creep/fatigue properties of this alloy to it's metallurgical properties and the fabrication and heat treatment of the alloy. In general it is concluded that titanium aluminides are strong low-density materials and their properties have been shown to be highly dependent on microstructure and fabrication processes. Lamellar microstructures have been found to offer superior creep resistance and attractive fracture toughness, particularly at elevated temperatures however it is weakened by the presence and development of multiple crack sites within the matrix. This work has highlighted the failure problems that exist in this γ-TiAl alloy. A potential problem with this class of material is its inherent low ductility and fracture toughness, particularly at lower temperatures in addition to its susceptibility to fatigue fracture at high temperatures.

*Keywords: Intermetallic;* γ-Titanium; fracture; creep; crack;

## Introduction

The evolution of the gas turbine has to a large extent been due to the development of new advanced materials capable of withstanding higher temperatures and thus allowing more efficient operation. An entire family of alloys, the so-called superalloys, based on Nickel and Cobalt has been developed almost exclusively for the gas turbine industry over the last 50 years [1]. These alloys are used for components operating above 550°C such as turbine blades, vanes, ducts, filters, cases, discs and combustion cans. The thermal efficiency of a gas turbine is only around 38 % with a gas entry temperature of 1200 °K, whilst at 1800 °K the efficiency rises to over 50 %, hence there is a strong demand for materials capable of operating at ever increasing temperatures. In response to this considerable research has and indeed is continuing to be performed on intermetallic materials such as γ-TiAl.

Lightweight Alloys for Aerospace Applications
Edited by Kumar Jata, Eui Whee Lee,
William Frazier and Nack J. Kim
TMS (The Minerals, Metals & Materials Society), 2001

The low-pressure turbine section, where blade temperatures up to 750°C and blade root temperatures of 500°C are commonplace, is seen as a typical environment for this material. γ-TiAl possesses some particularly attractive properties such as low density (approximately half that of nickel-based superalloys), and good creep resistance, however it presents the engineer with problems not associated with conventional materials.

An overview of the requirements for gas turbine materials

The aerofoil of a turbine blade is subjected to temperatures in the range 650-1100°C and stresses up to 150 MPa while the blade root experiences lower temperatures but stresses in the range 280-560 MPa. Materials for this application must possess creep resistance and high temperature strength, where creep is defined as the time dependent strain that accumulates upon loading a metallic component at elevated temperature. Additionally, the material must have sufficient ductility to accommodate the effects of stress concentrations, such as those at the fir-tree root of a turbine blade and be resistant to thermal and mechanical fatigue. High cycle fatigue (HCF) involves small variations in load but very high numbers of cycles, while low cycle fatigue (LCF) involves large variations in load and relatively few cycles and is usually associated with fixed strain amplitude. HCF is commonly caused by vibration whilst LCF is generated by the startup and shutdown of the turbine. Further requirements for such materials are a resistance to corrosion and oxidation due to the oxygen rich environment, which can contain many corrosive substances.

Non-rotating components, such as turbine guide vanes, encounter even greater temperatures of up to 1600°C but the stresses are lower at around 70 MPa [1]. The material requirements for these components are similar to those for turbine blades with the addition of high impact strength and occasionally weldability. To overcome the effects of temperature ceramic coatings are used as thermal barriers on the guide vanes, these can reduce the temperature the metal experiences by up to 300°C if air-cooling is employed.

A high tensile strength is required from materials used for turbine discs where the temperature is up to 800°C in the rim and the stresses are very high [1]. The effects of creep become less severe towards the centre of the disc since the temperature is lower but the stresses can be as high as 480 MPa. An important additional requirement for turbine disc materials is high fatigue resistance.

The use of γ-TiAl for non-rotating components and the high-pressure turbine blades is impossible due to the very high temperatures encountered in these areas. At the moment nickel-base alloys are the only materials used in these areas, however ceramics and other intermetallic alloys are being investigated for use in the future. γ-TiAl is intended for use in the low-pressure turbine section and for ducting associated with lower gas temperatures. A further requirement for any gas turbine alloy is low density. This has two effects namely a) to reduce centrifugal stresses in rotating components and b) to reduce the overall weight of the engine (of particular importance in aero-engines).

Background to the development of Titanium Aluminides

At present nickel-base superalloys are widely used in gas turbines; however the high density of nickel ($8.90 \times 10^3$ kg m$^{-3}$) is a major disadvantage. In response to this, much interest has been shown in titanium base alloys. Titanium has a density much lower than that of even the latest Ni-Al alloys ($4.54 \times 10^3$ kg m$^{-3}$ versus $8.30 \times 10^3$ kg m$^{-3}$), in addition titanium has a similar melting point to nickel. The use of conventional titanium alloys is limited by their poor oxidation and creep resistance at high temperatures, as a result the operating temperatures of these materials are generally below 600°C.

Improved properties are offered by intermetallic titanium aluminide alloys which are based on the $\alpha_2$ (Ti$_3$Al) and $\gamma$ (TiAl) phases. The development of these materials is comparable to that of NiAl and Ni$_3$Al. The titanium aluminides offer the following attractive properties, Low density, High modulus, good oxidation resistance and good creep resistance. These properties are not obtained without cost, in the case of $\gamma$-TiAl this is low ductility, particularly at room temperature. Four different microstructures have been identified in $\gamma$-TiAl depending on heat treatment, these are [2,3] a) Near-gamma, b) Duplex , c) Nearly lamellar , and d) Fully lamellar. see figure 1.

The development of titanium aluminides began in the 1950's, however it was not until the mid-1970's that major research began into $\gamma$-TiAl alloys. Pratt and Whitney and the USAF conducted this work. The findings of this work showed that a wrought Ti-48Al alloy, with minor additions and a fine duplex structure, offered the most attractive ductility and creep resistance. These values, although promising, were insufficient for the material to be used in aero-engines. Further research yielded a second-generation alloy, produced via rapid solidification wrought processing, with improved ductility, strength and oxidation resistance when compared to the wrought first-generation alloy.

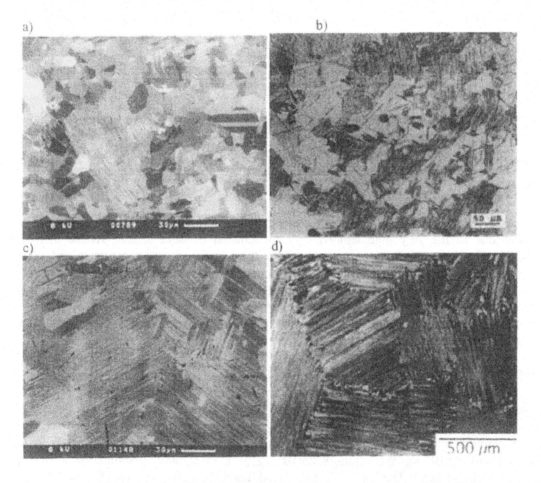

Figure1(a-d) – Examples of the four microstructures found in $\gamma$-TiAl

Unlike many typical engineering metals, e.g. steels, nickel-based alloys and titanium alloys, the ductility and toughness of γ-TiAl are not closely linked. Below the ductile-brittle transition temperature of 650-700°C, γ-TiAl alloys exhibit a ductility-toughness inverse relationship. This is illustrated by comparing a γ-TiAl alloy with a duplex microstructure to a lamellar material. The duplex material will typically have a high ductility but poor toughness while the lamellar material will exhibit the reverse properties. In simple terms ductility is defined as the ability of a material to resist crack initiation while toughness is regarded as a measure of a materials ability to resist crack growth. As a general rule, fine-grained γ-TiAl alloys offer attractive tensile properties but poor fracture toughness, while the reverse is true for large-grained materials because any crack will have to follow a tortuous path in order to grow. Fully lamellar materials have been found to offer the greatest resistance to creep and high cycle fatigue.

The production of γ-TiAl components has been achieved using investment casting. The second-generation Ti-48Al-2Cr-2Nb alloy in as-cast condition has a coarse, non-uniform lamellar microstructure which produces a material of low ductility and strength. A duplex microstructure can be obtained through heat treatment, such a material has suitable properties for use as an engine material.

Further refinement to the as-cast microstructure can be obtained via XD (exothermal dispersion) processing, the so-called XD alloys have a fine dispersion of second phase particles, typically $TiB_2$, which promote the formation of equiaxed grains. The properties of second-generation titanium aluminides are generally comparable to those of nickel-base superalloys; with the exception of fatigue crack growth resistance, ductility and impact resistance.

To demonstrate the potential of titanium aluminides General Electric conducted tests equivalent to 1500 flight cycles, during 1993 and 1994, on a full 98-blade low-pressure turbine disk. The blades were cast Ti-47Al-2Cr-2Nb and their performance suggested the material offered definite promise for aero-engine applications [5].

## 2. Material properties ti Ti-Al bases intermetallics

By virtue of their ordered structure, intermetallics based on TiAl and $Ti_3Al$ have attractive strength and modulus values at temperatures approaching those associated with Ni-base superalloys, unfortunately their use is limited by their brittle nature. The principal advantage of titanium aluminides over comparable materials is their low density, approximately half that of the Ni-based superalloys currently used in gas turbines. The large aluminium content of titanium aluminides gives them a greater oxidation resistance than pure titanium. The maximum operating temperatures of γ-TiAl and $\alpha_2$-$Ti_3Al$ are 870°C and 700°C respectively.

Both $\alpha_2$ and γ phases are ordered resulting in poor room temperature ductility. γ-TiAl has greater ductility and stiffness than $\alpha_2$-$Ti_3Al$ at both room temperature and elevated temperature. The ductile-brittle-transition-temperature of γ is approximately 700°C, below this temperature the dominant fracture mode is cleavage while above this temperature fracture is intergranular. At low temperature $\alpha_2$ is brittle yet retains its strength up to approximately 800°C. At 550°C $\alpha_2$ begins to show some ductility which increases with temperature. The room temperature ductility of γ-TiAl can be much improved by reducing the Al content and using heat treatment. This produces a two-phase 'duplex' microstructure giving much improved properties.

Table 1 - Some typical properties of various Ti alloys and comparable superalloys [6].
RT - room temperature, OT - operating temperature.

|  | Ti-base | TiAl-base | Ti$_3$Al-base | superalloys |
|---|---|---|---|---|
| Density (g cm$^{-3}$) | 4.5 | 3.7-3.9 | 4.1-4.7 | 8.3 |
| Modulus (GPa) | 96-115 | 160-176 | 120-145 | 206 |
| Yield strength (MPa) | 380-1150 | 400-630 | 700-990 | 800-1000 |
| Tensile strength (MPa) | 480-1200 | 450-700 | 800-1140 | 900-1200 |
| Max. temp. - creep ($^{\circ}$C) | 600 | 1000 | 750 | 1090 |
| Max. temp. - oxid. ($^{\circ}$C) | 600 | 900-1000 | 650 | 1090 |
| Ductility - RT (%) | ~20 | 1-3 | 2-7 | 3-5 |
| Ductility - OT (%) | high | 10-90 | 10-20 | 10-20 |
| Crystal structure | hcp/bcc | L1$_0$ | D0$_{19}$ | fcc/L1$_2$ |

As can be seen from table 1 above, both TiAl and Ti$_3$Al base alloys possess superior high temperature properties to commercially available titanium alloys. The modulus of both TiAl and Ti$_3$Al at high temperatures is greater than that of pure titanium at room temperature. As a replacement to Nickel-Base superalloys certain Ti Base aluminides show tendencies for possessing similar qualities and the most striking advantage is the density differential (see table 1). Summary of the effects of different alloying additions are shown in table 2. The relative changes in the material properties are taken as an initial indication of whether the alloy composition can be improved upon. However only a comprehensive testing programme under creep or creep/fatigue condition under different stress states will determine the eventual suitability of the alloy for use in the gas turbine industry.

Table 2 - Summary of the effects of different alloying additions [7].

| Alloying addition | Hardness | Yield Stress | Ductility |
|---|---|---|---|
| $\geq$ 50at.% Al | increase | increase | reduction |
| 1-5at.% Zr | increase | increase | reduction |
| 1-5at.% Hf | strong increase | Strong increase | slight reduction |
| V |  | Strong increase | slight reduction |
| Nb |  | increase | unclear |
| 0.5-2.5wt.% Mn |  |  | increase |
| $\leq$ 3at.% Be |  | slight increase | reduction |

Fracture Mechanisms in Titanium Aluminides

The preferred approach for improving the fracture properties of intermetallics is to reduce the ease of crack nucleation and propagation, thus increasing the critical stress. This can be achieved by removing grain boundary particles or other microstructural nuclei [8]. Improved understanding of brittle fracture in intermetallics can be achieved by [8], Attention to fracture micromechanisms and characterisation of fracture mode and determination of whether initiation or propagation controls fracture.

As the understanding of intermetallic fracture improves, modification of the microstructure to improve fracture behaviour can be attempted. Microstructural modifications may include [8]:

1. Refinement of the microstructure, either by processing or addition of extra phases;
2. Reduction of slip band size, by introduction of particulates or production of homogeneous microstructures;
3. Adding a ductile second phase.

Considerable insight has been gained into the effects of microstructure on the fracture toughness of TiAl from work on Ti-48Al alloys [9-12]. In the as-cast condition the microstructure consists of lamellar grains with small equiaxed grains along the lamellar grain boundaries. A duplex microstructure is produced by heat treatment at 1200°C for 1.7 hours. Heat treatment at 1200°C for 12 hours produces a fully $\gamma$ microstructure, while heat treatment at 1400°C for 0.1 hours produces a large grained fully lamellar microstructure. This highlights the sensitivity of the Ti-Al microstructure to different heat treatments.

The fatigue crack growth and fracture toughness of fully lamellar $\gamma/\alpha_2$ Ti-48Al alloys is heavily dependent on lamellar orientation, temperature and environment [10]. At high temperature the lamellar grains normal to the direction of crack growth, suffer from significant splitting along the lamellar plates [11]. It has also been found that large changes in composition do not appreciably affect fatigue crack growth rate but do affect fatigue threshold [13].

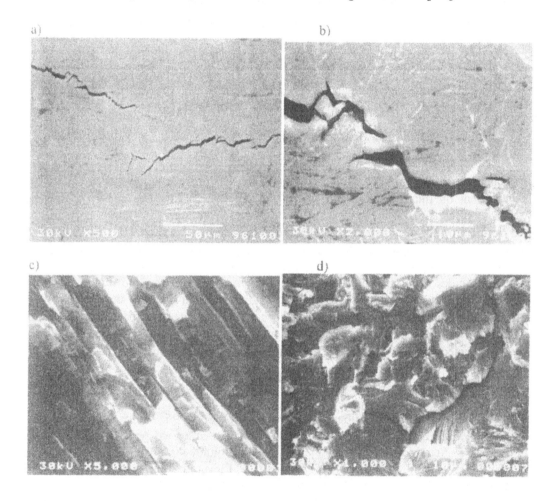

Figure 2: Intergranular crack at 700 °C a) X500, b) X2000, c) Interlamellar cracking (X5000) and d) cleavage fracture (X1000)

## Fracture modes for Ti-Al intermetallics

The fracture surface of as cast TiAl is characterised by transgranular cracks through lamellar grains; the cracks pass through the individual lamellae by translamellar cleavage. The crack direction changes violently as the interfaces between the individual lamellar are crossed. This has also been observed in TiAl crystals where the height of the steps on the fracture surface was measured to be equal to the average lamellar spacing (~10μm) [14]. In comparison the fracture surface of wrought XD TiAl is planar and exhibits signs of transgranular cleavage. Wrought and as-cast XD alloys suffer from secondary cracking along grain boundaries and lamellar interfaces respectively. Secondary cracks have also been observed at the interfaces of the $TiB_2$ particles and the matrix, suggesting an interaction occurs between the particles and the matrix [15].

As the aluminium content for most of these alloys at in the region of 30 weight % the main portion of the microstructure will consist of γ-TiAl phase. Etched specimens reveal a duplex structure with large lamellae colonies of γ-TiAl and $α_2$-Ti3Al with separate smaller grains of γ-TiAl and $α_2$-Ti3Al. The direction of the lamellas varies from colony to colony which is observed in the directional change of the crack propagation over the fracture surfaces. These aspects are shown in the micrographs in figure 2, which shows also the fracture behaviour at 700 °C. The white regions are the $α_2$-Ti3Al and the darker regions are the γ-TiAl. The figure also shows a more detailed high magnification image of the fracture surfaces showing respectively two forms of cracking consisting of fracture through the lamella and $α_2$-Ti3Al grains, with some intergranular separation at the γ-TiAl/ $α_2$-Ti3Al interfaces. Final fracture occurs when microcracks coalesce through a crack jumping process from lamella to lamella. Invariably multiple crack were found in the region of the crack tip [16].

## Conclusions

Titanium aluminides are strong low-density materials; their properties have been shown to be highly dependent on microstructure and fabrication processes. Lamellar microstructures have been found to offer superior creep resistance and attractive fracture toughness, particularly at elevated temperatures. A potential problem with this class of material is its inherent low ductility and fracture toughness, particularly at low temperatures. Whilst it has been found that the alloy has good potential for high temperature applications it has also been shown that the Titanium-Aluminides have unpredictable fatigue and creep deformation and fracture properties by exhibiting in many cases high stress sensitivity and cracking at multiple sites or fast fracture at a possible inclusion site. In addition it has been found that creep and fatigue life are very sensitive to changes in composition, heat treatment and batch to batch variation, and although this is to be expected for most engineering alloys it seems to be more critical in Ti-Al alloys.

## References

1. Nabarro, F.R.N. and H.L. de Villiers, The Physics of Creep. 1995: Taylor & Francis.
2. Hayes, R.W. and B. London, On the Creep Deformation of a Cast Near Gamma TiAl alloy Ti-48Al-1Nb. Acta Metallurgica et Materialia, 1992. Volume 40: p.2167-2175.
3. Worth, B.D., et al. Mechanisms of Ambient & Elevated Temperature Fatigue Crack Growth in Ti-46.5Al-3Nb-2Cr-0.2W. in Eighth World Conference on Titanium. 1995.
4. Kim, Y.W., Gamma Titanium Aluminides: Their Status and Future. Journal of Materials, 1995: p.39-41.
5. Ti-Aluminide Blades tested in GE engine, in Aviation Week & Space Technology. 1993.
6. Cam, G., The Alloying of Titanium Aluminides with Carbon, Imperial College. 1990, London University.
7. Whang, S.H., Temperature and Composition Dependent Deformation in γ-Titanium Aluminides, in Ordered Intermetallics: Physical Metallurgy and Mechanical Behaviour, C.R.W. Liu C.T., Sauthoff G., Editor. 1992, Kluwer Academic Publishers. p.279-298.

8.  Thompson, A.W., <u>Fracture of Intermetallics, in Structural Intermetallics,</u> L.J.J. Darolia R., Liu C.T., Martin P.L., Mirade D.B., Nathal M.V., Editor. 1993, The Minerals, Metals & Materials Society. p.879-884.

9.  Beaven, P.A., et al., Fracture and Ductilisation of $\gamma$-Titanium Aluminides, in Ordered Intermetallics: <u>Physical Metallurgy and Mechanical Behaviour,</u> C.T. Liu, Cahn, R.W., Sauthoff, G., Editor. 1992, Kluwer Academic Publishers. p.413-432.

10. James, A.W. and P Bowen, Fracture and Fatigue of TiAl Based Aluminides, in Titanium '92 Science and Technology. 1992: <u>The Minerals, Metals & Materials Society.,</u>1992

11. Rogers, N.J. and P. Bowen, Effects of Microstructure on Fracture Toughness in a Gamma-based Titanium Aluminide at Ambient and Elevated Temperatures, in Structural Intermetallics, L.J.J. Darolia R., Liu C.T., Martin P.L., Mirade D.B., Nathal M.V., Editor. 1993, <u>The Minerals, Metals & Materials Society.</u> p.231-240.

12. Balsdone, S., J., Jones, W. and Maxwell, D.C., 'Fatigue crack growth in Cast gamma-titanium aluminide between 25-954 $^{\circ}$C', Fatigue and Fracture of Ordered Intermetallic Materials:I, Edited by W. O. Soboyejo et. Al., <u>The Minerals ans Metals, Soc., TMS,</u> Materials Week, Oct.1993, pp.307-327

13. Soboyejo, W.O. and K. Lou, Grain Boundary Segregation and Intergranular Fracture in a Gamma-based Titanium Aluminide Intermetallic. <u>Scripta Metallurgica et Materialia,</u> 1993. Volume 29: p.1335-1339.

14. Yamaguchi, M. and H. Inui, Deformation Behaviour of TiAl compounds with the TiAl/Ti$_3$Al Lamellar Microstructure, in Ordered Intermetallics: <u>Physical Metallurgy and Mechanical Behaviour,</u> C.R.W. Liu C.T., Sauthoff G., Editor. 1992, Kluwer Academic Publishers. p.217-233.

15. Larsen, D.E., et al., Influence of Matrix Phase Morphology on Fracture Toughness in a Discontinuously Reinforced XD Titanium Aluminide Composite. <u>Scripta Metallurgica et Materialia,</u> 1990. Volume 24: p.851-856.

16. Nikbin, K. & Webster, G., 'Influence of state of stress on creep/fatigue failure of $\gamma$-titanium aluminide turbine blade material' <u>'Materials for Advanced Power Engineering 1998',</u> Eds J. Lecomte-Beckers et al, Proc 6th Liege Conf. Materials for Adv. Power Engineering, 1998.

# LIGHTWEIGHT ALLOYS FOR AEROSPACE APPLICATION

*Edited by:*
*Dr. Kumar Jata, Dr. Eui Whee Lee,*
*Dr. William Frazier and Dr. Nack J. Kim*

## TITANIUM ALLOYS

# Dwell-Fatigue Behavior of Ti-6Al-2Sn-4Zr-2Mo-0.1Si Alloy

*V. Sinha, M.J. Mills and J.C. Williams*

Pgs. 193-207

184 Thorn Hill Road
Warrendale, PA 15086-7514
(724) 776-9000

# Dwell-Fatigue Behavior of Ti-6Al-2Sn-4Zr-2Mo-0.1Si Alloy

V. Sinha, M.J. Mills, and J.C. Williams

Department of Materials Science and Engineering, The Ohio State University,
2041 College Road, Columbus, OH 43210

## Abstract

This paper presents the results of a recent study of normal-fatigue and dwell-fatigue response of an $\alpha/\beta$ forged Ti-6Al-2Sn-4Zr-2Mo alloy. The extent of micro-texture has been shown to be a key variable in dwell-fatigue susceptibility of these near-$\alpha$ titanium alloys. The micro-texture analysis of three different $\alpha/\beta$ forged pancakes has been carried out using electron backscatter diffraction (EBSD) techniques. By increasing the forging strain in the $\alpha/\beta$ phase field, the aligned $\alpha$ structure has been randomized and thereby the level of micro-texture is reduced. The failure modes and associated fractographic features under normal-fatigue and dwell-fatigue conditions are also discussed.

Lightweight Alloys for Aerospace Applications
Edited by Kumar Jata, Eui Whee Lee,
William Frazier and Nack J. Kim
TMS (The Minerals, Metals & Materials Society), 2001

# 1. Introduction

Ti-6Al-2Sn-4Zr-2Mo is a near-$\alpha$ titanium alloy, widely used for the hot portions of compressor rotor spools of aero-engines because of their superior creep properties. Aero engine rotating components are subjected to cyclic loads in service. The life of these components has traditionally been estimated based on the continuous cycling fatigue (normal-fatigue) response of this alloy. However, a significant life debit, of an order of magnitude or more, has been reported for various near-$\alpha$ titanium alloys when the specimens are held at maximum stress for 1 to 5 minutes in each cycle [1-7]. This effect has been called dwell-fatigue. This kind of life debit has been observed to usually be associated with a high peak stress [1] and a low temperature [8]. These conditions are also representative of the take-off transients of a typical flight. Thus, the life prediction of Ti-6-2-4-2 components based only on the normal-fatigue considerations can lead to an overestimation and can pose serious reliability problems. Therefore, there is a need to obtain a better understanding of the dwell-fatigue phenomena.

One of the models proposed by Evans and Bache [9] to explain the dwell-fatigue phenomena stresses the importance of having a weak (easy slip) region next to a hard (difficult slip) region in the testpiece in order to create a high local stress. This situation is suggested to promote a sub-surface faceted crack initiation on the basal planes; which is the feature commonly observed in dwell-fatigue tested specimens. The determination of the crystallographic orientations of neighboring grains/colonies in the microstructure is a very important step in understanding the dwell-fatigue response of near-$\alpha$ titanium alloys. A knowledge of the angles between loading axis and the important crystallographic directions (e.g. slip plane poles and slip directions) can give us an idea about the stress redistributions that may be occurring among different regions of the testpiece/component. Under these stress gradients, it is also logical to envision that hydrogen is getting concentrated in the high hydrostatic stress regions. These increased local hydrogen levels may also be detrimental with regards to the dwell-fatigue response of these alloys, as has been proposed by Hack and Leverant [3,4].

In the current study, we have examined the crystallographic orientation of large areas (1.5 mm X 1.5 mm) compared with the size of microstructural features. From this, we have obtained an idea about the size of regions containing similar-orientation $\alpha$ grains. Effects of thermo-mechanical processing (TMP) routes on the levels of micro-texture of $\alpha/\beta$ forged Ti-6-2-4-2 alloy have also been evaluated. The normal-fatigue and dwell-fatigue response of the material removed from the field-returned (retired) impellers are investigated. The crack initiation sites and crack growth modes under the two loading conditions are also studied using scanning electron microscopy (SEM) techniques.

# 2. Material and Microstructures

The Ti-6Al-2Sn-4Zr-2Mo-0.1Si alloy that was used in this study was supplied by Ladish Co., Cudahy, WI. This alloy was supplied in the form of three pancake forgings. The three forgings have undergone three different thermo-mechanical processing (TMP) routes, as described in detail in Table 1. The billet for the first pancake was $\alpha/\beta$ forged to a very high strain (~ 89%) in order to cause recrystallization and randomization of the aligned $\alpha$ (colony) structure of the starting material, thereby producing a randomly oriented primary $\alpha$ grains in the microstructure. This was subsequently solution heat treated at $T_\beta$ - 20 $^0$C to obtain a low volume fraction (~ 20-25%) of primary $\alpha$ (Fig. 1a) and oil quenched in order to suppress the re-growth of primary $\alpha$ grains during cooling. The resulting microstructure consisted of globular $\alpha$ grains and lamellar transformed $\beta$ regions. All these steps are representative of current industrial practice and are thought to reduce the dwell-fatigue susceptibility of this class of alloys. The billet for the second pancake was also $\alpha/\beta$ forged, however along the TMP route representative of earlier

Table 1: Thermo-mechanical Processing Routes

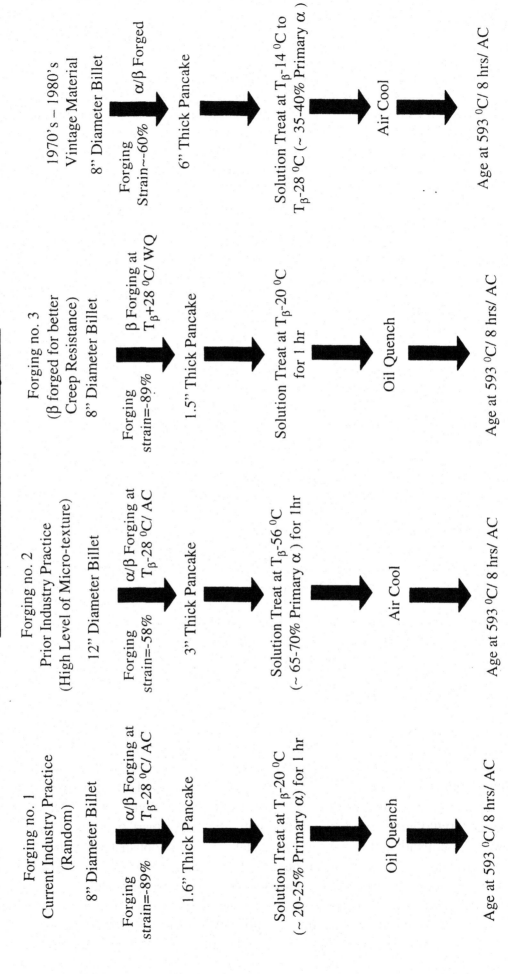

**Forging no. 1**
Current Industry Practice
(Random)

8" Diameter Billet

Forging strain=~89% → α/β Forging at $T_\beta$-28 °C/ AC → 1.6" Thick Pancake → Solution Treat at $T_\beta$-20 °C (~ 20-25% Primary α) for 1 hr → Oil Quench → Age at 593 °C/ 8 hrs/ AC

**Forging no. 2**
Prior Industry Practice
(High Level of Micro-texture)

12" Diameter Billet

Forging strain=~58% → α/β Forging at $T_\beta$-28 °C/ AC → 3" Thick Pancake → Solution Treat at $T_\beta$-56 °C (~ 65-70% Primary α ) for 1hr → Air Cool → Age at 593 °C/ 8 hrs/ AC

**Forging no. 3**
(β forged for better Creep Resistance)

8" Diameter Billet

Forging strain=~89% → β Forging at $T_\beta$+28 °C/ WQ → 1.5" Thick Pancake → Solution Treat at $T_\beta$-20 °C for 1 hr → Oil Quench → Age at 593 °C/ 8 hrs/ AC

1970's – 1980's Vintage Material

8" Diameter Billet

Forging Strain~-60% → α/β Forged → 6" Thick Pancake → Solution Treat at $T_\beta$-14 °C to $T_\beta$-28 °C (~ 35-40% Primary α ) → Air Cool → Age at 593 °C/ 8 hrs/ AC

Figure 1: Optical micrographs of the forgings being tested in this research. (a) Forging no. 1, α/β forged; (b) Forging no. 2, α/β forged; (c) Forging no. 3, β forged; and (d) 1970's - 1980's Vintage Material, α/β forged.

197

industrial practice. This was forged to a relatively lower strain level (~58%) and subsequently, solution heat treated at $T_\beta$ - 56 $^0$C to obtain a higher volume fraction (~ 65-70%) of primary $\alpha$ (Fig. 1b). It was then air cooled to obtain a bimodal microstructure (i.e. a microstructure consisting of primary $\alpha$ grains and transformed $\beta$ regions). The TMP parameters for forging 3 (Table 1) are consistent with current $\beta$-processing practice used by aero engine companies. The resulting microstructure (Fig. 1c) is used mostly because of the better creep resistance at high temperatures.

In addition to the three pancake forgings supplied by Ladish Co., specimen blanks from a few field-returned (retired) impellers were also electro-discharge machined (EDM) for an evaluation of their dwell-fatigue susceptibility. These vintage materials were processed in 1970's – 1980's. The approximate TMP parameters for the vintage materials examined in this study are described in Table1. This material, also Ti-6Al-2Sn-4Zr-2Mo alloy, has a bimodal microstructure as shown in Fig. 1d.

## 3. Experimental Procedures

### 3.1 Electron Back Scatter Diffraction (EBSD) experiments

Since the level of micro-texture is a key variable in the dwell-fatigue susceptibility of $\alpha/\beta$ forged near-$\alpha$ titanium alloys; the micro-texture analysis of the three $\alpha/\beta$ forged pancakes (forging 1, forging 2 and vintage material) was carried out using the EBSD techniques. Small rectangular coupons (~ 8 mm X 10 mm) were cut from forgings 1 and 2, with their sides along the circumferential and axial directions of the pancakes. Here, the axial direction corresponds to the thickness direction of the pancakes, which will get aligned with the axis of the aero-engines if these forgings were to be used as the compressor discs. The specimen blanks from the retired impellers were supplied in the form of cylinder with circular cross-sections. The axis of these cylindrical specimen blanks is along the circumferential direction of the pancake forgings from which they were machined out. A rectangular coupon from one of the cylindrical blanks was cut such that one of its sides is along the circumferential direction (i.e. the axis of the cylinder).

The rectangular coupons from the three $\alpha/\beta$ forged pancakes were mounted using conductive mounting powder. The mounted specimens were first ground with the emery paper and then polished to a 0.06 μm finish in a vibratory polisher using colloidal silica. The specimens were subsequently cleaned ultrasonically in distilled water, dried and baked in an oven (temperature ~ 70 $^0$C) before being loaded in the scanning electron microscope (SEM) specimen chamber.

The optimum conditions to obtain a good quality electron backscattered diffraction patterns (EBSP) for this material were found to be 70$^0$ tilt and 20 KV accelerating voltage. The background signal was collected by averaging 128 frames, and the subtraction of this background from raw pattern was found to give a high quality EBSP. The computer software (Supplier: HKL Technology, Burnt Hills, NY, USA) looks for a matching of the measured and calculated interplanar angles for a phase under investigation in order to index the EBSP from any grain. These interplanar angles are a function of c/a ratio in hcp structures. Consequently, the c and a lattice parameters for the $\alpha$-phase (hcp) in Ti-6-2-4-2 alloy were determined from a standard X-ray diffraction ($\theta$–$2\theta$) run. The lattice parameters were determined to be, $a_\alpha = 0.293$ nm and $c_\alpha = 0.468$ nm. The lattice parameter for $\beta$ phase (bcc) was determined to be, $a_\beta = 0.324$ nm. These lattice parameters were input into the crystal files of the computer package, which the software uses to calculate the various interplanar angles and compares those with the observed interplanar angles in the actual (experimental) EBSPs searching for a best fit.

In an order to obtain an idea about the area occupied by similarly oriented $\alpha$ grains, a relatively large area (1.5 mm X 1.5 mm) compared with the apparent size of microstructural features (Fig. 1) was analyzed for each specimen. For analyzing such a large area, the stage-scan mode was found to be more appropriate than the beam-scan mode. The step-size was selected to be 10 μm for all the three specimens, which is on the order of the primary α grain size (Fig. 1).

### 3.2 Normal-fatigue and Dwell-fatigue experiments

Both the normal-fatigue and dwell-fatigue experiments were carried out in lab air, at room temperature using a closed-loop servo-hydraulic test-frame. A schematic of the normal fatigue and dwell-fatigue cycles are depicted in Fig. 2. Both the tests are conducted at a stress ratio of R = 0, and the rate of loading and unloading is same in the two cases. The only difference is that we hold the load for 1 minute at the maximum value in each dwell-fatigue cycle. The loading and test conditions (R ratio, dwell-time and temperature) are chosen to correspond to the take-off transients of a typical flight with which the dwell-fatigue life debit is thought to be associated. Both types of tests are conducted under force-controlled conditions at a maximum stress of 828 MPa (~ 90% of yield strength).

The specimens for these tests have a circular cross-section with a gage section diameter of 5.08 mm (0.2") and the length of reduced section = 19.05 mm (0.75"). The other dimensions of the specimens are selected in accordance with ASTM E466 specifications. The threaded grip-section of the specimen was shot-peened in order to avoid any failure that may be caused due to fretting-fatigue in this region.

### 3.3 Fractography of failed specimens

The failed specimens were carefully taken out of the fixtures avoiding any damage to the fracture surfaces and subsequently cleaned ultrasonically in distilled water to remove any surface contaminants. Thereafter, the specimens were dried and baked in an oven (temperature ~ 70 $^0$C) before being loaded in the specimen chamber of SEM. The secondary electron images seem to provide a good idea about the topography of the fracture surfaces at an accelerating voltage of 10 KV.

## 4. Results and Discussion

### 4.1 Micro-texture analysis

The orientation maps for the three α/β forged pancakes are presented in Fig. 3a - 3c. Here, the points having the same color represent microstructural regions having similar crystallographic orientations. Forging 1 (which was forged to a strain of ~ 89%) has minimal micro-texture; whereas forging 2 and the vintage material (which were forged to a strain of ~ 58-60%) consist of several large regions (size ~ 500 μm) that have grains with similar crystallographic orientations. The $(0001)_\alpha$ poles of three similar-orientation regions (denoted by 1,2 and 3 in Fig. 3d) are depicted in a stereographic projection in Fig. 3e. The angle between the $(0001)_\alpha$ poles of any two same-color regions can be easily measured with the help of the software. For example, in Fig. 3e, the angle between poles 1 and 2 = $70^0$, the angle between poles 1 and 3 = $90^0$ and that between poles 2 and 3 = $66^0$. One can similarly obtain another pole figure (e.g. $(\bar{1}100)_\alpha$ pole-figure) for the three same-color regions and thus, the crystallographic orientation of these regions can be determined. This kind of analysis can be very useful in determining the crystallographic orientation of the dwell-fatigue initiation sites and their surrounding regions.

### 4.2 Normal-fatigue and Dwell-fatigue tests

The specimens made of the vintage material exhibit minimal dwell life debit, under the test conditions employed in this study. The normal-fatigue specimen failed in 35,077 cycles;

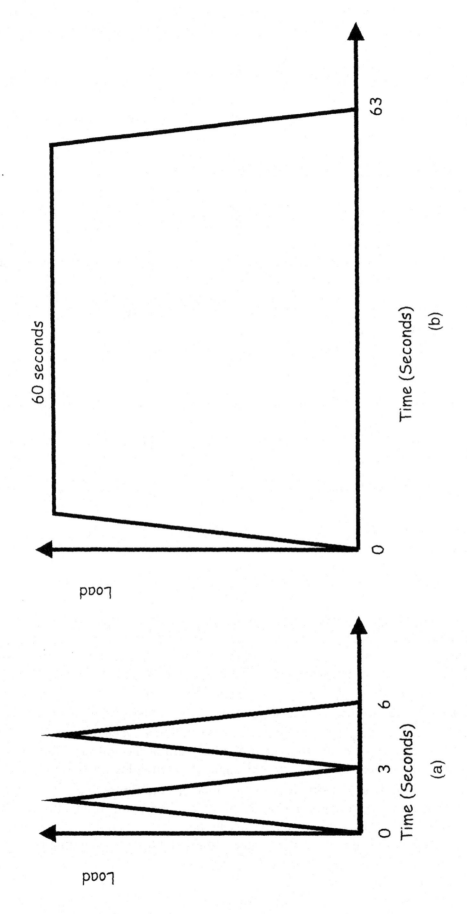

Figure 2: Schematic of loading Cycles. (a) Normal Fatigue (20 cycles per minute), and (b) Dwell-Fatigue cycle with 1 minute dwell at the maximum load

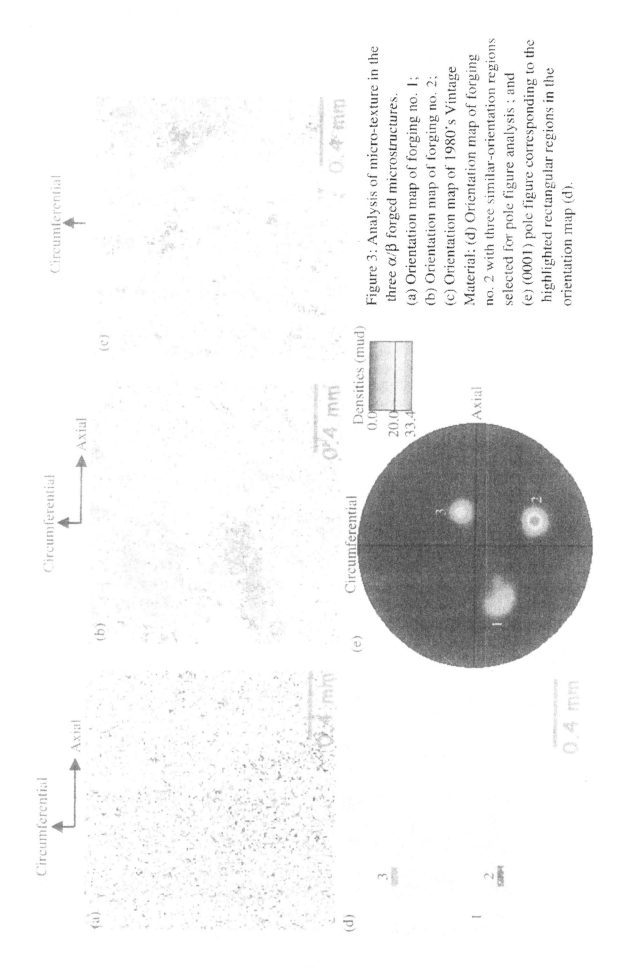

Figure 3: Analysis of micro-texture in the three α/β forged microstructures.
(a) Orientation map of forging no. 1;
(b) Orientation map of forging no. 2;
(c) Orientation map of 1980's Vintage Material; (d) Orientation map of forging no. 2 with three similar-orientation regions selected for pole figure analysis ; and (e) (0001) pole figure corresponding to the highlighted rectangular regions in the orientation map (d).

whereas the dwell-fatigue specimen failed in 17,877 cycles. In the literature, dwell life debit by about an order of magnitude (i.e. 10 times shorter life) has been reported [1,3,6]. In the present study, the relatively low dwell life debit (of a factor of 2) may be due to a lower applied stress-level and/or the basal poles of the neighboring same-color regions not being optimally oriented with respect to the loading axis to promote internal cracking along the basal planes. Furthermore, the total plastic strain to failure was similar within the margin of error of experimental measurements: it was 1.5% for the normal-fatigue test and 1.44% for the dwell-fatigue test. The levels of total accumulated strain under the two test conditions are currently being investigated further.

*4.3 Crack initiation and propagation mechanisms*

The fracture surface of the normal fatigue tested specimen was examined in detail under a scanning electron microscope. It revealed only one surface initiation site (Fig. 4a). River lines starting from this initiation site and moving towards the center of the testpiece are clearly visible (Fig. 4b). A cleavage-like fracture mode is observed in the vicinity of the initiation site (Fig. 4c); which is a characteristic of the near-threshold crack growth regime in titanium alloys. Further towards the center of specimen cross-section, fatigue striations are observed (Fig. 4d); which are the characteristics of the Paris regime of fatigue crack growth in titanium alloys having yield strength of 1050 MPa or lower. In the rest of the fracture surface, ductile-dimpled fracture mode is seen which represents the unstable fracture in these regions.

In contrast, for the dwell-fatigue tested specimen, three surface initiation sites are observed (Fig. 5). The river lines are seen to be emanating from all the three initiation sites. The area between each initiation site and the center of the testpiece was examined in detail at higher magnifications under SEM. Quite surprisingly, the cleavage-like fracture modes and the fatigue striations were seen to be emanating from only one of the three initiation sites (Fig. 6a-6c). The rest of the fracture surface shows ductile-dimpled fracture mode (Fig. 6d); which is representative of unstable failure due to overload.

In order to investigate if there are any other initiation sites away from the primary fracture surfaces; the gage-sections of normal-fatigue and dwell-fatigue specimens were examined using X-ray micro-radiography techniques. This examination clearly shows that there are no additional cracks in the normal-fatigue tested specimen; whereas there is one additional crack (about 300 μm long) in the gage-section of the dwell-fatigue tested specimen (Fig. 7). This additional crack in the dwell-fatigue testpiece is confirmed to be a surface crack by the fluorescent-type dye penetrant technique. Therefore, it appears that this crack has also initiated at the surface of the testpiece in the gage section. Preliminary results suggest that the multiplicity of cracks is associated with the ease of crack initiation under dwell-fatigue test conditions.

**5. Summary and Conclusions**

1. Dwell-fatigue life deficit can be a major reliability issue in the aero-engine compressor components made of Ti-6242 alloy.
2. Micro-texture appears to be a key variable in dwell-fatigue susceptibility of Ti-6242 forgings. The extent of micro-texture in different α/β forged pancakes has been analyzed with the help of Electron Back Scatter Diffraction (EBSD) techniques.
3. The level of micro-texture has been successfully changed by changing thermo-mechanical processing routes. By increasing the forging strain in α/β phase field the aligned α microstructure of the billet is randomized, causing a significant reduction in its level of micro-texture.
4. Minimal dwell life debit (of a factor of 2) has been observed for the vintage material tested to date. This may be attributed to a low level of applied stress or a non-optimal

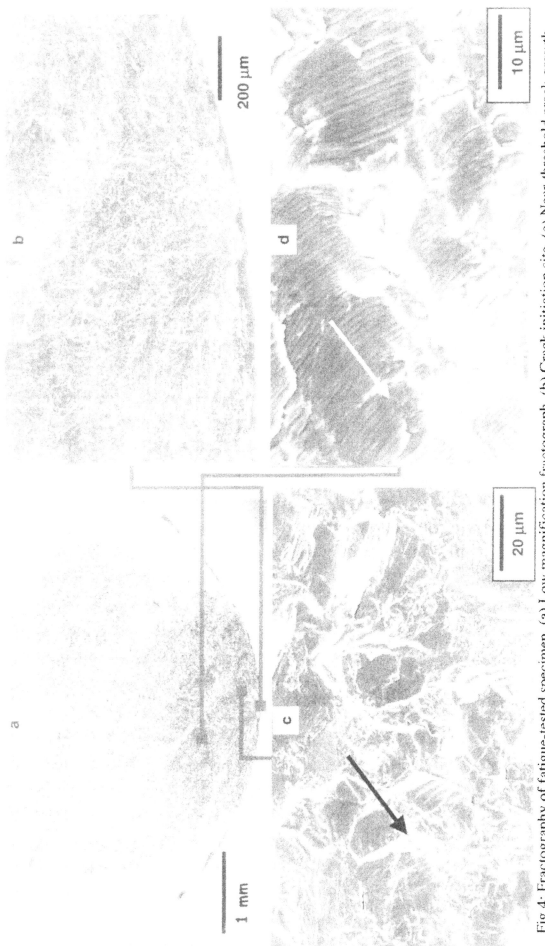

Fig 4: Fractography of fatigue-tested specimen. (a) Low magnification fractograph. (b) Crack initiation site. (c) Near-threshold crack growth regime, and (d) Paris regime of crack growth (fatigue striations). Crack growth direction is indicated by arrow in (c) and (d).

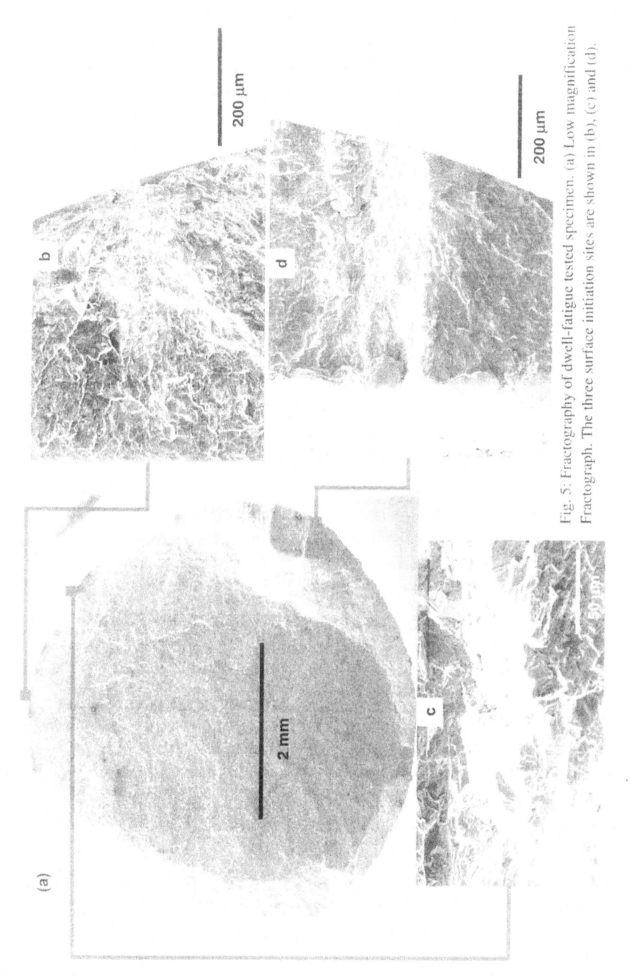

Fig. 5: Fractography of dwell-fatigue tested specimen. (a) Low magnification Fractograph. The three surface initiation sites are shown in (b), (c) and (d).

204

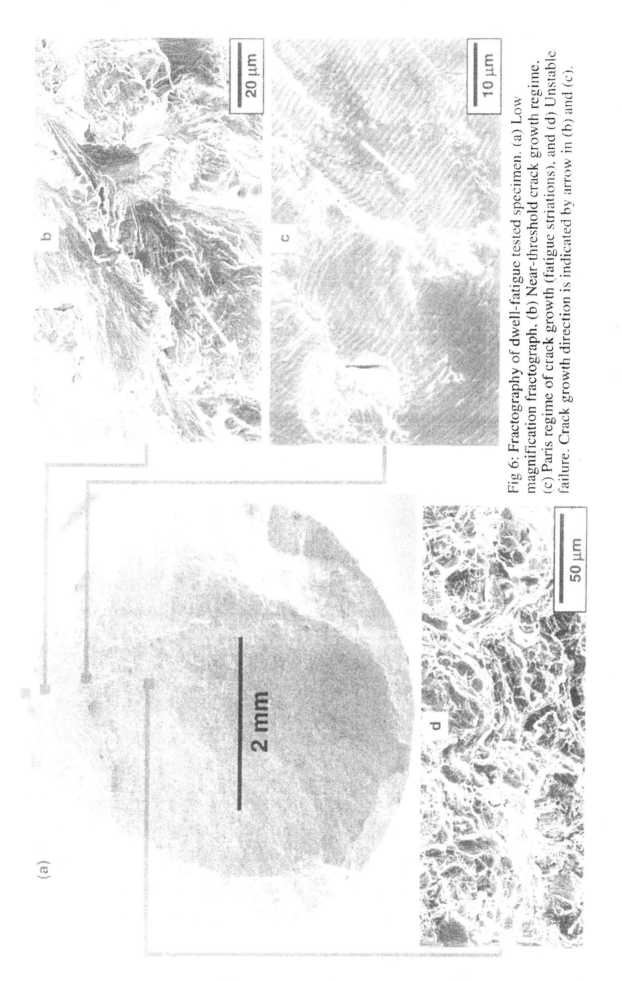

Fig 6: Fractography of dwell-fatigue tested specimen. (a) Low magnification fractograph. (b) Near-threshold crack growth regime. (c) Paris regime of crack growth (fatigue striations), and (d) Unstable failure. Crack growth direction is indicated by arrow in (b) and (c).

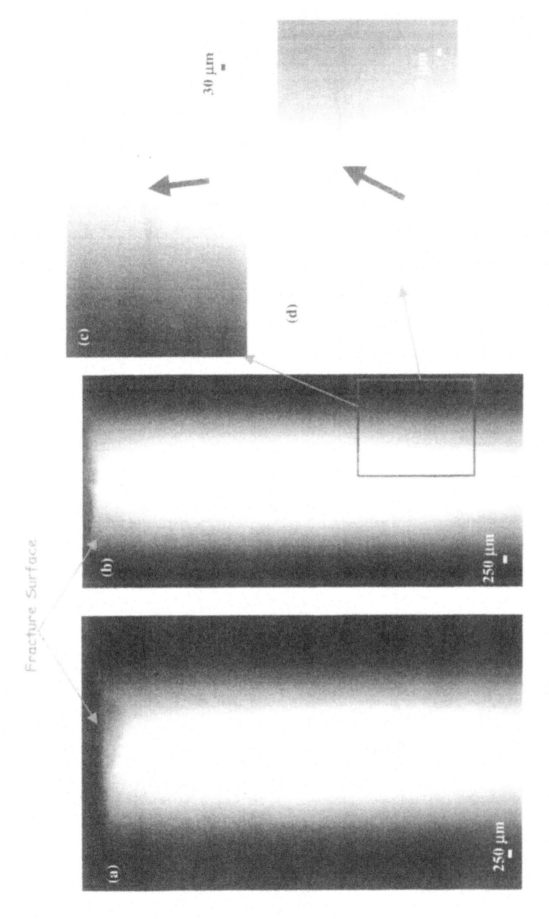

Fig. 7: X-ray micro-radiography of failed specimens. (a) Specimen tested under normal-fatigue condition, (b) Specimen tested under dwell-fatigue condition, (c) Higher magnification image of the area denoted by rectangle in (b), and (d) same as fig. (c) except the specimen is rotated by 90 degrees.

orientation of the basal poles of the neighboring same-orientation regions in the testpiece.

5.  Significant dwell life debit (about 10 times shorter dwell life) is usually associated with internal crack initiation sites; which have not been observed in the vintage material tested so far in this research program.

6.  Dwell-fatigue tested specimen had multiple crack-initiation sites, whereas normal-fatigue specimen had one crack-initiation site.

7.  Crack initiation and propagation mechanisms under normal-fatigue and dwell-fatigue appear to be similar. At low $\Delta K$ values, a faceted fracture mode is observed; whereas at higher $\Delta K$ values, fatigue striations are observed under both kinds of loading cycle.

8.  When tested under load control, similar levels of total plastic strain are observed under normal-fatigue and dwell-fatigue conditions. The levels of total accumulated strain under the two test conditions (normal-fatigue and dwell-fatigue) are currently being examined further for other peak stress levels and microstructural conditions.


## Acknowledgments

This research was supported by the Federal Aviation Administration. The authors would like to thank the Technical Monitor, Joseph Wilson, for his encouragement and support of this work. The authors would also like to thank Prof. S.I. Rokhlin, B. Zoofan and J.-Y. Kim for help with the non-destructive evaluation (X-ray micro-radiography and dye penetrant method) techniques. The supply of three pancake forgings by Ladish Co. (Cudahy, WI) is also gratefully acknowledged.

## References

1.  M.R. Bache, M. Cope, H.M. Davies, W.J. Evans and G. Harrison, International Journal of Fatigue, Vol. 19, Suppl. 1, pp. S83-S88, 1997.

2.  D. Eylon and J.A. Hall, Metallurgical Transactions, Vol. 8A, pp. 981-990, 1977.

3.  J.E. Hack and G.R. Leverant, Metallurgical Transactions, Vol. 13A, pp. 1729-1738, 1982.

4.  J.E. Hack and G.R. Leverant, Scripta Metallurgica, Vol. 14, pp. 437-441, 1980.

5.  A.P. Woodfield, M.D. Gorman, R.R. Corderman, J.A. Sutliff and B. Yamrom, Titanium '95: Science and Technology, pp. 1116-1124, 1995.

6.  W.J. Evans and C.R. Gostelow, Metallurgical Transactions, Vol. 10A, pp. 1837-1846, 1979.

7.  M.E. Kassner, Y. Kosaka and J.A. Hall, Metallurgical and Materials Transactions, Vol. 30A, pp. 2383-2389, 1999.

8.  D. F. Neal, Sixth World Conference on Titanium, Cannes, France, p. 175, 1988.

9.  W.J. Evans and M.R. Bache, International Journal of Fatigue, Vol. 16, pp. 443-452, 1994.

# LIGHTWEIGHT ALLOYS FOR AEROSPACE APPLICATION

*Edited by:*
*Dr. Kumar Jata, Dr. Eui Whee Lee,*
*Dr. William Frazier and Dr. Nack J. Kim*

## TITANIUM ALLOYS

# Characterization of Microstructural Evolution in a Ti-6Al-4V Friction Stir Weld

*Mary C. Juhas, G.B. Viswanathan and Hamish L. Fraser*

Pgs. 209-217

184 Thorn Hill Road
Warrendale, PA 15086-7514
(724) 776-9000

# CHARACTERIZATION OF MICROSTRUCTURAL EVOLUTION IN A Ti-6Al-4V FRICTION STIR WELD

Mary C. Juhas, G.B. Viswanathan and Hamish L. Fraser

The Ohio State University
Department of Materials Science & Engineering
2041 College Road, Columbus, OH 43210

## Abstract

The aim of this study was to characterize the microstructures present in a titanium alloy friction stir weld in an attempt to later relate them to mechanical properties and ultimately optimze weld performance. The allotropic phase transformation that occurs in titanium alloys, coupled with deformation and continuous cooling, produces complex stir zone microstructures compared with those observed in aluminum alloy friction stir welds. In this study, a mill annealed Ti-6Al-4V base material microstructure was completely transformed into a Widmanstätten microstructure during processing. Equiaxed $\alpha$ grains on the order of 1 $\mu$m in diameter were also present in the stir zone at triple points and $\alpha$ lath boundaries. The $\alpha$ laths contained basal faults similar to those observed previously in quenching experiments in the absence of deformation. There was no evidence of gross plastic deformation in the stir zone microstructure however curved misfit dislocations in the $\alpha/\beta$ interphase boundaries in the stir zone indicate grain boundary sliding. The thermomechanically affected zone was characterized by "fingers" of base material that appeared to be drawn into the stir zone. These "fingers" scaled closely with microtexture patterns observed the base material. Microhardness results showed no heat affected zone softening that is typically present in titanium alloy fusion welds.

Lightweight Alloys for Aerospace Applications
Edited by Kumar Jata, Eui Whee Lee,
William Frazier and Nack J. Kim
TMS (The Minerals, Metals & Materials Society), 2001

## Introduction

Titanium alloys have long been popular in the aerospace industry due to their superior strength to weight ratio, good fatigue properties, excellent corrosion resistance and manufacturability including forging, forming and joining. Recently, friction stir welding (FSW) has been extended from low temperature materials, such as those based on aluminum and copper, to titanium alloys. This solid state process for joining plate materials is achieved through the use of a non-consumable tool that provides both local frictional heat and physical stirring (plastic deformation) of the material to ultimately form a metallurgical bond. The development of FSW of titanium alloys is reliant upon extrinsic factors such as the identification of an appropriate tool material and optimization of process parameters. Additionally, intrinsic factors such as characterization of the weld microstructures and relating them to mechanical properties, must be addressed. Manufacturing processes such as forging and friction stir welding of $\alpha/\beta$ titanium alloys require control of process temperatures and cooling rates in order to optimize the microstructure for the intended application.

In this work, a microstructural characterization study has been performed on a FSW in Ti-6Al-4V, a popular aerospace alloy. The microstructural data obtained will be used as input to physically based models that will eventually predict microstructure and weld properties.

## Experimental Procedure

A Ti-6Al-4V FSW was produced for this study by TWI and provided in the as-welded condition. The base material was 6.4mm thick plate in the mill annealed condition. Several samples were sectioned transverse to the welding direction and prepared for metallographic analysis using standard polishing practices and etched with Kroll's reagent. Microstructural examination was performed using scanning (SEM) and transmission electron microscopy (TEM). The SEM analysis was performed on a Philips XL30 field emission gun (FEG) instrument. Orientation Imaging Microscopy (OIM) was performed to assess microtexture using a Philips XL30 FEG ESEM (Environmental SEM). Vickers microhardness data (500 g load) was obtained for the base material and various regions of the weld including thermomechanically affected zone (TMAZ), heat affected zone (HAZ) and stir zone (SZ). Thin foils were extracted from both the stir zone and the base material for examination on a Philips CM 200 transmission electron microscope. The foils were thinned and perforated by ion milling rather than jet electropolishing which has been associated with hydride formation along grain boundaries.

## Results

The base material microstructure, shown in Figure 1(a), has a significant amount of microtexture, as evidenced by the accompanying pole figure, Figure 1(b). The equiaxed $\alpha$ grains are bounded by $\beta$ ribs as shown more clearly in the TEM micrograph in Figure 2(a). Figure 2(b) is a TEM micrograph of an $\alpha/\beta$ interphase boundary which is defined by the Burgers orientation relationship $(101)_\beta//(0001)_\alpha$, $[1\bar{1}\bar{1}]_\beta//[2\bar{1}\bar{1}0]_\alpha$. It can be seen that the misfit dislocations, or intrinsic grain boundary dislocations, $\underline{b}//[0001]_\alpha$, are straight and evenly spaced. A comparison of this base material $\alpha/\beta$ interfacial structure will be made later with similar boundaries examined in the stir zone.

The stir zone microstructure, Figure 3(a), clearly shows that the process temperatures exceeded the $\beta$ transus temperature of ~995°C. The original $\alpha/\beta$ base material microstructure completely transformed to $\beta$ phase during processing. The Widmanstätten morphology and the relatively fine size of the transformed $\beta$ laths indicate a fairly rapid cooling rate from above the $\beta$ transus. The

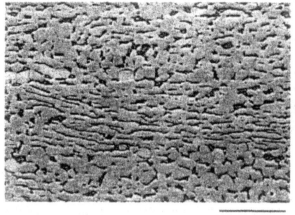

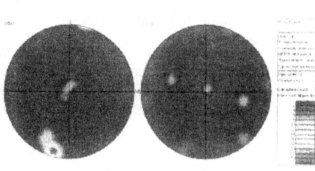

Figure 1(a). SEM micrograph of mill annealed Ti-6Al-4V base material

Figure 1(b). OIM pole figure of mill annealed base material showing microtexture.

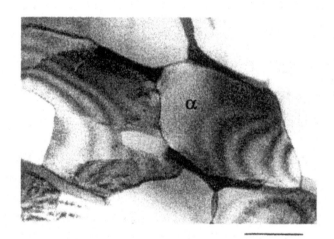

1μm

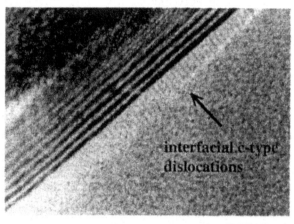

200nm

Figure 2(a). TEM micrograph showing equiaxed α grains bounded by β ribs.

Figure 2(b). TEM micrograph of α/β interface showing straight, evenly spaced c-type misfit dislocations.

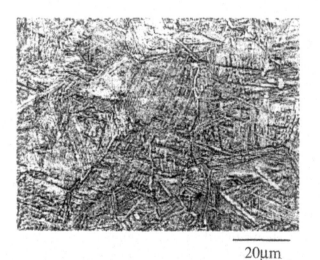

20μm

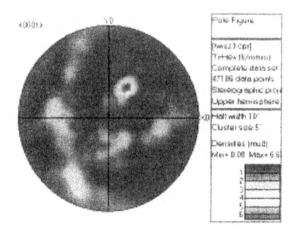

Figure 3(a). SEM micrograph of stir zone showing Widmanstätten microstructure.

Figure 3(b). OIM pole figure of stir zone showing moderate microtexture.

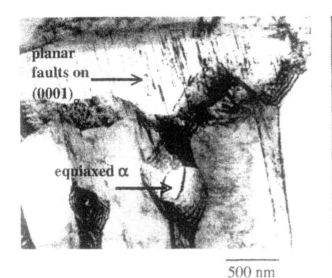

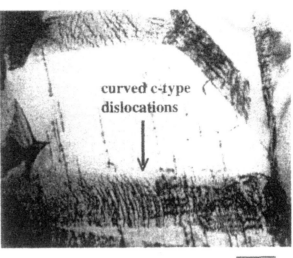

Figure 4(a). TEM micrograph of stir zone showing equiaxed α at triple point.

Figure 4(b). TEM micrograph of stir zone showing curved c-type dislocations at α/β interface.

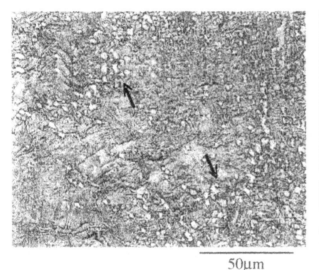

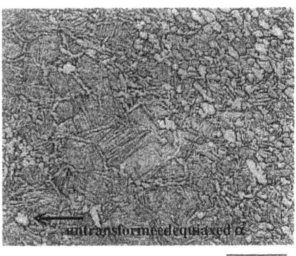

Figure 5(a). SEM micrograph of TMAZ showing fingers of the base material (arrows) that were drawn into the stir zone.

Figure 5(b). Higher magnification SEM micrograph of TMAZ showing untransformed equiaxed α on stir zone side.

pole figure shown in Figure 3(b) reveals a moderate amount of microtexture in the stir zone compared with that observed in the base material. Figure 3(a) shows that, in addition to lath morphology, the α phase is present along prior β grain boundaries. Examination of the stir zone microstructure using TEM, Figure 4(a), revealed a third morphology of α, namely equiaxed α, which is present at lath boundaries and triple points. Figure 4(a) also illustrates the presence of planar faults, shown edge-on, in the lath α. Further examination of these faults revealed that they lie on basal planes. Figure 4(b) is a TEM micrograph of an α/β interphase boundary in the stir zone. It can be seen that the misfit dislocations in the boundary are curved compared to those in the base material, shown in Figure 2(b).

The TMAZ was examined by optical metallography on both the leading and trailing sides of the

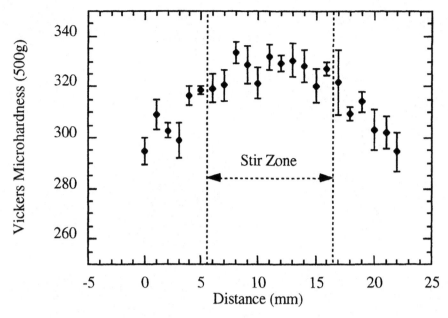

Figure 6. Microhardness profile showing increased hardness in stir zone compared with base material. Note the absense of HAZ softening.

weld. When examined at relatively low magnifications, e.g., 200X, it was noted that periodic "fingers" of the base material appeared to have been drawn into the stir zone. These fingers, which were more prevalent on the leading side of the weld than the trailing side, are considered to be too closely spaced to correspond to threads in the pin tool as shown in the SEM micrograph in Figure 5(a). Another region of the TMAZ microstructure on the leading side of the weld is shown at higher magnification in the SEM micrograph in Figure 5(b). This represents a region between the fingers described above where it can be seen that the base material microstructure (upper right) is distinct from that of the stir zone (lower left). The two-phase base material consisting of equiaxed α bounded by β ribs changes rather abruptly, giving rise to the Widmanstätten morphology that was observed in the stir zone. It should be noted however that some equiaxed α grains persist within the stir zone and similarly, small colonies of α laths appear with the equiaxed grains on the base material side. This distribution of morphologies was also noted in the periodic fingers described above.

The microhardness traverse, Figure 6, shows a distinct increase in hardness across the stir zone without the attendant hardness troughs in the HAZ. Local softening in the HAZ has been associated with aluminum alloy friction stir welds (1, 2) due to coarsening of the second phase particles. In addition, titanium alloy fusion welds are known to have softening in the HAZ due to localized growth of the prior β grains (3).

## Discussion

As illustrated in the previous section, there is a striking contrast between the base material and stir zone microstructures. It is clear that the process temperature under the tool exceeded β transus temperature however the peak temperature and cooling rate are unknown. There is no evidence of plastic deformation in the α+β phase field such as curved or broken Widmanstätten α laths in stir zone. Further, the dislocation density in the α laths is not excessively high, as would be expected in a worked microstructure. These observations support the hypothesis that metal transfer around the pin tool occurred entirely above the β transus and cooling occurred slowly enough to allow dislocations to annihilate or form sub-grains as observed in Figure 7, but rapidly enough to produce a refined Widmanstätten microstructure. As the temperature cooled from the β phase field into the α+β phase field, Widmanstätten α nucleated and grew. The mechanism associated with the evolution of the fine equiaxed α grains in the stir zone is unclear. The location of these α grains at triple points indicates that they may have formed via a nucleation and growth mechanism however this is not supported by the literature and cannot be substantiated in the present study. It is interesting to note the moderate microtexture in the stir zone even though this region had transformed to β phase during processing. The microtexture may have occurred during stirring in β phase and persisted during subsequent transformation. Microtexture has been reported in aluminum alloy friction stir welds (4).

The planar faults observed in the α laths on (0001) appear to be similar to those reported by Williams and Blackburn (5) and later by Woodfield et al (6) in quenched, undeformed material. The latter group reported these faults to be bounded by partial dislocations with Burgers vectors of the type $1/6<20\bar{2}3>$. It is unclear whether these faults formed as a result of cooling, local deformation or both. It is noted that the cooling rates generated during friction stir welding are obviously slower than those produced in quenching. Careful experiments that separate the effects of stresses due to cooling from those induced by deformation are required to understand the origin of these basal faults.

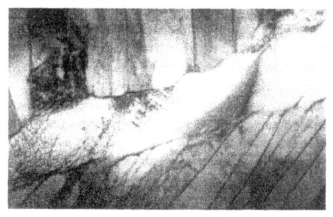

1μm

Figure 7. TEM micrograph of sub-grains in grain boundary α in stir zone.

It is assumed that the majority of the stirring process occurred at temperatures in excess of the β transus, however, there is evidence that some grain movement took place below that temperature. The curved c-type dislocations in the α/β interphase boundary indicate grain boundary sliding (7), in this case between the α laths and the β ribs. The ability of grains to slide or rotate relative to each other at high process temperatures may explain the relatively low dislocation density observed in the α laths. Grain boundary sliding may have occurred to relieve local stresses in favor of dislocation nucleation in the α laths and β ribs.

The fingers of base material that appear to have been drawn into the stir zone appear regularly from the top to the bottom of the plate. Although this pattern resembles the TMAZ of Al alloy FSWs when observed at low magnification, closer examination reveals that, rather than plastically deformed grains, these fingers represent alternating microstructural regions of base material and stir zone. It should be noted that these microstructures are referred to as "base material" and "stir zone" for clarity purposes. While there are most likely to be microscopic changes in the microstructures in the TMAZ, a careful TEM analysis has not yet been performed. The size of these fingers of base material scale closely with the size of the similarly oriented grains shown in the microtexture patterns in Figure 1(a). The regions of base material that were favorably oriented with respect to the local shear stresses would have been more easily swept by the tool into the stir zone. The material adjacent to the tool, which was cooler than the material under the tool, was mechanically stirred into the TMAZ but not heated above the β transus temperature. An SEM micrograph of the TMAZ is shown in Figure (5b). The persistence of equiaxed α grains among the Widmanstätten laths in the lower left of the micrograph indicates that the local temperature was not above the β transus for a long enough period of time to dissolve all of the equiaxed α.

The microhardness results indicate an increase in hardness in the stir zone compared with the base material. This hardness increase is explained by the finer scale of the Widmanstätten laths compared with the equiaxed α/β microstructure. There may also be a contribution from the basal faults observed in the lath α but further studies are required to support this statement. The lack of a local softening in the HAZ, as observed in Al alloy FSW, may be attributable to the relatively low thermal conductivity in Ti alloys compared with Al alloys. The HAZ softening observed in Ti alloy fusion welds is typically associated with coarsening of prior β grains which has not been observed in this study.

## Conclusions

The following conclusions have been drawn from this study.

1. The temperature of the stir zone exceeded the β transus producing a Widmanstätten microstructure upon cooling with increased hardness compared with the base material.
2. There was no apparent HAZ softening as often observed in Al alloy FSW and Ti fusion welds.
3. Equiaxed α grains, approximately 1μm in diameter, were observed in the stir zone.
4. Moderate microtexture was observed in the stir zone whereas significant microtexture was revealed in the base material. The latter may be associated with the irregularly shaped boundary between the stir zone and base material in the TMAZ.
5. Curved c dislocations were observed at α/β interphase boundaries in stir zone. This structure implies grain boundary sliding which may contribute to local strain accommodation below the β transus.
6. Planar faults were observed on basal planes of α laths in the stir zone. These are similar to faults previously reported in quenched, undeformed material.

## References

1. P. J. Ditzel, "Microstructure/Property Relationships in Aluminum Friction Stir Welds", (M.S. Thesis, The Ohio State University, 1997), 70.
2. M.J. Russell and H.R. Shercliff," Analytical Modelling of Microstructure Development in Friction Stir Welding", (Proc. of the Second International Symposium on Friction Stir Welding, Gothenburg, June, 2000).
3. W.A. Baeslack, J.R. Davis, and C.E. Cross, "Selection and Weldability of Conventional Titanium Alloys", Metals Handbook, vol. 6, Welding, Brazing and Soldering, (ASM International, 1993), 507 - 527.
4. T.W. Nelson, B. Hunsaker and D.P. Field, "Local Texture Charaterization of Friction Stir Welds in 1100 Aluminum", (Proc. of the First International Symposium on Friction Stir Welding, Thousand Oaks, CA, June, 1999).
5. J.C. Williams and M.J. Blackburn, "A Comparison of Phase Transformations in three Commercial Titanium Alloys", Transactions of the ASM, 60, (1967), 373 - 383.
6. A.P. Woodfield, J. White and M.H. Loretto, "The Nature of Faulted Defects in Ti Alloys", Scripta Metallurgica, 19, (1985), 33 - 36.

7.  S. Suri, G.B. Viswanathan, T. Neeraj, D.-H. Hou, and M.J. Mills, "Room Temperature Deformation and Mechanisms of Slip Transmission in Oriented Single-Colony Crystals of an α/β Titanium Alloy", <u>Acta Materialia</u>, 47, (3), (1999), 1019 - 1034.

**Acknowledgements**

The authors wish to thank Dr. Philip Threadgill from TWI for provision of the welded material and Ms. Su Meng and Mr. Peter Collins from the Department of Materials Science and Engineering at The Ohio State University for sample preparation. Special thanks are extended to Prof. Jim Williams, Honda Professor of Materials Science and Engineering at The Ohio State University for valuable technical discussions. This work is sponsored by the National Science Foundation, Dr. Bruce MacDonald, Contract Monitor (NSF grant DMR-0074267).

# LIGHTWEIGHT ALLOYS FOR AEROSPACE APPLICATION

*Edited by:*
*Dr. Kumar Jata, Dr. Eui Whee Lee,*
*Dr. William Frazier and Dr. Nack J. Kim*

## TITANIUM ALLOYS

## Microstructural Evolution During Hot Working of Ti-6Al-4V at High Strain Rates

*Seshacharyulu T., Steve C. Medeiros,*
*William G. Frazier and Prasad Y.V.R.K.*

Pgs. 219-228

184 Thorn Hill Road
Warrendale, PA 15086-7514
(724) 776-9000

# MICROSTRUCTURAL EVOLUTION DURING HOT WORKING OF Ti-6Al-4V AT HIGH STRAIN RATES

Seshacharyulu T.[1], Steve C. Medeiros[1], William G. Frazier[1], and Prasad Y.V. R.K.[2]

[1]AFRL/MLMR, Air Force Research Laboratory, Wright-Patterson AFB, OH 45433 USA

[2]Department of Metallurgy, Indian Institute of Science, Bangalore 560012 India

## Abstract

Microstructural conversion from lamellar to equiaxed is an important step in the thermomechanical processing sequence of Ti-6Al-4V and is conventionally achieved by cogging in the alpha-beta phase field. Since hot working at higher strain rates ($>0.1$ s$^{-1}$) in the two phase field produces microstructural defects such as adiabatic shear bands, cracking, and lamellae kinking, cogging is performed at slow speeds. Also, the occurrence of strain-induced porosity at lower strain rates and lower temperatures ($<850^{o}$C) demands higher temperature control during cogging to obtain defect-free products. In view of these difficulties, an effort has been made to find an alternative process for conversion. Isothermal compression tests conducted at high strain rates (1-100 s$^{-1}$) close to the beta transus revealed the evolution of an equiaxed microstructure consisting of alpha grains surrounded by a thin beta case. The new microstructure is found to be thermally stable and exhibited better mechanical properties over the conventional globularized structure. The equiaxed microstructure has been successfully reproduced under industrial manufacturing conditions using extrusion and sub-scale turbine-engine disk forging experiments at high speeds. The local variations in process parameters are estimated using finite element simulations of these operations and are correlated with microstructural observations.

Lightweight Alloys for Aerospace Applications
Edited by Kumar Jata, Eui Whee Lee,
William Frazier and Nack J. Kim
TMS (The Minerals, Metals & Materials Society), 2001

# Introduction

Among all titanium alloys, Ti-6Al-4V (Ti-6-4) is the most widely used and accounts for more than 50% of the aerospace applications [1]. Ti-6-4 is characterized as an $\alpha$ rich $\alpha+\beta$ alloy in which particular Al-V balance provides attractive mechanical properties required for structural applications [2]. The temperature at which $\alpha+\beta$ transforms to single phase $\beta$ ($\beta$ transus) in this alloy is important not only for mechanical working, but also for heat treatment [3]. The thermomechanical processing sequence for Ti-6-4 component manufacture is shown in Fig. 1, which consists of three steps, viz. ingot breakdown, conversion, and finishing [4]. Ingot breakdown is conducted in the $\beta$ phase field, which helps to achieving chemical homogeneity and breaking the as-cast microstructure. Beta processing followed by faster cooling rates including air cooling results in lamellar (transformed $\beta$) microstructure in which plates of $\alpha$ grow forming colonies according to the Burger's crystallographic orientation relationships between bcc and hcp structures. During conversion, the lamellar structure is broken down by extensive mechanical working in the $\alpha-\beta$ phase field to obtain fine equiaxed $\alpha+\beta$ microstructure. While lamellar structure possesses better fracture toughness and strength, equiaxed $\alpha+\beta$ structure provides excellent fatigue resistance and hence is preferred in rotating components such as turbine disks. Ti-6-4 semi-products are generally supplied in the equiaxed $\alpha+\beta$ microstructural condition, which are finish forged in the $\alpha-\beta$ range followed by appropriate heat treatment to obtain the required final microstructure in the component.

During processing of Ti-6-4, conversion is the most critical and time-consuming step due to large reductions and microstructural considerations. Conventionally conversion is achieved by cogging in the $\alpha-\beta$ range during which large ingots are reduced in cross-section by a series of bites along the length. In view of very large billet thickness, several steps of cogging and reheating are necessary to obtain final dimensions. The problem of prior beta boundary cracking, more popularly known in industry as 'strain induced porosity (SIP)', has been reported during microstructural conversion [5]. Detailed materials modeling conducted on Ti-6-4 by Seshacharyulu et al. [6] over wide temperature and strain rate ranges revealed that, precise temperature control is essential to avoid SIP, which is difficult to maintain during cogging of large billets.

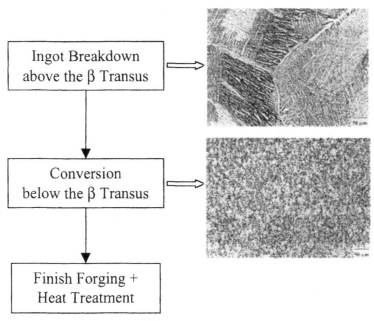

Figure 1: Multi-step thermomechanical processing sequence followed for Ti-6Al-4V component manufacture.

The mechanism of globularization is identified as the safe microstructural process for conversion, the lower temperature limit of which set by the onset of cracking at the prior β grain boundaries and the higher limit is set by the occurrence of transformed β structure when the β volume fraction exceeds about 60%. Also, a strain rate limit of about 0.1 s$^{-1}$ is set by the onset of flow instabilities. At higher strain rates, cracking along the localized flow bands caused by adiabatic heating occurs at low temperatures (below 900°C) while a geometric instability, viz. kinking of lamellae, predominates at higher temperatures. In view of the problems associated with the conventional microstructural conversion, an alternative process which is safer and faster than the conventional is highly sought after. Motivated by these requirements, this research study is conducted with the following objectives. Firstly, to 'look' for a microstructural process by which equiaxed α grains can be produced starting with a lamellar structure. The deformation behavior of Ti-6-4 closer to the β transus (β–α region) at different strain rates relevant to industrial metalworking processes has been considered for this purpose. Secondly, to reproduce the new microstructure under industrial processing conditions using scaled up validation experiments such as extrusion and forging.

## Experimental

Commercial grade Ti-6-4 with a Widmanstätten colony type starting microstructure consisting of lamellar α colonies in large prior β grains of about 200 μm and a grain boundary α layer of 5 μm thick is used. The β transus for this material is approximately 1010°C. Isothermal compression tests were conducted using a servohydraulic testing machine at a temperature of 1000°C and constant true strain rates $10^{-3}$, $10^{-2}$, $10^{-1}$, 1, 10, and 100 s$^{-1}$. Specimens of 22.5 mm height and 15 mm diameter were used. A borosilicate glass coating was applied to the specimens for lubrication and environmental protection. Concentric grooves of 0.5 mm depth were made on the top and bottom faces of the specimens to trap the lubricant. A 1-mm 45° chamfer was provided on the specimen edges to avoid fold-over in the initial stages of compression. Specimens were soaked for 15 minutes at the test temperature, deformed to half the height in each case to impose a true strain of about 0.6, and were air-cooled to room temperature after deformation. Deformed specimens were sectioned parallel to the compression axis and the cut surface was prepared for microstructural examination using standard metallographic techniques. Selected specimens were electropolished and observed in scanning electron microscope back-scatter mode.

## Results and Discussion

### Hot Deformation Behavior

Load-stroke data obtained from compression experiments are converted into true stress-true plastic strain (σ–ε) using standard method. The σ–ε curves at 1000°C and different strain rates are shown in Fig. 2. At strain rates slower than 0.1 s$^{-1}$, the curves are flat indicating steady-state behavior, while at higher strain rates (≥1 s$^{-1}$), softening/oscillatory behavior is observed. Although, the shapes of the σ–ε curves indicate a significant influence of strain rate on the deformation behavior, nothing conclusive can be said regarding the mechanisms of deformation since different mechanisms can exhibit similar features [7].

Detailed microstructural examination of specimens deformed at slow strain rates (≤ 0.1 s$^{-1}$) revealed fully transformed β structure which is normally expected when the deformation temperature is close to the β transus. A representative microstructure of specimen deformed at 1000°C/$10^{-3}$ s$^{-1}$ is shown in Fig. 3(a). Specimens deformed at higher strain rates (1-100 s$^{-1}$), on the other hand, exhibited fine equiaxed α grains. Representative microstructure of a specimen

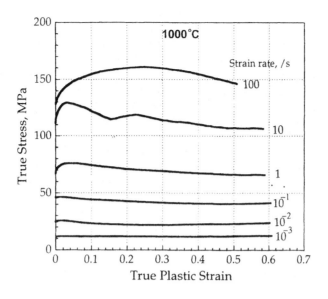

Figure 2: Flow curves of Ti-6Al-4V at 1000°C and different strain rates.

deformed at 1000°C/1 s$^{-1}$ is shown in Fig. 3(b). Back-scatter SEM images of a specimen deformed at 1000°C/100 s$^{-1}$ are shown in Fig. 4 which show a fine equiaxed α grains surrounded by a thin β case of approximately 1 μm thick. The new microstructure could be termed as *dissipative* since it evolves as a transient response to deformation at high strain rates close to the β transus. Additional experiments conducted at different temperatures and higher strain rates revealed that the structure reverts back to transformed β when the deformation temperature is above the β transus and the dissipative structure forms at temperatures up to about 30°C below the transus. Thermal stability of the microstructure, which is an important requirement during further processing, is evaluated by subjecting the material with dissipative starting structure to different heat tretments. Microstructure of a specimen subjected to stress-relief anneal at 705°C for 1 hour followed by air cooling is shown in Fig. 5(a), which as expected reveals no significant difference in features compared to the as-deformed structure. Microstructure of a specimen subjected to recrystallization anneal at 925°C for 1 hour followed by slow furnace cool is shown in Fig. 5(b). Completely recrystallized grains with slight grain growth can be observed in this microstructure. These microstructures confirm that the dissipative microstructure is thermally stable.

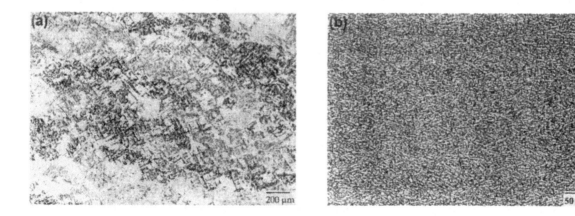

Figure 3: Microstructures of Ti-6Al-4V specimens deformed at 1000°C and (a) 10$^{-3}$ s$^{-1}$ (b) 1 s$^{-1}$.

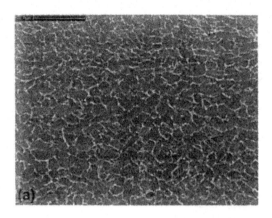
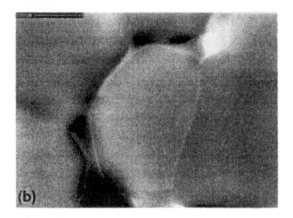

Figure 4: (a–b) Back scatter scanning electron micrographs of Ti-6Al-4V specimen deformed at 1000°C and 100 s⁻¹.

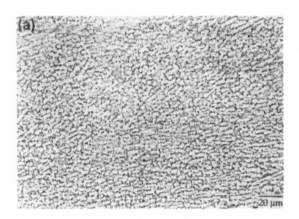
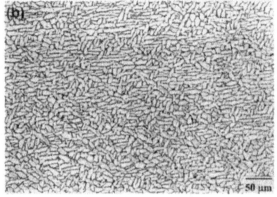

Figure 5: Microstructures of Ti-6Al-4V specimens deformed at $1000°C/100$ s⁻¹ and subjected to (a) stress-relief anneal at $705°C/1$ h/air-cool (b) recrystallization anneal at $925°C/1$ h/furnace-cool.

The mechanism of dissipative microstructural evolution may be qualitatively explained as an effect of strain rate on the phase transformation kinetics. Deformation at higher strain rates produces large amount of dislocations in the matrix, which act as heterogeneous nucleation sites for the $\alpha$ precipitation. The stress field associated with the dislocations compensates the strains developed due to phase transformation giving rise to equiaxed $\alpha$ structure. At lower strain rates, on the other hand, homogeneous precipitation of $\alpha$ phase occurs and the precipitate growth progresses by the movement of incoherent interface on a good plane of matching, which results in a typical Widmanstätten structure [8]. The experimental results indicate that there exists a threshold strain rate which changes precipitation reaction kinetics from homogeneous to heterogeneous. The development of a detailed model from the first principles explaining the dissipative microstructural evolution is in progress.

Validation Experiments
Extrusion and forging experiments were conducted to reproduce the dissipative microstructure under industrial processing conditions in larger components and to establish the influence of process variables on the microstructural evolution. These are explained in the following sections.

Extrusion: A high speed extrusion validation experiment was conducted using a 7000 kN capacity Lombard horizontal press located at the Wright-Patterson Air Force Base. The process parameters were selected in such a way that the average strain rate and temperature fall within the range for dissipative microstructure evolution. A round-to-round streamlined die made of

H-13 tool steel with an extrusion ratio of 6:1 was used. The die facing was coated with zirconia and a Necrolene™ lubricant was used to minimize the die-billet friction. The container was coated with Fiske 604™ lubricant. The billet dimensions were 75 mm diameter and 150 mm length with a 12.7 mm 45° chamfer at the nose. The billet was coated with Delta 151™ lubricant. A ram speed of 100 mm/s was used and the average strain rate during extrusion was calculated using the relation [9]

$$\dot{\varepsilon} = \frac{6v_0 D_b^2}{D_b^3 - D_p^3} \ln(\Gamma)$$ (1)

where $v_0$: ram speed, $D_b$: billet diameter, $D_p$: product diameter, and $\Gamma$: extrusion ratio. By substituting the values in Eq. (1), the $\dot{\bar{\varepsilon}}$ is estimated be about 15 s$^{-1}$. FEM simulations have revealed that the local variations in strain rate are between 1-20 s$^{-1}$. The temperature rise ($\Delta T$) due to deformation heating is estimated using the energy balance equation

$$\rho C_p \Delta T = \chi \bar{\sigma} \bar{\varepsilon}$$ (2)

where $\rho$ is the mass density, $C_p$ is the heat capacity per unit mass, $\chi$ is a coefficient which characterizes the fraction of plastic work that is converted into heat (usually taken constant and set equal to 0.95), $\bar{\sigma}$ is the effective stress, and $\bar{\varepsilon}$ is the effective strain. A billet temperature of 980°C was selected considering the estimated adiabatic temperature rise of about 30°C during deformation ($\varepsilon\sim1.8$). The die and the container were heated to 260°C. The billet was soaked at 980°C for 1 hour and quickly transferred to the press. Extrusion was performed using a follower carbon block and the product (extrudate) was air-cooled to room temperature. Extrudate was sectioned parallel to the extrusion direction and microstructural observations were made. Microstructures in the front end and middle of the extrusion are shown in Fig. 6 (a) and 6(b), which exhibit fine $\alpha$ grains elongated in the extrusion direction. This experiment demonstrated that the dissipative microstructure could be reproduced in larger components under complex manufacturing conditions.

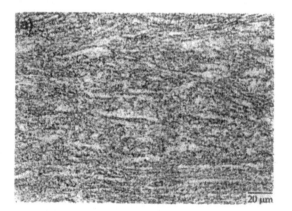 

Figure 6: Microstructures of Ti-6Al-4V extrudate specimens taken from (a) front end and (b) middle. The extrusion direction is horizontal.

Forging: Another validation experiment was conducted using a sub-scale turbine-engine disk forging. A hydraulic press of 10,000 kN capacity and shaped dies made of H-13 steel were used. The billet dimensions were 160 mm dia x 82.5 mm length. Billet was subjected to $\beta$ annealing to get fully transformed $\beta$ starting microstructure. A ram speed 105 mms$^{-1}$ was selected to achieve an average strain rate of 10 s$^{-1}$. Local variations in process parameters were estimated with the help of finite element analysis using a commercial software DEFORM™ [10]. The strain and strain rate distributions in one-half of the axi-symmetric disk are shown Figs. 7 (a) and 7(b) respectively. These simulations have shown that a large strain gradient exists from the center (~125%) to the outer rim region (~50%). The strain rate varies from 1-25 s$^{-1}$, with the flash and last-to-fill corner regions experiencing the highest strain rate. A billet temperature 985°C was selected in view of an expected adiabatic temperature rise of 25°C. Billet was coated with glass for lubrication and environmental protection. Forging was completed in about 0.5 s and the disk was air-cooled to room temperature. Forged disk was bisected and the cut surface was prepared for metallographic observations. The microstructures at two different locations are shown in Fig. 8, which exhibit fine equiaxed α structure. The α grain size is found to be extremely fine in the regions of higher strain rate and strain (e.g. corners) while slightly coarse grains were observed in the mid region. The disk forging experiment demonstrated that the equiaxed microstructure could be reproduced in components of complicated geometry using the new high strain rate microstructural process.

Tensile Properties: To evaluate the room temperature mechanical properties of the dissipative microstructure, tensile tests were conducted on the specimens machined from the extrudate. Cylindrical threaded specimens of 25 mm gage length and 4 mm gage diameter were used. Tensile properties of the dissipative microstructure averaged from four tests are shown in Fig. 9. The strength and ductility of mill-annealed microstructure from ref. [3] are also shown in Fig. 9 for the purpose of comparison. The dissipative microstructure exhibits higher strength as well as a 30% increase in ductility compared to the conventionally processed material. Examination of the fracture surfaces in SEM revealed flute type tearing fracture within the α grains, which is in contrast to the void nucleation and coalescence mechanism in the conventional microstructure. The increased strength and ductility in the dissipative microstructure suggest an increase in fracture toughness which is an important property for damage critical applications.

## Conclusions

1. An alternative process for Ti-6Al-4V microstructural conversion is found at higher strain rates closer to the β transus.
2. The new dissipative microstructure produced by the new process is found to be thermally stable and exhibits better mechanical properties than the conventionally converted microstructure.
3. The dissipative microstructure could be successfully reproduced in larger components using extrusion and shaped forging experiments.

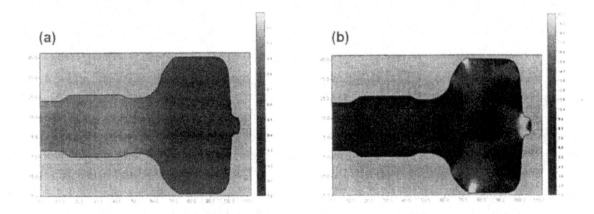

Figure 7: Local variations in one-half section of Ti-6Al-4V disk forging predicted from the finite element simulations: (a) strain and (b) strain rate.

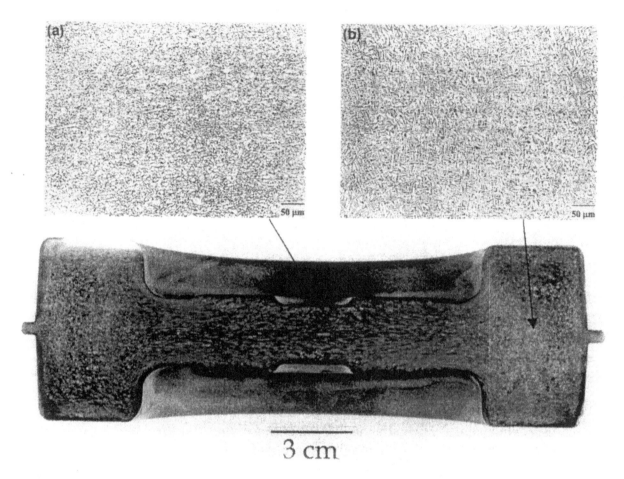

Figure 8: Macrostructure of the Ti-6Al-4V forged disk cross-section and microstructures at different locations (a) center and (b) outer rim.

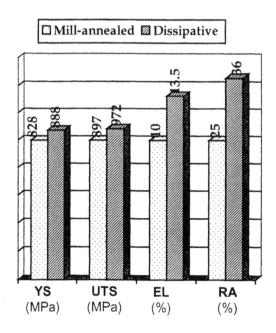

Figure 9: Tensile properties of Ti-6Al-4V with dissipative starting microstructure compared with those of mill-annealed structure.

## Acknowledgements

The authors would like to thank Dr. J.C. Malas for many stimulating discussions. TS is supported under the auspices of Air Force Contract No. F33615-96-D-5835-0018-14. YVRKP is thankful to the National Research Council, USA, for awarding him an associateship and to the Director of the Indian Institute of Science, Bangalore, for granting him a sabbatical leave. The assistance of Joe Brown and Travis Brown in extrusion and forging experiments is much appreciated.

## References

1. "Titanium Alloys for Aerospace", Advanced Materials and Processes, ASM International, (3) (1999), 39-41.

2. E.W. Collings, The Physical Metallurgy of Titanium Alloys (Metals Park: ASM, OH, 1984).

3. R. Boyer, G. Welsch, and E.W. Collings (eds.), Materials Properties Handbook: Titanium Alloys (Materials Park: ASM International, OH, 1994).

4. Y.V.R.K. Prasad et al., "Titanium Alloy Processing", Advanced Materials and Processes, ASM International, (6) (2000), 85-89.

5. S.R. Seagle, K.O. Yu, and S. Giangiordano, "Considerations in processing titanium," Materials Science and Engineering A, 263 (1999) 237-242.

6. T. Seshacharyulu et al., "Microstructural Mechanisms during Hot Working of Commercial Grade Ti-6Al-4V with Lamellar Starting Structure", unpublished research, WPAFB (2000).

7. Y.V.R.K. Prasad and T. Seshacharyulu, "Modelling of Hot Deformation for Microstructural Control", International Materials Reviews, 43 (1998) 243-258.

8. H.I. Aaronson, C. Laird, and K.K. Kinsman, Phase Transformations, Chapter 8, (Metals Park: ASM, OH, 1970).

9. G.E. Dieter, Mechanical Metallurgy, Third Edition, (McGraw-Hill, 1986) 629.

10. DEFORM User Manual, Scientific Forming Technologies Corporation, Columbus, OH, 1998.

# LIGHTWEIGHT ALLOYS FOR AEROSPACE APPLICATION

*Edited by:*
*Dr. Kumar Jata, Dr. Eui Whee Lee,*
*Dr. William Frazier and Dr. Nack J. Kim*

## TITANIUM ALLOYS

# The Application of a Novel Technique to Examine Sub-β Transus Isothermal Forging of Titanium Alloys

*Martin Jackson, Richard Dashwood,*
*Leo Christodoulou and Harvey Flower*

Pgs. 229-239

184 Thorn Hill Road
Warrendale, PA 15086-7514
(724) 776-9000

# THE APPLICATION OF A NOVEL TECHNIQUE TO EXAMINE SUB-β TRANSUS ISOTHERMAL FORGING OF TITANIUM ALLOYS

Martin Jackson, Richard Dashwood, Leo Christodoulou, and Harvey Flower

Department of Materials
Imperial College of Science, Technology and Medicine
Prince Consort Road
London SW7 2BP
UK

## Abstract

The paper reviews how a technique exploiting a novel specimen design and testing methodology can effectively yield a database of microstructural information for titanium alloys. By applying the technique, the effect of thermomechanical processing (TMP) parameters on microstructural development is elucidated.

Double truncated cone specimens are isothermal compressed at industrially comparable strain rates and sub-β transus temperatures to obtain microstructural information for a range of strains within a single test specimen. Rheological behaviour is fitted using a recently developed constitutive approach that incorporates an internal microstructural variable (lambda). A finite element modelling package is used to generate both strain and lambda profiles which complement the equivalent microstructural profile of the test specimen, providing a database of microstructural information for defined TMP conditions. An output of microstructural data can aid the implementation of finite element based isothermal forging simulations. The effectiveness of the technique is reviewed for the near-β alloy; Ti-10V-2Fe-3Al at 760°C and the α+β alloy; Ti-6Al-2Sn-4Zr-6Mo at 825°C, for a range of strain rates.

Lightweight Alloys for Aerospace Applications
Edited by Kumar Jata, Eui Whee Lee,
William Frazier and Nack J. Kim
TMS (The Minerals, Metals & Materials Society), 2001

# Introduction

In order to implement any finite element (FE) based process simulation programme to predict a final forged microstructure, a database of information relating final microstructure to thermomechanical processing (TMP) conditions has to be established. This can represent a significant work effort due to the fact that the key property determining microstructural features are strongly affected by the strain rate, temperature and strain pertaining to the deformation process. Even during a relatively simple hot die forging, significant through section variations in these variables are encountered. Using classical compression testing of cylinders, the test matrix required to evaluate the individual and combined effect of these variables, would be extremely large. The authors have overcome this problem by exploiting a novel double truncated cone specimen that permits the effect of a range of strains to be studied in one test.

The geometric benefits of the double truncated cone specimen are illustrated in fig. 1, an isothermally compressed specimen produces a controlled strain profile, designed to be compatible with strains in isothermal forging operations. The exact strain profile within the specimen is determined by mathematical simulation of the compression test, using FE modelling software. In previous work [1,2] the authors have also carried out a parametric study of the testing strategy to determine the effects and importance of friction and strain rate sensitivity on generated strain profiles.

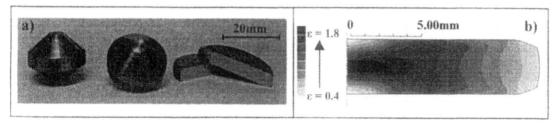

Figure 1: a) Photograph of double truncated cones, before and after deformation and; b) a finite element strain profile for the deformed double truncated cone.

Ti-10V-2Fe-3Al (Ti-10-2-3) is a high strength/toughness near-$\beta$ alloy, developed to provide weight savings over steels in airframe forging applications. Ti-6Al-2Sn-4Zr-6Mo (Ti-6246) is a high strength solid solution strengthened $\alpha+\beta$ alloy, used in elevated temperature applications, such as the intermediate pressure stage of a gas turbine engine compressor [3]. The $\beta$ transus of Ti-10-2-3 (~795-800°C) is relatively low compared to Ti-6246 (~940-960°C), allowing lower forging temperatures, enabling the use of cheaper tools/dies, within an overall costly production route [4].

# Experimental Procedure

Both the near-$\beta$ alloy, Ti-10-2-3 and $\alpha+\beta$ alloy, Ti-6246 originated from Timet UK, with compositions shown in Table I. A microstructural investigation found $\beta$ transus temperatures of 795±5°C and 950±5°C, for Ti-10-2-3 and Ti-6246, respectively.

Table I Compositions (in weight %) of alloys Ti-10-2-3 and Ti-6246 used in the present study.

| Alloy | Al | V | Mo | Fe | Sn | Zr | C | Si | $O_2$ | $H_2$ | $N_2$ |
|---|---|---|---|---|---|---|---|---|---|---|---|
| Ti-10-2-3 | 3.14 | 10.1 | - | 1.72 | - | - | 0.009 | - | 0.091 | 0.052 | 0.007 |
| Ti-6246 | 5.71 | - | 5.99 | 0.06 | 1.97 | 3.94 | 0.01 | 0.05 | 0.11 | 0.01 | 0.035 |

## Isothermal Forging Conditions

Isothermal compression was performed on a servo-hydraulic 100kN Mayes testing machine, with specimen heating provided by a Severn Furnaces 1000°C radiant furnace. Further details of the equipment are given in [2]. An isothermal forging test matrix, comprised of strain rates, 0.1, 0.01 and 0.001s⁻¹ and sub-β transus temperatures in the range 760-790°C and 825-925°C for Ti-10-2-3 and Ti-6246, respectively, was firstly carried out on uniaxial cylinder compression specimens (8mm Ø × 10mm height), to generate flow curves, providing rheology data for FE modelling. The double truncated cone specimens, were deformed to ~5mm thickness for the correlative microstructural information. All specimens were coated in a boron nitride release agent prior to compression, and received a water quench immediately after deformation.

## Flow Stress Algorithm

The constitutive equation is based on the approach outlined in [5] and incorporates an internal state variable (lambda), which can represent microstructural evolution. The model has been successfully applied to the forging of titanium aluminide aerofoils [6] and to the disc forging of nickel alloy, Nimonic AP1 [7], where the model is discussed in detail.

In brief, the Zener-Hollomon parameter ($Z$) is used to combine temperature ($T$) and strain rate ($d\varepsilon/dt$) effects where $Q$ is an apparent activation energy and $R$ is the universal gas constant,

i.e. $$Z = d\varepsilon/dt . \exp(Q/RT) \tag{1}$$

The model assumes that lambda ($\lambda$) evolves as an exponential function of strain ($\varepsilon$) towards a steady state condition ($\lambda_{ss}$) which is defined as a power function of $Z$:

$$\lambda_{ss} = \lambda_0 \times Z^q \tag{2}$$

Such that;
$$\lambda = \lambda_{ss} + (1 - \lambda_{ss}).\exp(-\alpha\varepsilon) \tag{3}$$

Then the flow stress ($\sigma$) is: $$\sigma = k \times Z^m (\lambda - \exp(-\beta\varepsilon)) \tag{4}$$

Where $k$, $\lambda_0$, $m$, $q$, $\alpha$ and $\beta$ are material constants.

## Finite Element Modelling

The flow stress algorithm was incorporated into the finite element code of the commercial package Forge2 [8], which was used to simulate the forging process. Fig. 2, shows a typical mesh used for the mathematical simulation along with the strain and lambda evolution during the test.

Input data included; geometric definition of the tooling and starting material, rheology data for the material, and friction behaviour at the tooling/material interface. Elastoplastic behaviour was modelled using 4-node linear elements. Output from the package includes strain, strain rate, temperature and user parameter (the internal variable, lambda) profiles.

The defined coefficient of friction for Forge2, is the interface friction factor, $\bar{m}$, where the shear stress at the interface of the die/sample, $\tau_i$, is directly proportional to the flow stress, $\sigma_0$.

$$\tau_i = \bar{m} \frac{\sigma_0}{\sqrt{3}} \tag{5}$$

For given conditions of lubrication and temperature, $\bar{m}$ is usually considered to have a constant value independent of the pressure at the interface, and vary from 0 (perfect sliding) to 1 (sticking).

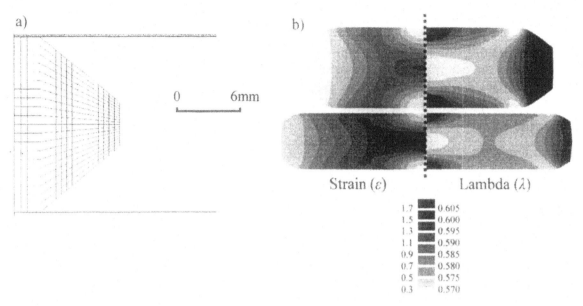

Figure 2:  a) Typical mesh used for simulation of isothermal forging and; b) Predicted strain and lambda profiles generated from a simulation of a specimen forged at 760°C at a strain rate of $0.1s^{-1}$.

## Results & Discussion

<u>As-Received Microstructures</u>

The as-received microstructure of Ti-10-2-3 (fig. 3a) is typical of a near-β alloy forged in the single phase β field, consisting of a transformed β grain structure, with a typical grain size of 300μm.  The β grains contain a large volume fraction of Widmanstattën α plates of high aspect ratio, and a near continuous distribution of α particles along the prior β grain boundaries.

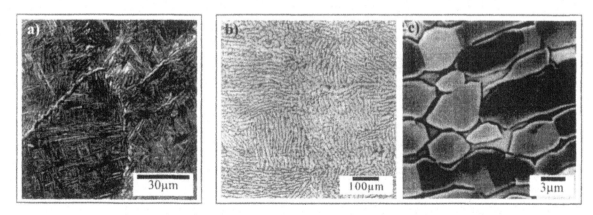

Figure 3:  The as-received microstructures of the materials used in the study; a) SEM backscattered electron image of Ti-10-2-3 illustrating a typical Widmanstatten α microstructure; b) Light micrograph illustrating the colonies of α-lamellae in Ti-6246; c) SEM backscattered electron image of Ti-6246 illustrating the sub-grain structure within the α-lamellae.

The as-received microstructure of Ti-6246, (figures 3b and 3c), exhibits colonies of relatively coarse α-lamellae, approximately 4μm in width, which reveal a sub-grain structure under SEM backscattered electron imaging, with β-phase between the α-lamellae (fig. 3c). Such a microstructure is typical of an α+β alloy billet material, which has received a secondary hot work at around 50°C below its β transus (~900°C), followed by an air cool. At the lower light microscopy magnifications (fig. 3b) continuous α-phase is observed at prior β grain boundaries, indicating that the material received a primary ingot breakdown in the single phase β field, prior to α+β secondary work. Phase analysis measurements reveal that the as-received microstructure contains 15% β-phase, with the balance being α-phase. This is common for α+β alloys at room temperature [9].

Compression Flow Curves

The stress-strain curves for Ti-10-2-3 at 760°C and Ti-6246 at 825°C for the range of strain rates are illustrated in fig. 4. Sub-β transus forging was characterised by rapid work hardening to a peak stress level, followed by flow softening towards a steady state flow stress. This characteristic peak stress, flow softening shape becomes less apparent with increasing temperature, and as shown in fig. 4, decreasing strain rate. The experimental curves were used to derive data (summarised in table II) for the constitutive model developed by Blackwell et al [7]. The predicted flow curves for the equivalent forging conditions are superimposed in fig. 4.

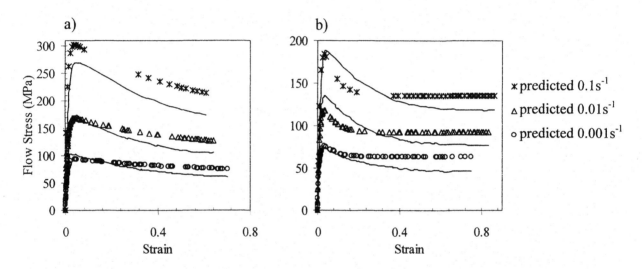

Figure 4: Comparison of experimental flow stress data with the model predictions of for; a) Ti-6246 at 825°C and; b) Ti-10-2-3 at 760°C. at strain rates of 0.1, 0.01 and $0.001s^{-1}$.

The fit between the calculated and measured flow curves for the whole data range was refined using numerical fitting methods. Both α (exponential damping constant relating strain and structure) and β (exponential damping constant relating stress and strain) were allowed to vary, generating best fit errors of 5.95 and 4.80 %, for Ti-10-2-3 and Ti-6246, respectively.

Table II  Constitutive data for Ti-10V-2Fe-3Al and Ti-6Al-2Sn-4Zr-6Mo.

| Alloy | k | $\lambda_0$ | m | q | α | β | Q/R |
|---|---|---|---|---|---|---|---|
| Ti-10-2-3 | 1.17E-06 | 23.32 | 0.21 | -0.04 | 17.00 | 101.12 | 98155 |
| Ti-6246 | 7.69E-05 | 22.86 | 0.26 | -0.06 | 1.83 | 123.87 | 67722 |

A die/specimen interface friction factor, m̄ of 0.6 was estimated from FEM analysis of the conventional ring test [1,2].

Microstructural Database Analysis

Mathematical simulations of the forging process were performed to generate both strain and lambda profiles, and were assigned to the equivalent backscattered image location on the forged specimen. Figures 5 & 7 illustrate a catalogue of microstructures at a range of strain rates for Ti-10-2-3 at 760°C and Ti-6246 at 825°C, respectively.

Ti-10V-2Fe-3Al at 760°C. Observations at the lowest strains in fig. 5, compared to the as-received microstructure (fig. 3a) indicate little change other than the reduction in α-phase imposed by the exposure to the elevated test temperature (~35°C below the β transus). Phase analysis of the images in fig. 5 indicates there is a negligible change in the percentage of α-phase as the strain rate decreases, and the time at temperature increases. It may, therefore, be assumed that near equilibrium phase conditions are reached at 760°C, prior to the forging.

At each strain rate microstructural development initiates from an equiaxed β structure with Widmanstattën α plates at low strains, evolving to a microstructure with greater backscattered electron contrast and high misorientation, with a dispersion of α particles at the β sub-grain boundaries. The orientation mismatch between the β sub-grains increases, due to local lattice rotation between constraining α plates.

As the strain rate decreases the Widmanstattën α plates initially present becomes spherodised at decreasing levels of strain. In the lower strain regions, decreasing the strain rate, results in less pronounced misorientation developing between regions of β partitioned by the α plates and a generally larger subgrain size. This suggests recovery processes anneal the dislocations introduced by the deformation more readily at the lower strain rates.

Measured values of the average α grain size at 760°C, indicate that the average α grain sizes increase as strain rate decreases. Fig. 6 reveals an approximately linear relationship between the average α grain size and lambda, with respect to an increase in both strain and strain rate.

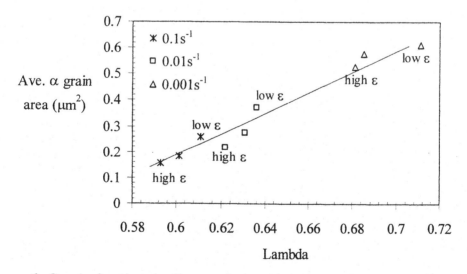

Figure 6: Graph showing the linear relationship of the average Ti-10-2-3 α grain size with the internal variable, lambda, for a range of strains and strain rates at the sub-β transus temperature of 760°C.

Ti-6246 at 825°C: Heating the as-received microstructure high in the α+β phase region (~125°C below the β transus), has the effect of partially dissolving the as-received α-lamellae, as shown in fig. 3c. The surviving equiaxed α grains (primary-α), retain grain diameters of ~4μm, equivalent to the original α-lamellae widths. Phase analysis results also presented in fig.

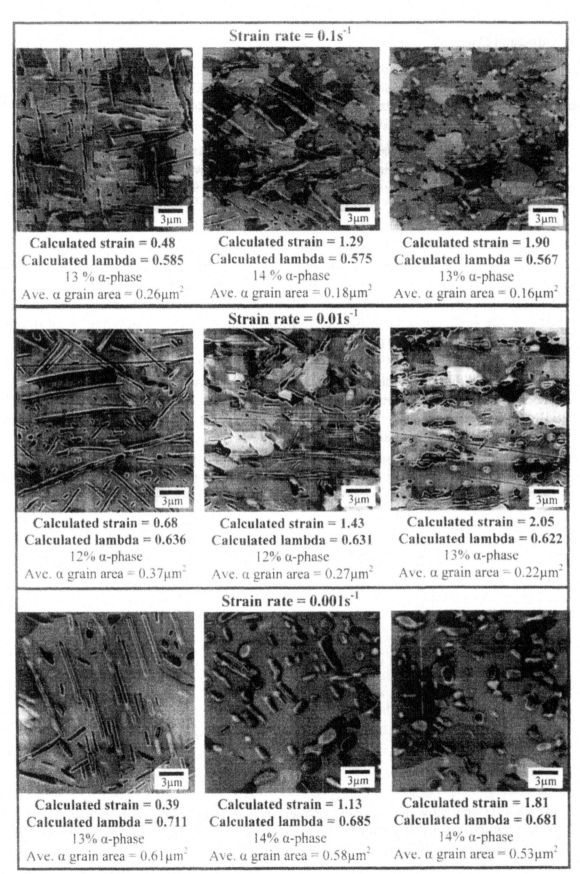

**Strain rate = 0.1s⁻¹**

| Calculated strain = 0.48 | Calculated strain = 1.29 | Calculated strain = 1.90 |
| Calculated lambda = 0.585 | Calculated lambda = 0.575 | Calculated lambda = 0.567 |
| 13 % α-phase | 14 % α-phase | 13% α-phase |
| Ave. α grain area = 0.26μm² | Ave. α grain area = 0.18μm² | Ave. α grain area = 0.16μm² |

**Strain rate = 0.01s⁻¹**

| Calculated strain = 0.68 | Calculated strain = 1.43 | Calculated strain = 2.05 |
| Calculated lambda = 0.636 | Calculated lambda = 0.631 | Calculated lambda = 0.622 |
| 12% α-phase | 12% α-phase | 13% α-phase |
| Ave. α grain area = 0.37μm² | Ave. α grain area = 0.27μm² | Ave. α grain area = 0.22μm² |

**Strain rate = 0.001s⁻¹**

| Calculated strain = 0.39 | Calculated strain = 1.13 | Calculated strain = 1.81 |
| Calculated lambda = 0.711 | Calculated lambda = 0.685 | Calculated lambda = 0.681 |
| 13% α-phase | 14% α-phase | 14% α-phase |
| Ave. α grain area = 0.61μm² | Ave. α grain area = 0.58μm² | Ave. α grain area = 0.53μm² |

Figure 5: Catalogue of Ti-10V-2Fe-3Al microstructures produced by the compression of truncated cone specimens, at strain rates; 0.1, 0.01 and 0.001s⁻¹ at 760°C, along with the corresponding calculated strain and lambda value, generated from the mathematical simulation, plus α-phase percentages and average α grain area measurements.

7 reveal a negligible variation in the percentage of primary-α with strain and strain rate. Interestingly, it has been shown in previous studies [10] that the primary-α accommodates a significant percentage of the deformation, indicated by the shape change of the α grains which is observed.

Fig. 7 effectively illustrates how microstructural development in the β matrix initiates with the break-up of α-plates (secondary-α), at the low strains. With increasing strain, the α-plates fragment forming a 'necklace' morphology at high strains. The necklace α morphology effectively pins sub-grain boundaries of the softer β-phase. Thus, with increasing strain the local orientation mismatch across the constricted β sub-grains increases. At higher strains the β sub-grain rotations eventually lead to the development of high angle grain boundaries, giving the appearance of a recrystallised microstructure.

Decreasing the strain rate provides more time for coarsening of the α-plates and β sub-grains. Fig. 8a, illustrates, with the aid of microstructural analysis from previous work, at 825°C and zero strain [10], the reduction of volume fraction of secondary-α at the slower strain rates of 0.01 and 0.001 s$^{-1}$. The combination of longer times at temperature and applied strain is sufficient to promote faster diffusional rates of the dissolution of secondary-α. This observation is also displayed in fig. 8b, where at 0.1 s$^{-1}$, the percentage of secondary-α is insensitive to lambda, compared to slower strain rates at 825°C, where a sensitivity relationship seems to be evident, with increasing strain.

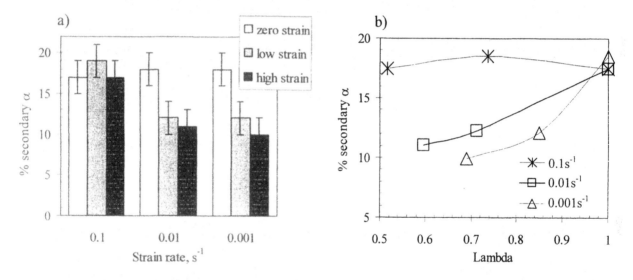

Figure 8: a) Bar chart illustrating the variation of secondary-α volume fraction at different strains, for the range of strain rates; b) Graph showing the relationship between secondary-α percentage and lambda, for the range of strain rates.

## Conclusions

The novel testing technique and specimen geometry were effective in examining the microstructural development of the near-β alloy, Ti-10-2-3 and α+β alloy, Ti-6246 for a range of strains and strain rates at the sub β transus deformation temperatures of 760 and 825°C, respectively.

The technique revealed an analogous microstructural development in both Ti-10-2-3 and the β matrix of the Ti-6246. α-plates are broken with increasing deformation and they, or their spherodised remnants effectively pin sub-grain boundaries of the softer β-phase with increasing

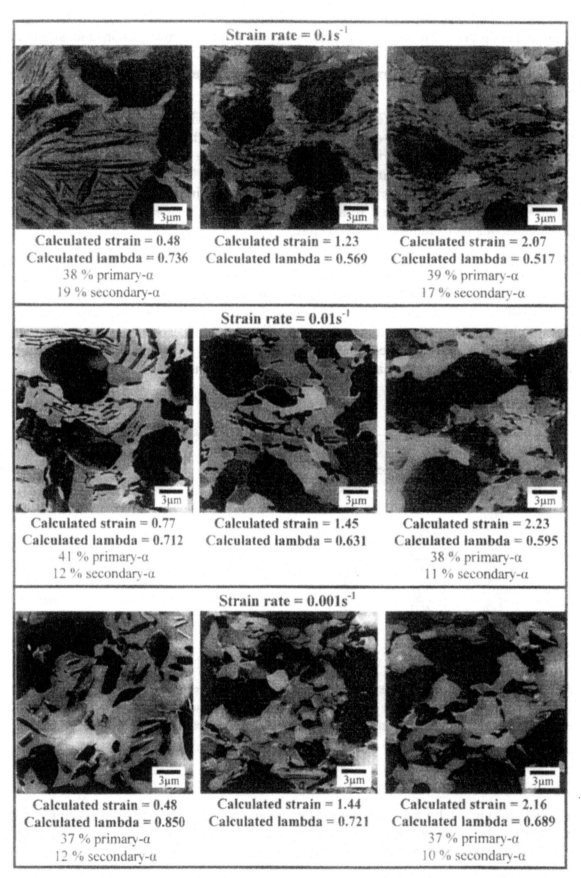

Figure 7: Catalogue of Ti-6Al-2Sn-4Zr-6Mo microstructures produced by the compression of truncated cone specimens, at strain rates; 0.1, 0.01 and 0.001s⁻¹ at 825°C, along with the corresponding calculated strain and lambda value, generated from the mathematical simulation, plus primary-α and secondary-α phase percentages.

238

strain. The local orientation mismatch between the constricted β sub-grains increases, thus leading to the development of high angle grain boundaries, giving the appearance of a recrystallised microstructure, in the case of the Ti-6246 β matrix.

Preliminary microstructural prediction results indicate a distinct relationship between the microstructure of Ti-10-2-3 and the corresponding state variable (lambda). The average α grain size of Ti-10-2-3 at 760°C, exhibits an approximate linear relationship with lambda, with respect to an increase in both strain and strain rate. However, further analysis is required to generate an equally well-defined relationship for Ti-6246. The difficulties associated with the microstructural prediction of Ti-6246 are due to the existence of two morphological and chemically different forms of α-phase, which respond differently to TMP parameters.

The paper has further demonstrated both the effectiveness of the technique to yield a microstructural database to assess TMP parameters for titanium alloys, and the potential to execute a full investigation for a range of alloys and TMP conditions, which will aid isothermal forging predictions.

## Acknowledgements

The authors gratefully acknowledge the financial support of the EPSRC and the computer facilities provided under EPSRC grant GR/L86821.

Partial support and provision of materials from DERA is acknowledged, as well as the technical support from Drs. J.W. Brooks, H.S. Ubhi, and A. Wisbey at DERA Farnborough, UK.

## References

1. L. Christodoulou, R.J. Dashwood and H.M. Flower, 'Techniques For Thermomechanical Property Evaluation In Titanium Based Materials', Third Pacific Rim International Conference on Advanced Materials and Processing, ed. M. Ashraf Imam, (Honolulu, Hawaii, July 1998).

2. M. Jackson et al., 'The Application of a Novel Technique to Examine Thermomechanical Processing of Near-β Titanium Alloy Ti-10V-2Fe-3Al', Materials Science and Technology, Vol. 16 (Nov-Dec 2000).1437-1444.

3. G.W. Kuhlman, 'Ti-6Al-2Sn-4Zr-6Mo', Materials Properties Handbook–Titanium Alloys, (ASM International, 1994) 479.

4. S. Shah, 'Isothermal Forging and Hot-Die Forging', Metals Handbook 9th Edition, vol. 14, (ASM International, 1988) 150-175.

5. P.S. Bate, AGARD Conf. Proc., Vol. 426, NATO (1987) p22.1.

6. J.W. Brooks, T.A. Dean, Z.M. Hu and E. Wey, 'Three-Dimensional Finite Element Modelling Of A Titanium Aluminide Aerofoil Forging', Journal of Materials Processing Technology 80-81 (1998) 149-155.

7. P.L. Blackwell, J.W. Brooks and P.S. Bate, 'Development Of Microstructure In Isothermally Forged Nimonic Alloy AP1', Materials Science and Technology, 14 (1998), 1181-1188.

8. Forge2®, version 2.9.04, Transvalor S.A., Sophia Antipolis, France (2000)

9. S.L. Semiatin, V. Seetharaman and I. Weiss, 'The Thermomechanical Processing of Alpha/Beta Titanium Alloys' Journal of Materials, (June 1997), 33-39.

10. M. Jackson et al., 'The Application of a Novel Technique to Examine Sub-β Transus Isothermal Forging of Ti-6Al-2Sn-4Zr-6Mo' (Paper presented at the IoM meeting, Titanium Alloys at Elevated Temperature, University of Birmingham, UK, 11th September 2000) 10.

# LIGHTWEIGHT ALLOYS FOR AEROSPACE APPLICATION

*Edited by:*
*Dr. Kumar Jata, Dr. Eui Whee Lee,*
*Dr. William Frazier and Dr. Nack J. Kim*

## TITANIUM ALLOYS

## Stress Corrosion Cracking of α-Ti in a Methanol Solution

*X.Z. Guo, K.W. Gao, L.J. Qiao and W.Y. Chu*

Pgs. 241-249

184 Thorn Hill Road
Warrendale, PA 15086-7514
(724) 776-9000

# STRESS CORROSION CRACKING OF α-TI
# IN A METHANOL SOLUTION

X.Z. Guo, K.W. Gao, L.J. Qiao and W.Y. Chu

Department of Materials Physics, University of Science and Technology Beijing
Key Laboratory for Environment Fracture of Ministry of Education
Beijing 100083, China

## Abstract

The flow stress of a specimen of α-Ti before unloading was different with the yield stress of the same specimen after unloading and forming a passive film through immersing in a methanol solution at various constant potentials. The difference is the film-induced stress. The film-induced stress and susceptibility to SCC in the methanol solution at various potentials were measured. At open-circuit potential and under anodic polarization, both film-induced tensile stress $\sigma_p$ and susceptibility to SCC $I_\sigma$ had a maximum value. The film-induced stress and SCC susceptibility, however, decreased steeply with decrease in potential under cathodic polarization. When the potential was less than $-250mV_{SCE}$, the film-induced stress became compressive, correspondingly, susceptibility to SCC was zero. Therefore, the variation of film-induced stress with potential was consistent with that of SCC susceptibility with potential. A large film-induced tensile stress is the necessary condition for SCC of α-Ti in the methanol solution.

Lightweight Alloys for Aerospace Applications
Edited by Kumar Jata, Eui Whee Lee,
William Frazier and Nack J. Kim
TMS (The Minerals, Metals & Materials Society), 2001

# Introduction

Transmission electron microscopy (TEM) observations show[1-4] that the corrosion process can facilitate dislocation emission and motion during stress corrosion cracking (SCC) of brass, Type310 stainless steel, $\alpha$-Ti and Ti$_3$Al+Nb. Nanocracks of SCC will nucleate in a dislocation-free zone (DFZ) only when the corrosion-enhanced dislocation emission and motion develop to a critical condition. How can the corrosion process facilitate dislocation emission and motion? It has been proposed that divacancies induced by anodic dissolution can facilitate the climb of edge dislocations, which results in anodic polarization-enhanced ambient creep.[5-6] TEM observations, however, show that corrosion process during SCC enhances dislocation emission, multiplication and motion and these dislocations tend to glide over slip planes instead of undergoing climb.[1-4] So, there must exist another mechanism for the corrosion-enhanced dislocation emission and motion in addition to the divacancies enhanced-dislocation climb.

The *in situ* tensile tests in TEM show that the dislocation emission begins when the applied stress intensity $K_{Ia}$ reaches a critical value $K_{Ie}$, thereafter dislocation emission and motion cease under keeping constant load for some time. Furthermore, if increasing applied stress intensity from $K_{Ie}$ to $K_{Ie}+\Delta K_I$, dislocation emission, multiplication and motion will reappear.[7-8] If a constant deflection device is attached to TEM, the sample is stressed after the dislocation configuration ahead of a loaded crack tip has reached steady state. After exposed to a corrosion solution for some period of time, it appears that the dislocation emission, multiplication and motion are active again.[1-3] Comparing *in situ* SCC using the constant deflection device[1-3] with *in situ* tensile test in TEM,[7-8] it can be seen that corrosion-enhanced dislocation emission and motion is analogous to stress-enhanced dislocation emission and motion. It seems that there is an additive tensile stress induced during SCC even under constant displacement condition, which assists the applied stress to promote dislocation emission and motion. The additive tensile stress generated by SCC which is controlled by the anodic dissolution instead of hydrogen, should be related to the passive film or de-alloyed layer during immersion.

Many experiments show that metal foils with a protective layer formed on one side are concave or convex during anodic polarization using a potentiostat because of a tensile or compressive stress, respectively, generated at or near the passive film interface.[9] The stress is in direct proportion to the deflection and in inverse to the thickness of the passive film. Our experiments show that the brass and $\alpha$-Ti foils with a protective layer formed on one side are concave during original corrosion in an ammonia and methanol solution, respectively.[4, 10] This is due to corrosion-induced tensile stress developed at the dezincification layer interface for brass[10] and at the passive film interface for $\alpha$-Ti.[4] The variation of susceptibility to SCC of brass in the ammonia solution with potential is consistent with the variation of corrosion-induced stress.[11] At open-circuit potential, both corrosion-induced tensile stress($\sigma_p$) and susceptibility to SCC($I_\sigma$) had a maximum value. Both tensile stress $\sigma_p$ and $I_\sigma$ decrease slightly under anodic polarization, but reduce steeply with the decrease in potential under cathodic potential. At the cathodic potential of $-500\text{mV}_{SCE}$, corrosion-induced stress becomes compressive because of copper-platting layer, correspondingly, susceptibility to SCC is zero.[11] The film-induced stress of $\alpha$-Ti in the methanol solution containing 0.6mol/L KCl decreases with increase in water content and is equal to zero at 10% H$_2$O, and the susceptibility to SCC decreases correspondingly from 98% to 0.[4] Therefore, a large tensile stress induced by the passive film or a dealloyed layer is a necessary condition for SCC.

The deflection during corrosion of $\alpha$-Ti in the methanol solution can be measured exactly, but the thickness of the passive film does not. Therefore the film-induced stress calculated is too large to believe. In this paper, the film-induced stress will be measure using a direct method. The passive film-induced stress and susceptibility to SCC for $\alpha$-Ti in the methanol solution under various applied potentials will be measured simultaneously, and then the relationship between the corrosion-induced stress and SCC susceptibility is looked for.

## Experimental Procedure

The material used was ASTM Gr.2 Ti of 99.2% purity, with yield strength of 362 MPa. The specimens included flat strips with dimensions of $0.2\times4\times20mm^3$ for the stress measurements and single edge notched specimens with dimensions of $0.2\times10\times95mm^3$, containing a notch of 5mm depth and root radius of 0.1mm for slow strain rate tests (SSRT). All specimens were polished down to 1000 grit finish, and then annealed in vacuum at 650°C for 1.5h.

The susceptibility to SCC was evaluated in terms of the percent strength loss, $I_\sigma$, and the relative decrease of time to fracture, $I_t$, measured through SSRT. $I_\sigma=(1-\sigma_F^*/\sigma_F)\times100\%$, $I_t=(1-t_F^*/t_F)\times100\%$, where $\sigma_F$, $t_F$ and $\sigma_F^*$, $t_F^*$ are the strength and time to fracture of the notched tensile specimens during tensioning in air and the methanol solution containing 0.6 mol/L KCl, respectively, at a displacement rate of $1.89\times10^{-5}mm/s$. The specimens were etched in a 1% $HF+2\%HNO_3$ solution for 1min in order to remove the air-formed oxide[9]. After cleaning in alcohol and acetone, the specimens in the solution were kept at various constant potentials from $-100mV_{SCE}$ to $-400mV_{SCE}$ during SSRT. The stable open-circuit potential is $-210mV_{SCE}$.

The flat tensile specimens were extended in air to a plastic strain of $\varepsilon_P\geq1\%$ with a displacement rate of $1.89\times10^{-5}mm/s$. After unloaded, the specimens were immersed in the $1\%HF+2\%HNO_3$ solution for 1min to remove the oxide and then put into the methanol solution at various constant potentials for 1h to form various passive films. After that, the specimens with the passive film were again extended in air to yield. The yield stress of a reloaded specimen without passive film must be equal to the flow stress of the specimen before unloaded. The yield stress of the specimen with passive film, $\sigma_f^*$, however, was different with the flow stress of the specimen before unloaded, $\sigma_f$. The difference between $\sigma_f$ and $\sigma_f^*$ is the film-induced stress, i.e. $\sigma_P=\sigma_f-\sigma_f^*$.

## Experimental Results

The average stress and time to fracture of the notched specimen during extending in air were $\sigma_F=397Mpa$ and $t_F=49.5h$, respectively. The stress and time to fracture of the notched specimen during SSRT in the methanol solution are listed in Table 1. The susceptibilities to SCC, $I_\sigma$ and $I_t$, are also listed in Table 1. The variation of SCC susceptibility with potential is shown in Fig.1.

Table 1. SCC susceptibility and film-induced stress at various potentials

| V, $mV_{SCE}$ | $t_F$, h | $\sigma_F$, MPa | $I_t$, % | $I_\sigma$, % | $\sigma_P=\sigma_f-\sigma_f^*$, MPa |
|---|---|---|---|---|---|
| -100 | 3.0 | 72 | 94 | 82 | 68 |
| -200 | 3.2 | 88 | 93 | 78 | 55 |
| -210* | 3.5 | 88 | 93 | 78 | 51 |
| -220 | 4.0 | 93 | 92 | 77 | -- |
| -230 | 40.0 | 221 | 19 | 44 | -- |
| -240 | 57.5 | 342 | 0 | 14 | 12 |
| -280 | -- | -- | -- | -- | -7 |
| -300 | 60.5 | 379 | 0 | 0 | -17 |
| -400 | 49.0 | 401 | 0 | 0 | -- |

* Steady open-circuit potential

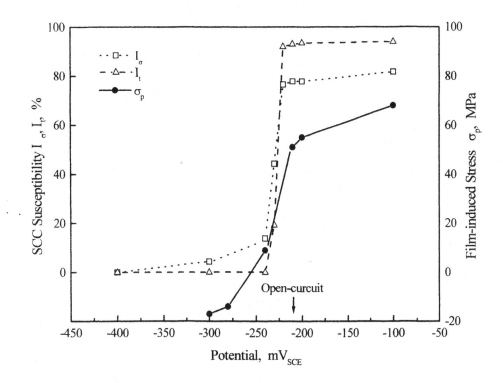

Figure1: Susceptibility to SCC and film-induced stress of α-Ti in the methanol solution vs potential.

The stress-strain curves of the specimens extended in air before unloaded and after unloaded and then forming passive films at various potentials are shown in Fig. 2. The dotted lines shown in Fig. 2 were the curves extended in air. After unloaded at the point of A, the specimen was immersed in the methanol solution containing 0.6mol/L KCl for 1h under various constant potentials in order to form a passive film. After dried, the specimens with various passive films were again extended in air, as shown by the solid lines of Fig. 2. The difference between the flow stress at A and the yield stress of the specimen with passive film is the passive film-induced stress and listed in Table 1, too.

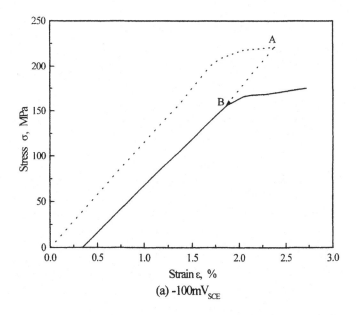

(a) -100mV$_{SCE}$

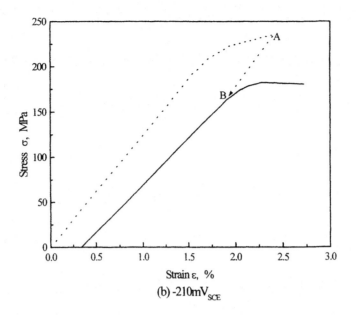

(b) -210mV$_{SCE}$

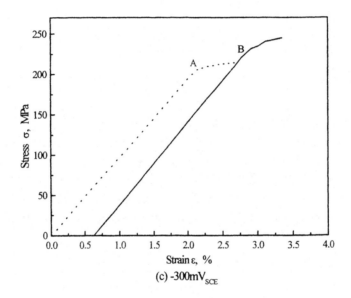

(c) -300mV$_{SCE}$

Figure 2: Stress vs strain curves for α-Ti. The dotted line is the curve
extending in air and the solid line is that extending again in air
after unloading and forming a passive film through immersing in
the methanol solution at various constant potentials for 1h.
(a) at constant potential of –100mV$_{SCE}$    (b) at –210mV $_{SCE}$
(steady open-circuit potential)    (c) at –300mV$_{SCE}$.

The variation of the film-induced stress with potential is also shown in Fig. 1. Fig. 1
indicates that at the steady open-circuit potential of –210mV$_{SCE}$ and under anodic polarization,
both film-induced tensile stress and susceptibility to SCC have a maximum value. Both the
film-induced stress and SCC susceptibility, however, decrease steeply with decrease in potential
under cathodic polarization. When the cathodic potential was less than –250mV$_{SCE}$, the film-
induced stress became compressive. Correspondingly, susceptibility to SCC was zero.
Therefore, the variation of SCC susceptibility with potential was consistent with that of the
film-induced stress. A large film-induced tensile stress is a necessary condition for SCC of α-Ti
in the methanol solution.

During original corrosion of α-Ti in the methanol solution, the potential increases with corrosion time and reaches a steady value of –210mV$_{SCE}$, as shown in Fig. 3. If the unloaded specimen was corroded for 1h, a film-induced compressive stress of –7Mpa was obtained because the potential forming passive film was less than –280mV$_{SCE}$. If corrosion time of the unloaded specimen was longer than 4h, the film-induced stress became tensile because the potential was higher than –240mV$_{SCE}$. XPS of the passive films formed at constant potential of –280mV$_{SCE}$ and –210mV$_{SCE}$ are shown by curve (a) and (b) in Fig. 4, respectively. Fig. 4 indicates that the compositions of the passive films formed at different potential are different.

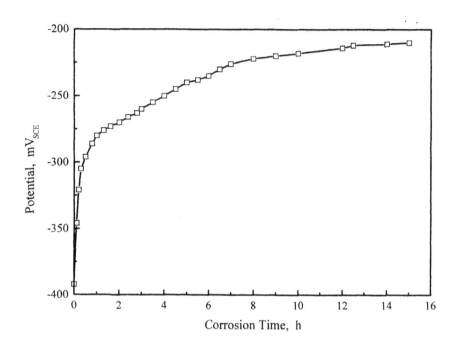

Figure 3: Potential vs corrosion time of α-Ti in the methanol solution.

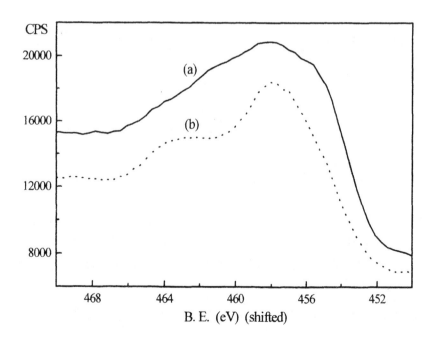

Figure 4: XPS of passive films formed at constant potential of
–280mV$_{SCE}$ (a) and –210mV$_{SCE}$ (b)

247

# Discussion

The passive film formation involves either the inward diffusion of anions or outward immigration of cations.[7] If only the anions move, the anion transport number $\beta=1$, new passive film is generated at the metal/passive film interface, i.e., the passive film grows inward. Since the volume of the passive film is greater than the volume made available by the ionization, the lattice of the surface layer expands. The metal matrix of the sample hinders the surface layer consisting of the passive film to elongate, resulting in a compressive stress. If only cations move outward, the anion transport number $\beta=0$, new passive film is generated only at the passive film-solution interface, and there are many free spaces (vacancies) left at the metal side of the inner interface produced by the oxidation of $\alpha$-Ti. The other part of the sample hinders the metal layer containing vacancies to contract, resulting in a tensile stress at the inner interface. Actually, $0<\beta<1$. When $\beta=\beta_c$, no additive stress is generated which is corresponding to the case of $\alpha$-Ti in the methanol solution at potential of $-250m$ $V_{SCE}$. When the potential was higher than $-250mV_{SCE}$, $\beta<\beta_c$, a tensile stress was generated, and a compressive stress corresponded to $V<-240mV_{SCE}$.

During SCC, the tensile stress generated by forming the passive film during corrosion may assist the applied stress to facilitate SCC. For a pre-cracked sample, the additive tensile stress generates an additive stress intensity factor $K_{IP}$. During tensile tests in vacuum (or air), when an applied stress intensity factor $K_{Ia}$ reaches a critical value for dislocation emission, $K_{Ie}$, dislocations emit from a loaded crack tip. During SCC, however, there exists $K_{IP}$, so that when $K_{Ia}(SCC)=K_{Ie}-K_{IP}$, dislocations emit. That is to say, during SCC, when the applied stress intensity $K_{Ia}$ is lower than $K_{Ie}$, which is the critical stress intensity for dislocation emission in air, dislocations begin to emit. Therefore, the corrosion process itself can generate an additive tensile stress through forming passive film and then facilitate dislocation emission and motion, which has been proved by a series of observations in TEM.[1-4]

# Summary

Passive film formed during corrosion of $\alpha$-Ti in the methanol solution at various potentials can generate an additive stress. The film-induced stress was tensile if the constant potential was higher than $-250mV_{SCE}$, and a film-induced compressive stress corresponded to $V<-240mV_{SCE}$. The variation of the film-induced stress with potential was consistent with that of susceptibility to SCC in the same solution. Therefore, a large film-induced tensile stress is a necessary condition for SCC of $\alpha$-Ti in the methanol solution.

# Acknowledgments

This project was supported by the Special Funds for the Major State Basic Research Projects, G19990650, and the NNSF of China (No. 50071010)

# References

1. B. Gu et al., "The In-situ TEM Observation of Corrosion Facilitating Dislocation Emission, Multiplication and Motion for Brass," Sripta Metall, 32 (4) (1995), 637-640.

2. K.W. Gao et al., "In-situ TEM Observation of Dissolution-enhanced Dislocation Emission, Motion and The Nucleation of SCC for Ti-24Al-11Nb Alloy in Methanol," Scripta Metall, 36 (2) (1997), 259-264.

3. K.W. Gao et al., "In-Situ Transmission Electron Microscopic Observation of Corrosion-Enhanced Dislocation Emission and Crack Initiation of Stress Corrosion", Corrosion, 56 (5) (2000), 515-522.

4. H. Lu et al., "Stress Corrosion Cracking Caused by Passive Film-Induced Tensile Stress," Corrosion, 56 (11) (2000), 1112-1118.

5. H.H. Uhlig, "Effect of Surface Dissolution on Plastic Deformation of Iron and Steel," J. Electrochem. Soc., 123 (11) (1976), 1699-1701.

6. B. Gu et al., "The Effect of Anodic Polarization on The Ambient Creep of Brass," Corrosion Sci., 36 (8) (1994), 1437-1445.

7. Q.Z. Chen et al., "In Situ TEM Observation of Nucleation and Bluntness of Nanocracks in Thin Crystals of 310 Stainless Steel," Acta Metall. Mater, 43 (12) (1995), 4371-4376.

8. Q.Z. Chen et al., "Nucleation, Blunting or Propagation of a Nanocrack in Dislocation-Free Zone of Thin Crystals," Fatig Frac Eng Mater Struc, 21(1998), 1415-1423.

9. J.C. Nelson, and R.A. Oriani, "Stress Generation During Anodic Oxidation of Titanium and Aluminum," Corrosion Sci., 34 (2) (1993), 307-326.

10. Lu H, Gao K W, and Chu W Y, "Investigation of Tensile Stress Induced by Dezincification Layer During Corrosion for Brass," Corrosion Sci., 40 (10) (1998), 1663-1670.

11. X.Z. Guo et al., "Stress Corrosion Cracking Relation with Dezincification Layer-Induced Stress," Metall Mater Trans A, in press.

## Figure Caption

Figure 1: Susceptibility to SCC and film-induced stress of $\alpha$-Ti in the methanol solution vs potential

Figure 2: Stress vs strain curves for $\alpha$-Ti. The dotted line is the curve extending in air and the solid line is that extending again in air after unloading and forming a passive film through immersing in the methanol solution at various constant potentials for 1h.
(a) at constant potential of $-100mV_{SCE}$ (b) at $-210mV_{SCE}$ (steady open-circuit potential (c) at $-300mV_{SCE}$.

Figure 3: Potential vs corrosion time of $\alpha$-Ti in the methanol solution.

Figure 4: XPS of passive film formed at constant potential of $-280mV_{SCE}$ (a) and at $-210 mV_{SCE}$ (b)

# LIGHTWEIGHT ALLOYS FOR AEROSPACE APPLICATION

*Edited by:*
*Dr. Kumar Jata, Dr. Eui Whee Lee,*
*Dr. William Frazier and Dr. Nack J. Kim*

## TITANIUM ALLOYS

## Investment Casting of Titanium Alloys with CaO Crucible and CaZrO$_3$ Mold

*S.K. Kim, T.K. Kim, M.G. Kim, T.W. Hong and Y.J. Kim*

Pgs. 251-260

184 Thorn Hill Road
Warrendale, PA 15086-7514
(724) 776-9000

# INVESTMENT CASTING OF TITANIUM ALLOYS WITH CaO CRUCIBLE AND CaZrO$_3$ MOLD

S.K. Kim, T.K. Kim, M.G. Kim, T.W. Hong and Y.J. Kim

School of Metall. & Mater. Eng., Sungkyunkwan Univ.,
300 Chunchun-dong, Jangan-gu,
Suwon, Gyungki-do 440-746, Korea

## Abstract

A cost-saving process has been developed to produce cast parts of titanium alloys using CaO melting crucibles and CaZrO$_3$ investment casting molds. An assessment was made of the suitability of CaO crucible and the thermal stability of Al$_2$O$_3$, ZrO$_2$, CaO, and CaZrO$_3$ molds against molten titanium alloys. Induction melting of sponge and scrap titanium was possible in the CaO crucible under argon atmosphere without excessive overheating and holding time of the melt. The CaZrO$_3$ mold could be used in a steam autoclaves and was easy to remove from castings, and cast parts had very limited reaction layer with enough surface finish quality. The thermal stability of the molds evaluated has also been established for pure titanium, Ti-6Al-4V alloy, and $\gamma$-TiAl.

## Introduction

Titanium alloys have long been attractive for advanced aerospace structural applications due to their many attractive features. However, the use of these alloys has been limited to forging materials, in which the limitations of components in complexity and size exist. The need to increase component complexity and size as well as reduce manufacturing costs and lead times has resulted in new opportunities and added expectations of investment casting.

Lightweight Alloys for Aerospace Applications
Edited by Kumar Jata, Eui Whee Lee,
William Frazier and Nack J. Kim
TMS (The Minerals, Metals & Materials Society), 2001

Almost all titanium investment castings have been produced in vacuum-arc and induction-skull remelting using rammed graphite molds with a minimal potential for contamination. However, these processes have very low energy efficiency and it is difficult to obtain sufficient superheat in the molten metal to be able to fill a mold satisfactory. The various aspects of titanium melting processes are summarized in Table I.[1,2] Furthermore, the relatively high thermal conductivity of the graphite mold promotes premature freezing and the mold ramming is a labor-intensive process, which cannot be easily automated.[3]

If a refractory material were capable of melting and casting the highly reactive titanium alloys without excessive contamination, thereby allowing induction melting and conventional molding to be used, a significant reduction in the cost and an improvement in the quality of castings could be achieved. CaO possesses a greater thermodynamic stability than most of the more common refractory oxides as shown by free energy data in Fig. 1,[4] which is calculated by using the thermochemical data of Ref. [5]. CaO is also much less expensive than $ZrO_2$ and $Y_2O_3$. Therefore, CaO based crucible and mold materials can show promise for melting and investment casting of titanium alloys.

This paper describes the work conducted for the purpose of developing a cost–saving process for induction melting and investment casting of titanium alloys using CaO melting crucibles and $CaZrO_3$ investment casting molds, including the relative thermal stability of the molds evaluated for pure titanium, Ti-6Al-4V alloy, and $\gamma$-TiAl.

Table I  Comparison of various titanium melting processes

| Melting Method | | | | Pressure, Pa | Power Effectivity | Magnetic Effect | Composition Change | Charges |
|---|---|---|---|---|---|---|---|---|
| Cold Crucible Melting | VAR | Non-Consumable Electrode | Graphite Tungsten | 0.1~$10^5$ | Low (DC) High Current, Low Voltage | Middle | Low | Uneasy |
| | | | Water-Cooled | | | | | |
| | | Consumable Electrode (Skull Melting) | | | | | | |
| | Electron Beam Melting | | | $10^{-4}$~10 | Low (DC) LC, HV | Low | High (Al, Cr) | Easy |
| | Plasma Arc Melting | | | > $10^3$ | Low (DC) LC, LV | Middle | Low | Easy |
| | Induction Melting | Skull | | < $10^5$ | Low (AC) | High | Low | Uneasy |
| | | Magnetic Suspension | | | Middle (AC) | High | | |
| Refractory Crucible Melting | Graphite Induction Melting | | | < $10^5$ | Middle (AC) | Low | Low | Easy |
| | Lime Induction Melting | | | ~ | High (AC) | High | | |

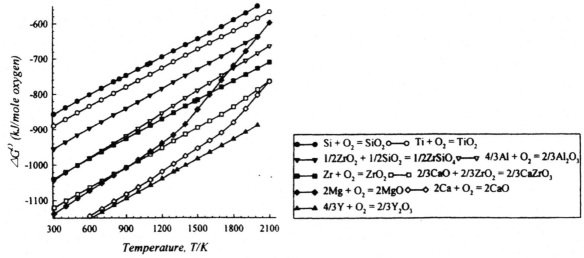

Figure 1: Standard free energy for the formation of oxides calculated using the data in Ref. [5].

## Experimental procedure

<u>Investment casting mold preparation</u>

The wax pattern for evaluating the thermal stability of the molds evaluated pure titanium, Ti-6Al-4V alloy, and $\gamma$-TiAl was made by pouring wax into a simple cylindrical metal mold ($\varphi$ 9 mm $\times$ 30 mm) with an integral pouring basin. The wax pattern was then inspected and dressed to eliminate any imperfection resulting pouring. After that, the pattern was dip coated with a slurry containing desired oxide powders (about 45 $\mu$m.) and binders. It was then stoccoed by using under 0.1 mm oxide powders after draining the excess slurry, and drying, dipping, and stuccoing procedures were repeated by dipping with the same slurry and stuccoing with gradually large oxide powder from 0.1 mm to 0.5 mm. The mold was dried at controlled temperature (300 $\pm$ 2K) and relative humidity (50 $\pm$ 10%) for least 4 h. Finally, the mold thickness of ~5 mm was built. This was followed by dewaxing in an autoclave at about 440K at a pressure of about 0.7 MPa and then by firing at about 1273K. With some difficulties to handle a CaO slurry, a CaO mold was prepared by ramming CaO powders into the annular space between the rod ($\varphi$ 9 mm) and the cylinder (20 mm in inner diameter). The oxides and binders used are given in Table II.

Table II  Types of oxides and binders examined and their mixing ratio

| Oxide | Binder | Mixing ratio |
|---|---|---|
| $Al_2O_3$ | Colloidal silica | first dipping 3 : 1 |
| $ZrO_2$ | $ZrOCl_2 \cdot 8H_2O$ : Ethanol = 20g : 100cc | |
| $CaZrO_3$ | $ZrCl_2O \cdot 8H_2O$ 1mol + $(CH_3COO)_2CaH_2O$ 1mol | after second dipping 2.5 : 1 |
| CaO | Ethanol 100g + $CaCl_2$ 4g | |

## Metal-mold reaction

The materials used for the metal-mold reaction analysis were sponge pure titanium, vacuum-arc remelted Ti-6Al-4V, and plasma-arc melted $\gamma$-TiAl. To prevent contamination from refractory crucibles, the metal-mold reaction was evaluated by non-consumable tungsten electrode plasma-arc melting and drop casting procedure. The photograph of the apparatus for the metal-mold reaction analysis is shown in Fig. 2. The charge was placed on the drop cast hearth. The vacuum furnace was evacuated to 0.133 Pa ($10^{-3}$ torr) and refilled with argon to 1.5 $\times$ $10^5$ Pa (1150 torr). This cycle of evacuation and flushing with argon was repeated at least three times. Melting was initiated by striking an arc against the charge, and when the charge was fully molten condition, the arc was directed at the center of the hearth and the current was then raised to allow the molten metal to flow into the mold positioned below the hearth, as shown in Fig. 2. After sufficient cooling, the as-cast rods were taken out of the molds, sectioned, and polished by using a Keller solution. The relative thermal stability of the molds was evaluated with regard to microstructural characterization of regions below the surface of the rods and microhardness profile. The microstructural characterization was evaluated by an optical microscope and the microhardness profile was established by measuring a microhardness at 0.2 ㎜ intervals from the surface on the vertical cross section of the rods.

Figure 2: Photograph of a plasma-arc melting furnace for evaluating thermal stability.

Figure 3: Photograph of a vacuum induction melting furnace for investment casting.

## CaO crucible preparation and investment casting procedure

CaO crucibles were prepared by a laboratory ramming technique. 4 g of calcium chloride was dissolved in 100 g of ethanol and 600g of dry calcium oxide powders and 30g of calcium chloride-ethanol solution were kneaded for 20 min. The mixture was then hand-rammed and the resultant crucible was fired in an electric resistance furnace at 1223K for 2 h.

Investment casting was carried out using 30 kW and 35 kHz vacuum induction furnace (Fig. 3). A CaO crucible charged with approximately 800 g of sponge pure titanium, vacuum-arc remelted Ti-6Al-4V, or plasma-arc melted $\gamma$-TiAl and a $CaZrO_3$ mold were placed in the furnace. The furnace was closed and evacuated to a vacuum of 0.133 Pa and flushed with argon three times, and then backfilled with argon at a pressure of $1.5 \times 10^5$ Pa. The furnace was switched on and the temperature was gradually increased until the charge was melted. The power was turned off 60 s after complete melting as judged by the visual inspection and the melt was poured into the static or centrifuged mold by tilting the crucible.

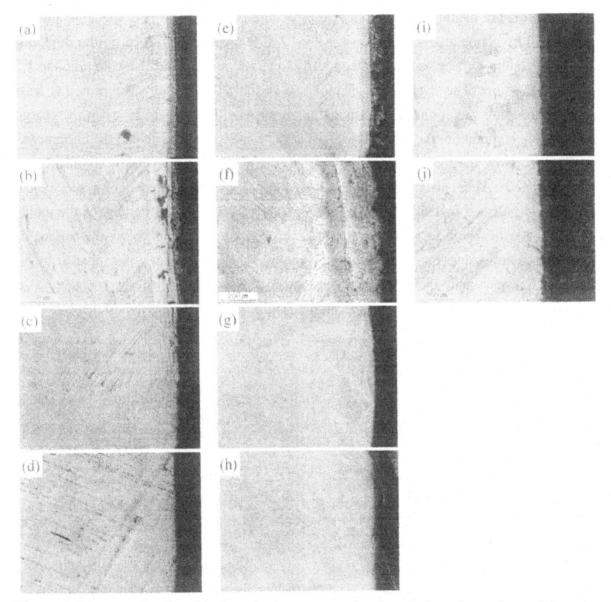

Figure 4: Photographs showing the microstructures of regions below the surface of the as-cast (a) cpTi in $Al_2O_3$ mold, (b) cpTi in $ZrO_2$ mold, (c) cpTi in CaO mold, (d) cpTi in $CaZrO_3$ mold, (e) Ti-6Al-4V in $Al_2O_3$ mold, (f) Ti-6Al-4V in $ZrO_2$ mold, (g) Ti-6Al-4V in CaO mold, (h) Ti-6Al-4V in $CaZrO_3$ mold, (i) $\gamma$-TiAl in $Al_2O_3$ mold, and (j) $\gamma$-TiAl in $ZrO_2$ mold.

## Results and discussion

Thermal stability of mold materials

The basic requirements of mold materials can be summarized as formability, strength, permeability, and collapse-ability. However, mold materials for titanium investment casting should have a greater thermodynamic stability because of the high melting point and extreme chemical reactivity of titanium and its alloys in the molten state. The development of fundamental data regarding the reactivity of these alloys with a wide variety of mold materials has been investigated.[6] Rammed graphite was the earliest commercial mold-making technique for producing titanium castings and is still used today. However, it is sometimes difficult to control the precise shape of the graphite mold during the drying and firing stages, which limits the dimensional accuracy of the final product.[3] The relatively high thermal conductivity of the graphite mold also promotes premature freezing and the mold ramming is a labor-intensive process, which cannot be easily automated. $Y_2O_3$ or CaO stabilized $ZrO_2$ and $Y_2O_3$ based molds have been received considerable attention and are generally used. However, these materials are expensive and $Y_2O_3$ is fairly expensive; therefore it would be desirable to use them as an additive or only in selected mold sections.

Although the use of CaO as a refractory has been tempered by the fact that it is the prime slag former at low temperature with other oxides, it has high hydration susceptibility, and it is easy to be deoxidized, the work has been concentrated on the study of CaO based mold material for the purpose of developing a cost–saving process for induction melting and investment casting of titanium alloys. This is because (1) the problems mentioned above can be reduced or solved if the ramming and firing are carefully and shortly processed and the melting and casting is carried out in a vacuum atmosphere, (2) CaO is cheaper than $ZrO_2$ and $Y_2O_3$, (3) its melting point is high (2888K), and (4) CaO possesses a greater thermodynamic stability than most of the more common refractory oxides as shown by free energy data in Fig. 1. The thermal stability of CaO against molten titanium alloys has been also confirmed by Sato who have studied the suitability of CaO as crucibles for titanium and reactive alloy castings.[7-10] Figure 4 shows the as-cast microstructures of the regions below the surface of the castings with different mold materials for pure titanium (Fig. 4a, b, c, d), Ti-6Al-4V alloy (Fig. 4e, f, g, h), and $\gamma$-TiAl (Fig. 4i, j). The pure titanium and Ti-6Al-4V alloy castings in $Al_2O_3$ and $ZrO_2$ molds have clear reaction layers, whereas negligible reaction layers are found in the pure titanium and Ti-6Al-4V alloy castings when cast in CaO mold.

However, the work done has revealed the other problems CaO still has as a refractory. It is difficult to control the CaO slurry viscosity during slurry making and dipping and storage and pinhole is found on the surface of the castings because the ethanol based binder should be used. The optimum strength of the mold cannot also be obtained with this CaO powder and ethanol based slurry system. Therefore, a new mold material, $CaZrO_3$, has been received considerable attention because (1) the viscosity of water-based $CaZrO_3$ slurry is not changed, (2) therefore, dewaxing procedure can be done in a steam autoclave, (3) the optimum strength can be obtained, and (4) the thermal stability of $CaZrO_3$ mold is not less than that of CaO. As shown in Fig. 4, the pure titanium and Ti-6Al-4V alloy castings in $CaZrO_3$ mold have negligible reaction layers. These results are in good accordance with the microhardness profiles established by measuring a microhardness at 0.2 mm intervals from the surface on the vertical cross section of the castings, as shown in Fig. 5.

For pure titanium and Ti-6Al-4V alloy castings, the relative thermal stability of the examined oxides can be graded in the order of increasing stability in the following sequence; $Al_2O_3 \Rightarrow ZrO_2 \Rightarrow CaO \Leftrightarrow CaZrO_3$. This grading follows the free energy data for the formation of these oxides.

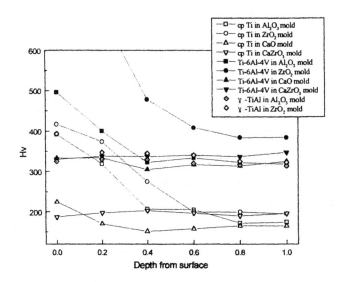

Figure 5: Effect of mold materials on microhardness profile of the pure titanium, Ti-6Al-4V, and $\gamma$-TiAl castings.

For $\gamma$-TiAl castings in $Al_2O_3$ and $ZrO_2$ molds, Fig. 4 shows negligible reaction layers. The result can be explained by the study of Misra.[11] It shows that $Al_2O_3$ would be chemically stable in TiAl-based alloys with Al concentrations higher than those corresponding to the $\gamma$-TiAl phase. However, the recent study of Campbell,[12] which is about the suitability of magnesia, calcia, alumina, and yttria for manufacturing melting crucibles and investment casting molds for $\gamma$-TiAl, shows that magnesia and silica containing alumina are unsuitable but both pure calcia and yttria-coated magnesia are acceptable for use with the $\gamma$-TiAl alloy, although the experiment condition cannot be compared with that of Misra. Notwithstanding the unclear explanation about the result and the potential reactivity of molten titanium alloys with conventional colloidal silica bonded $Al_2O_3$ and $ZrO_2$ molds, it is worth noting that the condition involved in investment casting does not lead to the extensive breakdown of the chemical constituents of these molds.

Suitability of CaO crucible

The suitability of the CaO crucible was tested by melting and casting experiments of pure titanium and T-6Al-4V alloy. Considerable difficulty has been frequently encountered in melting titanium under reduced pressure of argon, because the charge tended to form a bridge in the crucible. This is probably due to the boiling of CaO generated from titanium reduction of the CaO crucible at high temperature.[13] The oxygen contamination is also a major problem in induction melting of titanium in the CaO crucible. Therefore, the charge material is carefully placed for a smooth meltdown without bridging and the careful melting procedure is carried out. The metal charge is heated to about 1100K and held for 1000 s at this temperature for degassing. The furnace is backfilled with argon at a pressure of $1.5 \times 10^5$ Pa and temperature is gradually increased until the charge is melted.

## Conclusion

The major points that this paper attempts to show are the followings:

1. Good results are obtained by using CaO melting crucibles and $CaZrO_3$ investment casting molds for induction melting and investment casting of titanium and its alloys. These materials can be processed by the standard foundry equipment and the relative ease and inexpensiveness of the process is established.

2. Conventional vacuum induction furnaces can be readily adaptable to produce cast parts of pure titanium, Ti-6Al-4V alloy, and $\gamma$-TiAl without highly skilled techniques.

3. For pure titanium and Ti-6Al-4V alloy castings, the relative thermal stability of the examined oxides can be graded in the order of increasing stability in the following sequence; $Al_2O_3 \Rightarrow ZrO_2 \Rightarrow CaO \Leftrightarrow CaZrO_3$. This grading follows the free energy data for the formation of these oxides.

4. Although it is uncertain, it is worth noting that the condition involved in investment casting of $\gamma$-TiAl does not lead to the extensive breakdown of the chemical constituents of conventional colloidal silica bonded $Al_2O_3$ and $ZrO_2$ molds.

## References

1. H.B. Bomberger and F.H. Froes, "The Melting of Titanium," JOM, December (1984), 39-47.

2. S.K. Kim, T.W. Hong, S.H. Lee, and Y.J. Kim, "Melting and Investment Casting of Titanium Alloys," J. Kor. Foundrymen's Society, 19 (3) (1999), 210-215.

3. D. Eylon, F.H. Froes, and R.W. Gardiner, "Development in Titanium Alloy Casting Technology," JOM, February, (1983), 35-47.

4. S.K. Kim, T.W. Hong, and Y.J. Kim, "Evaluation of Thermal Stability of Mold Materials for Magnesium Investment Casting," Mater. Trans. JIM, 42 (2001), in press.

5. I. Barin, Thermochemical Data of Pure Substances (Weinheim: VCH, 1993), 49-1736.

6. R.L. Saha, T.K. Nandy, R.D.K. Misra, and K.T. Jacob, "On the Evaluation of Stability of Rare Earth Oxides as Face Coats for Investment Casting of Titanium," Met. Trans. B, 21B (1990), 559-566.

7. T. Sato, Y. Yoneda, H. Iwanabe, Y.J. Kim, and R. Sakagami, "Vacuum Melting and Casting of Some Intermetallic Compounds with Lime Refractory," Imono, 61 (11) (1989), 803-807.

8. T. Sato, N. Matsumoto, Y. Yoneda, T. Takahashi, H. Iwanabe, and R. Sakagami, "Production of Titanium Castings with Lime Refractory," Imono, 62 (9) (1990), 732-737.

9 T. Sato, N. Matsumoto, and Y. Yoneda, "Induction Melting and Casting of Chromium," Imono, 63 (4) (1991), 358-363.

10. H. Ito, S. Nakazawa, M. Hanada, T. Takahashi, and T. Sato, "Service Life of Electrically Fused Lime Crucible for Vacuum Induction Melting of Reactive Alloys," Imono, 68 (4) (1996), 343-347.

11. A.K. Misra, "Reaction of Ti and Ti-Al Alloys with Alumina," Met. Trans. A, 22A (1991), 715-721.

12. J.P.Kuang, R.A.Harding, and J.Campbell, "Investigation into Refractories as crucible and mold materials for melting and casting $\gamma$-TiAl alloys," Meter. Sci. Tech., 16 (2000), 1007-1016.

13. T. Sato, N. Matsumoto, Y. Yoneda, H. Iwanabe, and R. Sakagami, "Improvement of the Internal Soundness of Titanium Castings prepared by Induction Melting in Lime Crucibles," Imono, 64 (5) (1992), 312-317.

# LIGHTWEIGHT ALLOYS FOR AEROSPACE APPLICATION

*Edited by:*
*Dr. Kumar Jata, Dr. Eui Whee Lee,*
*Dr. William Frazier and Dr. Nack J. Kim*

## COMPOSITES

# Compressive Behavior of Ti-6Al-4V/TiC Layered Composites: Experiments and Modeling

*A.J. Wagoner Johnson, C.L. Briant, C.W. Bull and K.S. Kumar*

Pgs. 261-271

184 Thorn Hill Road
Warrendale, PA 15086-7514
(724) 776-9000

# Compressive Behavior of Ti-6Al-4V/TiC Layered Composites: Experiments and Modeling

A.J. Wagoner Johnson, C.L. Briant, C.W. Bull, and K.S. Kumar

Brown University
Division of Engineering, Box D
Providence, RI 02912

## Abstract

The Ti-6Al-4V/TiC composite system was studied in compression at strain rates of $0.1s^{-1}$ and $1000s^{-1}$ at room temperature. Symmetric three-layered structures were successfully fabricated with equal layer thickness by diffusion bonding individual layers. Each layer consisted of either monolithic Ti-6Al-4V or a particulate reinforced composite of Ti-6Al-4V and 10 volume %TiC. Layer-interfaces were macroscopically flat and free of voids; the matrix was continuous through the structure. The strength of the layered structures was significantly improved over the homogeneous Ti64. In addition, the structures were more damage resistant than the homogeneous particulate composite. Cracks were blunted and deflected within the composite layer, approximately parallel to the hard/soft material interface, thereby arresting the crack before catastrophic failure. A simple finite element model accurately represented the engineering stress-strain behavior as well as other macroscopic features observed in the deformed specimens. The model also served as a tool to understand the influence of friction on the deformation behavior of the layered structures.

Lightweight Alloys for Aerospace Applications
Edited by Kumar Jata, Eui Whee Lee,
William Frazier and Nack J. Kim
TMS (The Minerals, Metals & Materials Society), 2001

# Introduction

Particulate composite materials are frequently considered for high performance applications in the defense and aerospace industries to enhance material properties such as specific strength and modulus [1,2]. Ti-6Al-4V/TiC, which will be referred to as Ti64/TiC, is an example of such a composite. Ti64 has a high specific strength and is therefore a reasonable candidate for the matrix material. The carbide reinforcement has almost four times the stiffness of the matrix and a comparable density. Other reinforcements react with the matrix forming an interface product that degrades mechanical properties [3,4]. Our own research plus that of others [3,4] show no evidence that any reaction product degrades the mechanical properties of the Ti64/TiC composite system. However, other studies have shown that the TiC particles become C deficient during high temperature processing [5,6]. The Ti64/TiC interaction will not be discussed in this paper.

Studies in our laboratory have shown that this composite is prone to fracture in compression through shear cracks that form at quasi-static strain rates, and through failure along adiabatic shear bands at high strain rates. At both strain rates, these failure mechanisms are oriented at 45 degrees to the loading axis.

Structures that combine alternating layers of hard and soft materials have the potential to improve specific strength and damage resistance. For this reason we have fabricated layered structures of the Ti64/TiC particulate composite with un-reinforced Ti64 and studied its deformation in compression. Our results show that this layered material has a higher fracture resistance than the Ti64/TiC composite material and that cracks that form in the particulate composite layers are deflected at the interface between the layers. We have also found that a simple finite element model can be used to describe the engineering stress – engineering strain behavior and shape changes in the sample that occur during compression tests.

# Experimental Procedure

Symmetric, three-layered structures of monolithic Ti64 and a composite of Ti64 with 10 vol. % of particulate TiC were fabricated by diffusion bonding the individual layers. The starting materials were pre-alloyed Ti-6Al-4V powder (-325 mesh) and particulate TiC (1-5μm). These constituents were blended in the appropriate amounts, hot pressed at $1000^{\circ}C$, and isothermally forged in vacuum ($5 \times 10^{-5}$ Torr maximum) at $875^{\circ}C$. The individual compacts were sectioned perpendicular to the forging axis to make layers 2mm +/- 0.09mm thick. These were then diffusion bonded in a graphite die at $925^{\circ}C$ and 40MPa for four hours to make the layered structures. Two structures were studied. The 10/0/10 structure consisted of a layer of Ti64 sandwiched between two layers of the 10%TiC composite. The other structure, referred to as the 0/10/0 structure, was the inverse.

Compression tests were conducted at strain rates of $0.1s^{-1}$ and $1000s^{-1}$ using a servo-hydraulic test machine and a Split-Hopkinson Bar [7], respectively. Tool steel platens were used and specimen ends were lubricated before each test. Sample dimensions were approximately 6.35mm in diameter by 5mm in height. Specimens were deformed up to engineering strains of 25% in the slow strain rate tests and 13% in the high strain rate tests. Deformed samples were sectioned and examined by optical metallography and scanning electron microscopy.

# Experimental Results

Figure 1 is a representative micrograph showing the interface between the monolithic Ti64 and the Ti64/10TiC composite. The interface was free of voids and there was evidence of grain growth in the Ti64 matrix across the original interface. Thus the compacts can be described as having a continuous phase of Ti64 throughout with regions containing particulate TiC.

The engineering stress-strain results are presented in Figure 2a and 2b for the two different strain rates used in this study. These figures show the data for the layered structures as well as results for the monolithic Ti64 and the Ti64/10TiC composite. The yield stress and flow stress for the layered structures were higher than those for monolithic Ti64 but lower than those for the Ti64/10TiC composite. All four materials showed approximately the same hardening behavior at $0.1s^{-1}$. The oscillations in data from the high strain rate tests made comparison of the work hardening rate in these samples more difficult.

Previous work in our laboratory showed that the Ti64/10TiC material failed at approximately 18-19% plastic strain when tested at a strain rate of $0.1s^{-1}$. Both layered structures were deformed to 25% plastic strain at this strain rate and did not fail. The layered structures tested at $1000s^{-1}$ were deformed to 13% plastic strain for the 10/0/10 structure and 14% for the 0/10/0 structure without failure. In contrast, the Ti64/10TiC composite failed along adiabatic shear bands at 10-12% plastic strain when tested at this rate. Due to limitations of the experimental set-up, neither the monolithic Ti64 nor the layered materials could be strained to failure at the high strain rate. However, these results demonstrate that the layered structures could be deformed to higher strains than the Ti64/10TiC composite.

Figure 3a shows low magnification images of the 10/0/10 structure deformed at $0.1s^{-1}$ to 25% plastic strain. Cracks had formed in opposite corners at 45 degrees to the loading axis. The cracks are visible at a higher magnification in Figures 3b and 3c. The two large cracks abruptly changed direction just before reaching the Ti64-Ti64/10TiC interface. The crack-path continued in the Ti64/10TiC layer, running roughly parallel to the interface, before stopping. Figure 4 shows micrographs of the 0/10/0 structure also deformed to 25% strain at $0.1s^{-1}$. No cracks are visible at low magnification. The specimen had several crack-like voids in the corner of the monolithic layer closest to the platens, oriented at 45 degrees to the loading axis. No cracks were found in the middle layers of either structure.

The initially flat interfaces in each structure changed shape during deformation. Figures 3a and 4a show macroscopic images of both structures. The 10/0/0 structure (Figure 3a) adopted a shallow, U-shaped interface and the 0/0/0 structure (Figure 4a) formed a buckled or wavy interface. The interface shape for each deformed specimen appeared to be independent of both strain and strain rate. There was no evidence of shear deformation along the interfaces.

# Model

A simple finite element model was used to predict the compressive deformation behavior of the layered structures deformed at $0.1s^{-1}$ to 25% engineering plastic strain [8]. The goal was to reproduce the engineering stress-strain behavior and to understand the role of friction in deformation and failure. The experimental coefficient of friction was unknown; therefore the two structures were modeled using two different boundary conditions on the specimen ends. The top and bottom surfaces were either completely unconstrained or fully constrained from undergoing radial expansion. Boundary conditions at the interface required each node at the interface in one layer to displace with its initial counterpart in the adjacent layer. Each layer was given the plastic properties from the corresponding experimental data.

Several assumptions were made in order to simplify the model. The particle-matrix interaction was not considered; the properties throughout each layer were assumed to be homogeneous. Consequently, material damage was not accounted for within the layers or at the layer interfaces. No failure criterion was implemented. In addition, experimental engineering stress-strain curves from the homogeneous materials were assumed to be perfectly linear and were extrapolated beyond the actual measured values. The hardening behavior of Ti64 tested at $0.1s^{-1}$ was nearly linear throughout the test while that of the 10%TiC composite remained linear through most of the experiment, although it failed at a strain of 18-19%. Thus this assumption should not greatly affect the calculated results.

Results from the experiment and the simulations are plotted together in Figure 5. The model adequately represents the engineering stress-strain behavior of both layered structures. The model also captured the general shape change of the sample. Consider the result for the 10/0/10 sample. Figures 6a and b show the shape of the sample predicted by the model after deformation to 25% plastic strain with constrained and unconstrained boundary conditions, respectively. Note that in both calculations the softer Ti64 layer is extruded out during the deformation. Superimposed on the figures is the shape of the sample experimentally tested to this strain. For both calculations there is good agreement between the calculated and measured profiles of this extruded layer. In the experiment, the material directly under the platens extended more than the amount predicted by the fully constrained calculation but less than that predicted by the unconstrained calculation. Both calculations produced the U-shape in the interface that was observed experimentally. The effect is more pronounced in the sample that was fully constrained and the experimental observation lies between the two calculated predictions.

Also shown in the figure are the shear stress contours for the two different calculations. In the sample that was fully constrained, the highest shear stresses occurred at the corner of the sample and extended at approximately 45 degrees in toward the center of the sample. The shear stresses at the interface are significantly lower. In the unconstrained calculation the shear stresses at the interface are similar to those observed in the constrained calculations, but the shear stresses at the corner of the sample are absent. Since we observed cracks emanating from the corners of the tested sample, we conclude that these were caused by the friction-induced shear stresses.

Figures 6c and d show similar results for the 0/10/0 sample. In this case the stiffer Ti64/10TiC middle layer was not extruded as much as the softer layers on either side. Both calculations captured this shape change, as shown by the superimposed experimental sample profile. Again, friction caused shear stresses to develop at 45 degrees to the loading axis. The contour plots for both of the constrained structures (Figure 6a and c) show that for the same applied strain, the shear stress in the 10/0/10 structure is greater than in the 0/10/0 structure at a given material point. Recall that the model does not account for the presence of individual particles and so this difference is related only to the relative stiffness of the Ti64/10TiC and Ti64 materials. Particle cracking and particle/matrix debonding would also contribute to the crack initiation in the 10/0/10 structure.

Thus we conclude that this model was able to reproduce the engineering stress – engineering strain curves for the layered structures and to also predict the shape changes that occurred. A more detailed model would be required to predict the microscopic damage observed.

# Summary and Conclusions

1. Two symmetric, three-layered structures were successfully fabricated with macroscopically flat interfaces and a continuous matrix.
2. The layered structures had improved properties over their homogeneous constituents. Both layered structures had higher strength than the monolithic Ti64, and more damage resistance than the particulate Ti64/10TiC composite.
3. The layer-interface remained intact during deformation at both strain rates and large cracks were deflected away from the interface in the 10/0/10 structure deformed $0.1s^{-1}$. Neither large cracks nor adiabatic shear bands were observed in samples deformed at $1000s^{-1}$.
4. A simple finite element model accounted for the engineering stress-strain behavior.
5. The model showed that constraining the specimen ends determines the macroscopic shape change in the specimen and at the layer-interfaces and contributed to the crack formation observed in the sample.

# Acknowledgments

The research was supported primarily by the MRSEC Program of the National Science Foundation under Award Number DMR-9632524. AWJ gratefully acknowledges the support of the Army Research Office through an AASERT award, DAAG55-98-1-0453 and wishes to thank L.B. Freund for several helpful discussions throughout this investigation.

# References

1. S. Ranganath, "A Review of Particulate-Reinforced Titanium Composites," J. Mater. Sci., 32 (1997), 1-16.
2. E.S.C. Chin,"Army Focused Research Team on Functionally Graded Armor Composites," Mat. Sci. Eng., A259, (1999), 155-161.
3. M.H. Loretto and D.G. Konitzer, "The Effect of Matrix Reinforcement Reaction on Fracture in Ti-6Al-4V-Base Composites," Met. Trans. A., 21A (1990), 1579-1587.
4. J.H. Zhu, P.K. Liaw, J.M. Corum, M. Ruggles, and H.E. McCoy, Jr., "Mechanical Behavior and Damage Mechanisms of Titanium Alloy and Composite," Micromechanics of Advanced Materials, eds. S.N.G. Chu et al., (Minerals, Metals and Materials Society/AIME, USA, 1995), 459-468.
5. D.G. Konitzer and M.H. Loretto, "Microstructural Assessment of Ti-6Al-4V-TiC Metal-Matrix Composite," Acta. Metall., 37 (1989), 397-406.
6. P. Wanjara, R.A.L. Drew, J. Root, and S. Yue, "Evidence for Stable Stoichiometric $Ti_2C$ at the Interface in TiC Particulate Reinforced Ti Alloy Composites," Acta. Mater., 48 (1994), 1443-1450.
7. P.S. Follansbee, Metals Handbook, vol. 8, (Metals Park, OH: American Society for Metals, 1985), 198.
8. ABAQUS 5.8 Online Documentation, Hibbitt, Karlsson, and Sorensen, Inc., Pawtucket, RI.

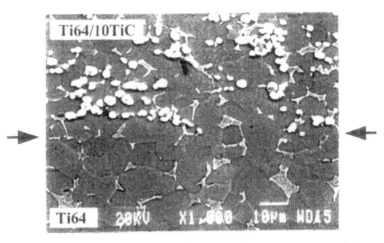

Figure 1. Scanning electron micrograph of a Ti64-Ti64/10TiC interface in the 10/0/10 layered composite. Arrows indicate the location of the layer-interface. Note the matrix grains spanning the interface.

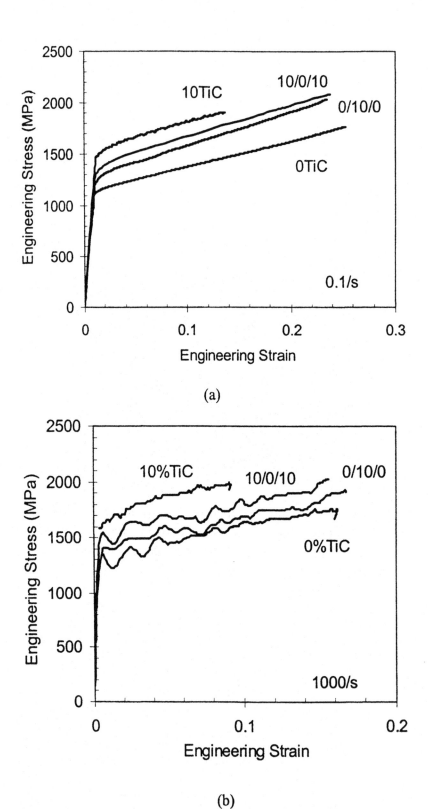

(a)

(b)

Figure 2. Engineering stress-strain data for layered and Ti64/10TiC composites and monolithic Ti64 tested (a) at 0.1s$^{-1}$ and (b) 1000s$^{-1}$.

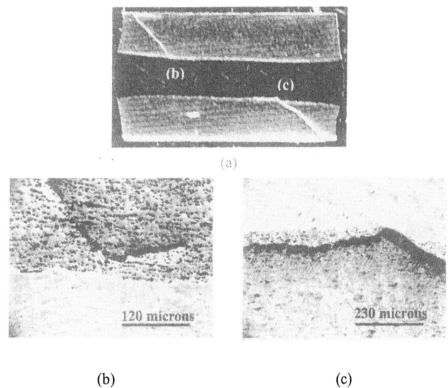

(b)                     (c)

Figure 3. (a) Optical macrograph of the 10/0/10 sample deformed to 25% plastic strain at $0.1s^{-1}$ showing the U-shaped interface and cracks in specimen corners. (b and c) Micrographs of the cracks deflecting near the layer interface.

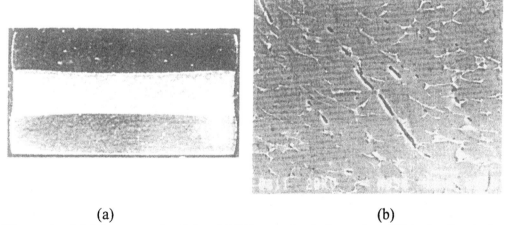

(a)                     (b)

Figure 4. (a) Macrograph of the 0/10/0 sample deformed to 25% plastic strain at $0.1s^{-1}$ showing the buckled interface. (b) Crack-like voids oriented at 45 degrees to the loading axis observed in the specimen corners closest to the platens in the Ti64 layer.

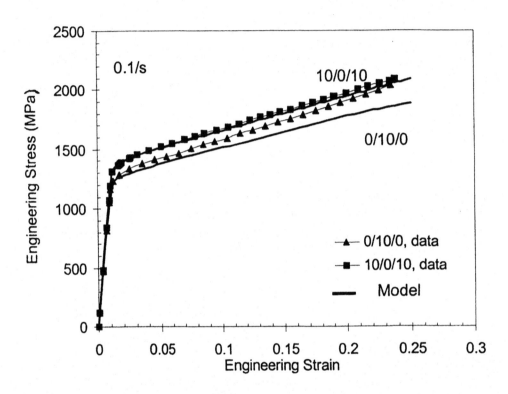

Figure 5. Engineering stress-strain curves from simulations compared to data from experiments for samples deformed to 25% plastic strain at $0.1s^{-1}$.

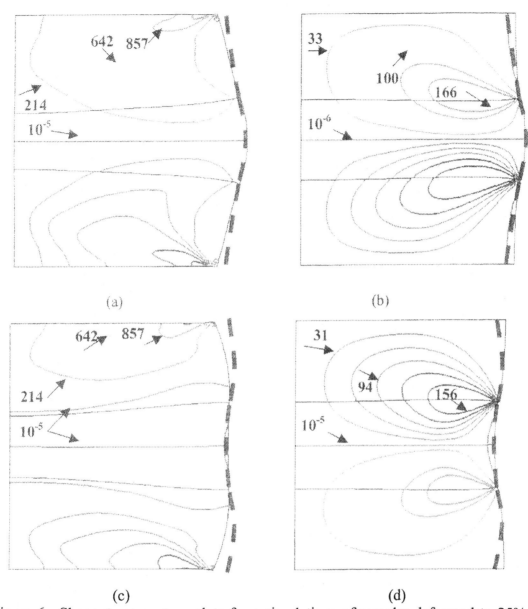

Figure 6. Shear stress contour plots from simulations of samples deformed to 25% plastic strain at $0.1s^{-1}$ for the constrained and unconstrained boundary conditions. (a) The 10/0/10-constrained (b), 10/0/10-unconstrained (c), 0/10/0-constrained and (d) 0/10/0-unconstrained simulations.

# LIGHTWEIGHT ALLOYS FOR AEROSPACE APPLICATION

*Edited by:*
*Dr. Kumar Jata, Dr. Eui Whee Lee,*
*Dr. William Frazier and Dr. Nack J. Kim*

## COMPOSITES

## High Ductility Cast
## Aluminum Beryllium Alloys

*Nancy F. Levoy and William T. Nachtrab*

Pgs. 273-281

184 Thorn Hill Road
Warrendale, PA 15086-7514
(724) 776-9000

# HIGH DUCTILITY CAST ALUMINUM BERYLLIUM ALLOYS

Nancy F. Levoy and William T. Nachtrab

Starmet Corp., Research & Development
2229 Main St., Concord, MA 01742

## Abstract

Cast aluminum beryllium alloys have recently been introduced in the market for high performance aerospace applications where high stiffness and low density are critical properties. During solidification of cast aluminum beryllium alloys, a two phase composite microstructure develops in which the primary beryllium phase forms within an aluminum matrix. These alloys typically contain more than 60 weight percent beryllium and are approximately 20% lighter and 3 times stiffer than conventional aluminum alloys. However, these high beryllium alloys are limited by low ductility and the high cost of beryllium. New alloys with lower beryllium content have now been developed, which optimize mechanical properties such as strength and ductility while still providing high specific stiffness. This paper will describe how alloy composition can be tailored to provide different combinations of properties based on application requirements. Results will be discussed in the context of property development in metal matrix composites.

Lightweight Alloys for Aerospace Applications
Edited by Kumar Jata, Eui Whee Lee,
William Frazier and Nack J. Kim
TMS (The Minerals, Metals & Materials Society), 2001

# Introduction

Aluminum-beryllium alloys made by various casting processes have recently been developed for use in applications requiring a combination of light weight and high stiffness such as aircraft, satellites and high speed positioning devices. These alloys are based on compositions having greater than 50 percent by weight beryllium with the balance of the alloy being aluminum and other minor alloying elements. For example, an alloy containing 65% by weight beryllium has a density of 2150 $Kg/m^3$ and an elastic modulus of 205 GPa, which is approximately 20% lighter and 3 times stiffer than conventional aluminum alloys. While current aluminum-beryllium alloys are well suited for their intended applications, they are not well suited for applications where it is necessary to have high ductility. In an alloy produced by casting, current aluminum-beryllium alloys typically exhibit less than 10% elongation, which restricts their use in applications that require cold forming as a means of fabrication or in components that must be damage tolerant. Applications of Al-Be alloys containing greater than 50 weight % Be have also been limited due to the high cost of Be.

The goal of the work presented in this paper was to develop new castable Al-Be alloys for applications that do not require the high Be content of current alloys and that would benefit from higher ductility and toughness. A series of nine new Al-Be alloys with reduced Be content (20 to 40 wt%) were produced by investment casting. Results of this work show that, through alloying, it is possible to achieve useful strength in aluminum-beryllium alloys with less than 50% by weight beryllium. In addition, the ductility of these alloys exceeds 20% elongation in the as-cast condition. This combination of properties compares favorably with other aluminum matrix composites, such as cast Al-SiC.

# Experimental Procedures

Al-Be alloys containing 20, 30 and 40 % Be were investigated. All compositions are given in weight percent unless otherwise indicated. Material for this study was produced by investment casting procedures previously described.[1] Each alloy was vacuum induction melted and poured into a preheated investment mold incorporating twenty cylindrical net-shape tensile specimens.

Alloy formulations included in this study are shown in Table I. Ge was added to each alloy at the level of 1 % to improve castability and decrease porosity in the product. The beneficial effects of Ge in Al-Be alloys have previously been described.[1,2] Levels of Ag were varied to optimize strength while still achieving high levels of ductility. Ag was added at levels from 1 to 4 %.

A second set of experiments were performed to investigate effects of Co on strength and ductility. Co was added at levels from 0.5 to 2%.

Tensile tests were performed on samples in both the as-cast and heat treated conditions. The heat treatment used included solutionizing 2 hours in vacuum at 550°C, quenching in water, then aging 4 hours at 190°C.

Table I  Alloy Formulations (in weight %; balance Al)

| %Be | %Ge | %Ag | %Co |
|-----|-----|-----|-----|
| 20 | 1 | 4 | 0 |
| 30 | 1 | 3 | 0 |
| 40 | 1 | 1 | 0 |
| 40 | 1 | 2 | 0 |
| 40 | 1 | 3 | 0 |
| 40 | 1 | 4 | 0 |
| 40 | 1 | 3 | 0.5 |
| 40 | 1 | 3 | 1 |
| 40 | 1 | 3 | 2 |

In addition to tensile properties, Young's modulus was determined by ultrasonic methods. Microstructure was characterized in a scanning electron microscope (Cambridge S360) equipped with an energy dispersive spectroscopy system (Link eXL).

**Results and Discussion**

Properties

Cast Al-Be alloys are a type of in-situ metal matrix composite. During solidification, a two phase composite microstructure develops in which the primary Be phase forms within an Al matrix. As has been shown in previous studies of both powder metallurgy and cast Al-Be alloys,[3] the density and elastic modulus of the alloys of the current study are a direct function of the %Be in the alloy, and closely follow rule-of-mixtures behavior. Other alloy additions have only a minor effect on these properties (Figure 1).

Conversely, tensile properties (yield strength, tensile strength, and %elongation) are strongly affected by other alloy additions. As will be shown below, tensile strength in alloys of the present study containing 20, 30, and 40% Be is comparable to the tensile strength of a commercial cast Al-Be alloy containing 65% Be (Beralcast® 191).[1] This level of strength was consistently achieved in combination with greater than 20% elongation.

Table II shows the properties of the alloys at varying levels of Be and Ag additions. Ag segregates to the Al matrix and does not directly interact with the Be phase, although it does contribute to overall microstructural refinement and formation of equiaxed microstructures.[4] Ag acts to strengthen the alloy by forming a precipitation strengthening phase with Al in the Al matrix. Ag has a significant effect on tensile properties in the as-cast condition, and Ag containing alloys can be further strengthened by heat treatment. Optimum as-cast results were achieved for the combination of 3% Ag and 1% Ge as minor alloying additions. Heat treatment of this alloy (solutionize, quench and age) led to an increase in tensile strength without a loss of ductility. The best properties for heat treated material containing 40% Be were 138 MPa yield strength, 248 MPa ultimate tensile strength, and 21% elongation.

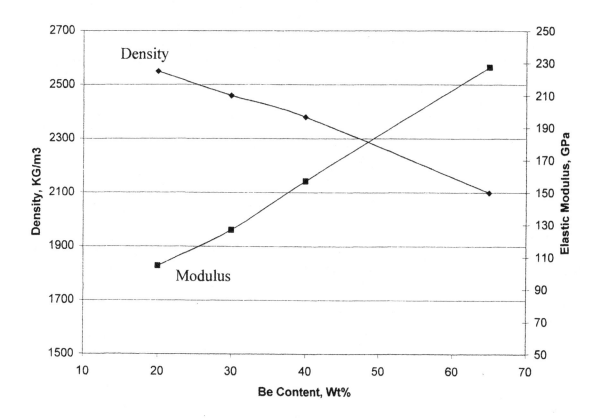

Figure 1: Density and elastic modulus as a function of Be content.

Table II Tensile Properties (As-cast)

| Composition (wt%) | 0.2% Yield Strength (MPa) | Ultimate Tensile Strength (MPa) | % Elongation (in 25 mm) |
|---|---|---|---|
| 40Be-58Al-1Ag-1Ge | 102.7 | 180.5 | 24.9 |
| 40Be-57Al-2Ag-1Ge | 120.6 | 209.5 | 21.1 |
| 40Be-56Al-3Ag-1Ge | 137.8 | 228.7 | 21.1 |
| 40Be-55Al-4Ag-1Ge | 126.1 | 216.3 | 20.8 |
| 30Be-66Al-3Ag-1Ge | 127.5 | 208.1 | 22.5 |
| 20Be-75Al-4Ag-1Ge | 121.3 | 198.4 | 20.0 |
| 65Be-31Al-2Si-2Ag* | 138.5 | 190.2 | 2.3 |

* Beralcast® 191

Although the main goal of this study was to develop Al-Be alloys with maximized ductility, higher strength alloys with good ductility were also investigated. Alloy formulations were selected to look at the strengthening effects of cobalt on lower Be alloys. Table III shows the effect of Co additions on the tensile properties of a 40% Be alloy. Significant strengthening occurs as the % Co increases, and ductility remains three to five times the ductility of alloys containing greater than 50% Be.

In cast alloys, Co segregates to the Be phase and strengthens this phase predominantly by a solid solution strengthening effect.[1,2,5] By SEM/EDS, it was confirmed that essentially all of the Co segregates to the Be phase in these alloys. The effect of Co is to increase the hardness and strength of the Be reinforcing phase of the Be-Al composite. The yield and tensile strength of the Al-Be composite increase directly with the strengthening of the Be phase. The Al matrix remains essentially unchanged by the Co additions, allowing overall composite strengthening to occur while still maintaining good ductility.

Table III  Effect of Co Additions on Tensile Properties (As-cast)

| Composition (wt%) | 0.2% Yield Strength (MPa) | Ultimate Tensile Strength (MPa) | % Elongation (in 25 mm) |
|---|---|---|---|
| 40Be-56Al-3Ag-1Ge | 137.8 | 228.7 | 21.1 |
| 40Be-55.5Al-3Ag-1Ge-0.5Co | 148.8 | 234.9 | 16.1 |
| 40Be-55Al-3Ag-1Ge-1Co | 168.1 | 252.9 | 13.9 |
| 40Be-54Al-3Ag-1Ge-2Co | 178.4 | 272.8 | 13.6 |

Microstructure

The microstructure for alloys containing varying levels of Be are shown in Figures 2 and 3. In these back-scattered electron images taken in the SEM, the dark imaging phase in the microstructure is the beryllium phase; the light imaging matrix is the aluminum phase. Figure 2 shows a typical cast alloy containing 65% Be. For this alloy, note that although the aluminum phase completely fills the interdendritic space around the beryllium phase, the volume of aluminum phase separating the beryllium dendrites is small. Figure 3 shows microstructures for alloys containing 20, 30 and 40% Be. In figure 3, note how the separation between the beryllium phase increases with the increasing amount of aluminum in the alloy. A minimum interdendritic spacing of the Be phase appears to be required to produce alloys with high ductility.

Figure 3D shows an alloy containing 40% Be with an addition of 2%Co. Note that the microstructure is coarser than for the 40% Be alloy without Co (Figure 3A). The coarsening of the microstructure may contribute to the reduced ductility of this alloy.

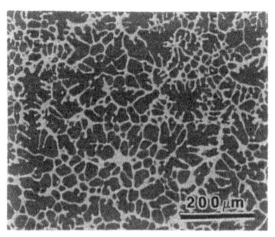

Figure 2: Back-scattered electron image showing microstructure of cast alloy with
composition 65Be-31Al-2Si-2Ag

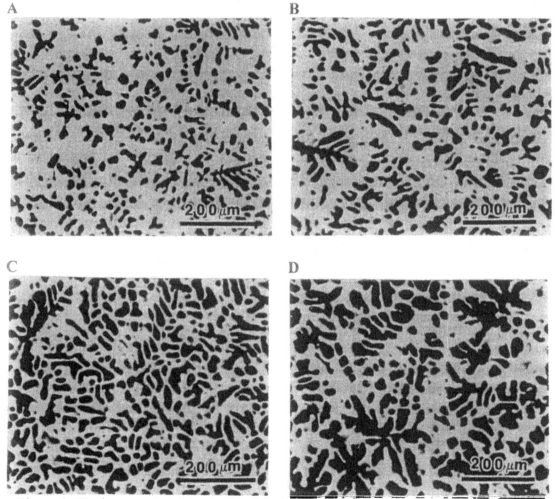

Figure 3: Back-scattered electron images showing microstructures of cast alloys with
compositions (a) 20Be-75Al-4Ag-1Ge; (b) 30Be-66Al-3Ag-1Ge; (c) 40Be-56Al-3Ag-
1Ge; and (d) 40Be-54Al-3Ag-1Ge-2Co.

## Discussion

It has long been assumed that, in Al-Be alloys, the Be phase provides the composite with high strength and stiffness, and low density, while the Al phase is responsible for the ductility of the composite. The present study is in agreement with past studies that have shown that the density and stiffness of Al-Be alloys are direct functions of the Be content,[6,7,8] while other alloy additions have only small effects on these properties.

Strength and ductility, however, are more complex phenomenon. As has been shown before, alloying to increase the strength of the Be phase (e.g. by Co additions) does in fact increase the strength of the overall alloy. However, simply increasing the %Be does not uniformly increase the strength of the alloy. Likewise, alloy strength can be enhanced by the addition of alloy elements that segregate to the Al phase.

Al is necessary for ductility in these alloys; Be by itself is brittle, and methods for increasing the ductility of Be without creating a distinct ductile phase (e.g. by alloying with Al, Mg, or Ag) have largely failed.[3,9] However, ductility does not increase steadily with increasing amounts of Al.

There appear to be several distinct regimes for ductility in these alloys. Above 40 weight % Be (equivalent to approximately 50 volume % Be) a limited amount of ductility (i.e. < 10% elongation) is imparted to these alloys by the Al phase. There is a sharp increase in the % elongation for alloys with 40 weight % or less Be, but the present study shows that the % elongation does not vary much for alloys containing between 20 and 40 weight % Be. A minimum interdendritic spacing of the Be phase appears to be required to produce alloys with high ductility. Once this minimum spacing is exceeded, increasing aluminum content does not contribute to additional increases in ductility.

Furthermore, some data indicate that the % elongation drops off slightly when %Be is reduced to 20 % or less. A possible explanation is that grain growth in the Al phase may account for the decrease in % elongation as the Be content drops below a critical value; this requires further investigation.

Quantitative microstructural analysis would be required for improved understanding of the relationship between specific microstructural features and ductility for these alloys. Quantitative microstructural analysis done by Underwood[10] in the 1970s yielded predictive correlations between specific quantitative microstructural features and properties for extruded and rolled binary Al-Be alloys. These correlations do not apply to the cast, higher order alloys of the present study, but they do show the usefulness of this type of analysis for Al-Be alloys.

## Summary

Cast Al-Be alloys containing 20 to 40% Be were investigated, with the goal of optimizing mechanical properties such as strength and ductility while still providing high specific stiffness. Results of this work show that, through alloying, it is possible to achieve useful strength combined with ductility exceeding 20% elongation in Al-Be alloys with less than 50% Be. Thus, it is possible to manufacture cast Al-Be alloys that are suitable for many engineering applications not restricted by the limited ductility of current Al-Be alloy compositions.

# References

1.  N.F. Levoy and B.J. Smith, "Beryllium Aluminum Alloy Development For Investment Casting," in *Synthesis/Processing of Lightweight Metallic Materials II,* C.M. Ward-Close, F.H. Froes, D.J. Chellman, and S.S. Cho, eds., The Minerals, Metals and Materials Society, 1997, pp. 363-374.

2.  W.T. Nachtrab and N.F. Levoy, "Beryllium-Aluminum Alloys for Investment Castings," Advanced Materials and Processes, Vol. 151, No. 5, May 1997, pp. 23-25.

3.  G.J. London, "Alloys and Composites," in Beryllium Science and Technology, Volume 2, D.R. Floyd and J.N. Lowe, eds., Plenum Press, NY, 1979, pp. 297-308.

4.  W.T. Nachtrab, N.F. Levoy, and K.R. Raftery, "Light Weight, High Strength Beryllium-Aluminum Alloy," US patent No. 5,603,780, Feb. 18, 1997.

5.  W.T. Nachtrab, N.F. Levoy, and R.L. White, "Ductile, Light Weight, High Strength Beryllium-Aluminum Cast Composite Alloy," US Patent No. 5,417,778, May 23, 1995.

6.  J.M. Marder, "Aluminum-Beryllium Alloys," Advanced Materials and Processes, 1997, vol. 152, pp. 37-40.

7.  D. Hashiguchi and F.C. Grensing, Adv. Powder Metall. Particulate Mater., 1995, vol. 3, pp. 12-13.

8.  D.H. Carter et. al., "Age Hardening in Beryllium-Aluminum-Silver Alloys," Acta Mater., 1996, vol. 44, p. 4311.

9.  X.D. Zhang et. al., "Microstructural Characterization of Novel *In-Situ* Al-Be Composites," Metallurgical and Materials Transactions A, Vol. 31A, November 2000, pp. 2963-2971.

10. E.E. Underwood, "Quantitative Shape Parameters for Microstructural Features," Microscope, 1976, vol. 24 (1), pp. 49-64.

# LIGHTWEIGHT ALLOYS FOR AEROSPACE APPLICATION

*Edited by:*
*Dr. Kumar Jata, Dr. Eui Whee Lee,*
*Dr. William Frazier and Dr. Nack J. Kim*

## COMPOSITES

# Characterization of Reinforcing Particle Size Distribution in a Friction Stir Welded Al-SiC Extrusion

*S.C. Baxter and A.P. Reynolds*

Pgs. 283-293

184 Thorn Hill Road
Warrendale, PA 15086-7514
(724) 776-9000

# CHARACTERIZATION OF REINFORCING PARTICLE SIZE DISTRIBUTION IN A FRICTION STIR WELDED Al-SiC EXTRUSION

S. C. Baxter and A. P. Reynolds
Department of Mechanical Engineering
University of South Carolina
Columbia, SC 29208, U.S.A.

## Abstract

Friction stir welding (FSW) is a new technique that shows great promise for improving the quality of welds in high strength aluminums. Relative motion between a rotating, non-consumable tool and the work-piece produces a solid-state weld via *in situ* extrusion and forging. In this work, friction stir welds were made on a discontinuously reinforced aluminum, specifically, a 7093-25% SiC + 15% Al extrusion. Statistical image analysis was applied to metallographic sections of the as extruded and as welded material, to investigate and characterize the material microstructure within the weld. Of particular interest is the change in particle size distribution across/through the weld.

Lightweight Alloys for Aerospace Applications
Edited by Kumar Jata, Eui Whee Lee,
William Frazier and Nack J. Kim
TMS (The Minerals, Metals & Materials Society), 2001

## Introduction

Friction stir welding (FSW) is a new technique that shows great promise for improving the quality of welds in high strength aluminums, which have traditionally been poor candidates for conventional welding techniques. Relative motion between a rotating, non-consumable tool and the work-piece produce a solid-state weld via *in situ* extrusion and forging.

While it is clear that during the FSW process material in the weld is transported and displaced due to the combination of frictional heating and mechanical deformation, the exact nature of the movement around the tool during welding is not fully understood. Welds in homogeneous metals are currently under investigation, (see, e.g., [1]), however less has been done to investigate welds in engineered heterogeneous materials. Welding particulate reinforced metal matrix composites presents additional challenges; in particular the potential difficulty of welding an effectively stiffer material containing abrasive particles, but it also offers a built-in mechanism, in the form of the spatial arrangement of the particles, which can assist in providing insight into the associated material displacements. Ultimately, accurate information on material microstructure at the composite level will also be necessary to make predictions of the effective behavior and properties of the welded component, and thus the effectiveness and reliability of the weld. In this work, friction stir welds were made on a 7093-25% SiC + 15% Al extrusion. Statistical image analysis was applied to metallographic sections of the as-extruded and as-welded material, to investigate and characterize the material microstructure within the weld. Of particular interest were volume fraction and particle size distributions of the particulate within and out of the weld.

## Material System

The material system used in this work was a discontinuously reinforced aluminum (DRA). The base alloy was Aluminum 7093, a precipitation hardened aluminum alloy containing zinc and magnesium as the major alloying elements. The alloy was produced by powder metallurgy, i.e., the alloy was produced by spraying liquid 7093 in a jet, breaking it up into droplets. The droplets, solidifying rapidly, freeze into a highly non-equilibrium composition and structure. Traditionally the metal matrix composite would be fabricated by combining the alloy particles with particles of the reinforcing phase, silicon carbide, and formed into a solid part by combinations of various thermo-mechanical treatments like Hot Isostatic Pressing (HIP), forging, and/or extrusion. DRAs are of interest because of typically higher specific modulus and superior specific strength, as compared to unreinforced aluminum alloys, and their ability to be processed using conventional metalworking techniques such as extrusion, rolling and/or forging. Commercially pure aluminum was also introduced into this composite during a subsequent extrusion process, to increase the fracture toughness, and resulted in a tube-like swirl of the pure material along the length of the extrusion [2]. The DRA was provided in the as-extruded condition so the heat treatment of the matrix is not well defined nor necessarily optimal. The focus of this work is the distribution of the particle reinforcement in the aluminum alloy matrix.

## Experimental procedures

Using a Supermill vertical milling machine, friction stir butt welds were made of extruded lengths of the composite. The weld was made in two passes, one from each side (top and bottom).

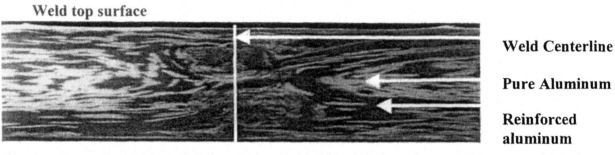

**Weld top surface**

**Weld Centerline**

**Pure Aluminum**

**Reinforced aluminum**

Figure 1: Transverse view of welded, extruded samples. Vertical line is drawn at the center of the weld, light areas are extruded pure aluminum, and dark areas are aluminum alloy matrix reinforced with silicon carbide particulate. The welding direction extends into the plane of the paper.

The welded sample consists of two extruded sections joined along sides parallel to the axis of the extrusion. The weld, and the swirl of pure aluminum, extend along the length of the sample, the longitudinal direction; the height is measured from the weld top to bottom surfaces. The width of the welded sample is defined by the horizontal distance from the weld centerline, defined as the transverse direction. (See Figure 1) .

Digital images were made of sections cut from the welded samples and labeled according to the schematic shown in Figure 2. The samples were mounted in epoxy resin and polished in order to be viewed and photographed using a metallographic microscope. Images taken along longitudinal faces, parallel to the centerline of the weld, were examined from the sections labeled

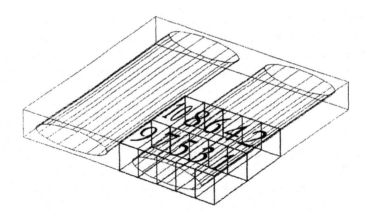

Figure 2: Schematic of sections of the welded samples. Images of the longitudinal planes were taken from the even numbered sections and images were made of the transverse faces from the odd numbered sections. Elliptical tubes approximate the location of the pure aluminum swirls.

10, 8, 6, 4 and 2, where the plane from section 10 is within the weld. Images of transverse planes were examined from the odd numbered sections, 9, 7, 5, 3 and 1, where section 9 extends from within the weld to outside of the **Heat Affected Zone** (HAZ). Table 1 lists the measured distances from the weld centerline of each of the longitudinal sections and from the weld centerline to the mid-point of each of the transverse sections.

Table I  Distance of the sample sections from the Weld centerline.

| Section # | Distance from Weld Centerline (mm) |
|---|---|
| 1 | 47.80 |
| 2 | 45.75 |
| 3 | 36.50 |
| 4 | 32.26 |
| 5 | 28.36 |
| 6 | 26.72 |
| 7 | 22.90 |
| 8 | 18.57 |
| 9 | 13.93 |
| 10 | 6.02 |

## Statistical Characterization

### Volume Fraction

The statistical software package, Simple PC, provided a semi-automated method of stereological analysis of the digital images. The local volume fraction of silicon carbide particles within the matrix aluminum was estimated by scanning ten sub-areas of each section, exclusive of the pure aluminum, and calculating the area fraction of the particles in each. Average values were then calculated for each section. Locally, the volume fraction of silicon carbide particles in the reinforced, unwelded matrix was approximately 28%.

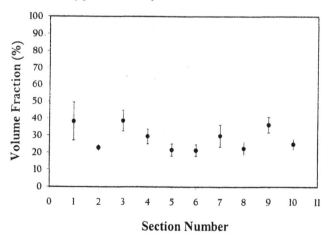

Figure 3: Local volume fraction of silicon carbide reinforcing particles for longitudinal and transverse sections. Error bars indicate +/- 1 standard deviation.

Figure 3 plots the average volume fraction for longitudinal sections (2, 4, 6, 8 and 10) and transverse sections (1, 3, 5, 7,and 9, measured from the center of the section) versus section number. Error bars indicate +/- 1 standard deviation. The correlation factor between volume fraction and distance from the weld centerline is 0.252 and suggests no relationship between the two quantities. The variations shown in the plots are likely due to a combination of the inherent variation in the original particle distribution, due to fabrication and extrusion, and the number of sub-areas examined.

Particle Size and Distribution of Relative Area

The distribution of particles within sub-fields of each section was also analyzed using Simple PC. The software calculated individual particle size by assuming an elliptical shape for each particle. Five sub-fields were examined from each section. Two sections are compared; Section 10, within the weld, and Section 2, the section furthest from the weld center line.

More than twice as many particles were counted in the samples taken from Section 10 than in the samples taken from Section 2; 716 and 341 respectively. However, the volume fraction of the particulate calculated over the sections was comparable: 24.8% in Section 10 and 22.8% in Section 2. The average particle size from Section 10 was 9.01 $\mu m^2$, roughly half the average size of the particles found in sampling Section 2, of 17.43 $\mu m^2$. The minimum particle size observed in both sections was 0.02 $\mu m^2$, the maximum particle sizes found in Section 10 and 2 respectively were 141.27 $\mu m^2$ and 122.15$\mu m^2$. Figure 4 plots particles versus particle area for the two sections. The graph on the left shows all particle areas and the one on the right has been expanded to show particles less than 20 $\mu m^2$ in area. Mean and median lines are drawn on the expanded graph. The most dramatic effect seems to be the extreme change in the median particle size. The median particle size from Section 10 was 0.32 $\mu m^2$, i.e., half of the particles had areas below this size, this is in contrast to the median particle size in Section 2 of 11.2 $\mu m^2$.

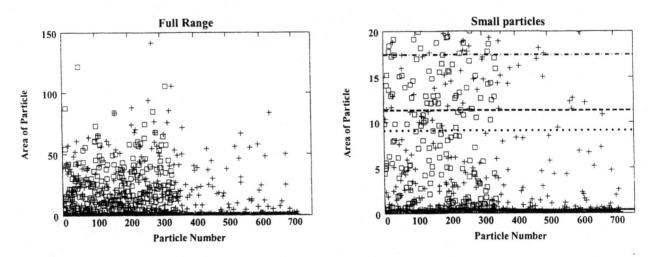

Figure 4: Particle Counts from Sections 10 marked with crosses and from Section 2 with squares. Full range of particle sizes in $\mu m^2$ are plotted in graph on left, enlarged region of particles less than 20 $\mu m^2$ is plotted on the right. Vertical lines on left are, from top to bottom, mean particle size Section 2, Median value Section 2, Mean particle size Section 10 and Median value Section 10.

Micrographs of sub-areas of each of the sections are shown in Figure 5; dark areas are the silicon carbide particles and light areas the aluminum alloy. On the left, the reinforced aluminum within the weld area clearly shows an increase in the number of small particles. Lower magnification

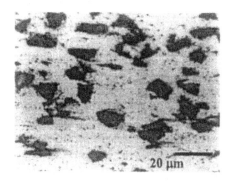 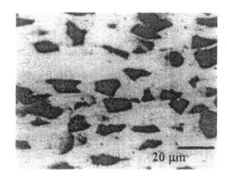

Figure 5: Micrographs of SiC reinforced Aluminum in the weld, Section 10 (left) and outside of the weld, Section 2 (right). The image at the left clearly indicates the presence of a larger number of small particles.

micrographs, shown in Figure 6, show particle spatial distribution on a larger scale. Qualitatively, there appears to be no increase in particle clustering due to the flow mechanisms within the weld.

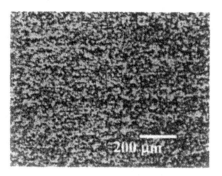 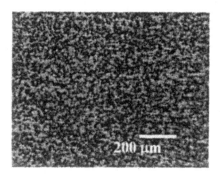

Figure 6: Micrographs of larger scale particle distribution. a) Section 10, within the weld. b) Section 2, outside of the weld.

While the number of particles changes dramatically between the two sections, with the net result being a large increase in the number of small particles, the shift in relative volume fraction is also of importance. To assess the relative influence of the shift in particle number/size distribution, statistical data was also calculated to illustrate the shift in the distribution of mass. This was done by normalizing particle size by the total particle area recorded from each sample.

Table II lists the statistical parameters associated with normalized volume fractions related to mean particle sizes, (particle area divided by total area of particulate over all sub-areas). Particles below the mean particle size in the unwelded sample make up approximately 63% of

Table II  Statistics of Relative Volume Fractions and Particle Sizes.

| Section # | Mean ($\mu m^2$) | % Rel. Vol. Frac. below mean (2) | % Rel. Vol. Frac below mean (10) |
|---|---|---|---|
| 2 | 17.43 | 62.65 % | - |
| 10 | 9.02 | 75.70 % | 81.56 % |

| Section # | Median ($\mu m^2$) | % Rel. Vol. Frac. below median (2) | % Rel. Vol. Frac below median (10) |
|---|---|---|---|
| 2 | 11.26 | 8.58 % | - |
| 10 | .32 | 8.48 % | .59 % |

the relative volume fraction. Particles below this same mean particle size (from Section 2) make up approximately 76% of the relative volume fraction in Section 10. As a comparison, particles in Section 10 below the mean particle size, calculated from Section 10, make up approximately 82% of the relative volume fraction. Thus, while there are many more small particles in the weld sample, the change in the 'mass' contributed by 'small' particles is not large. Similarly, half the particles in Section 2 lie below its median particle size, by definition, and these particles make up approximately 8.6% of the relative volume fraction. Roughly 77% of the number of particles in Section 10 lie below the median value from Section 2, and this number makes up approximately 8.5% of the total relative volume fraction.

Histograms were also constructed for the two samples over the full range of relative volume fractions and are shown in Figure 5 for Section 10 and Section 2. The histograms are divided into ranges of approximately 0.05%. These plots show very little change in the distribution of volume fraction; the mean values of the histogram heights are the same, 0.025, for both sections. This suggests that more corners and sharp edges are being knocked off large particles, than large particles being shattered.

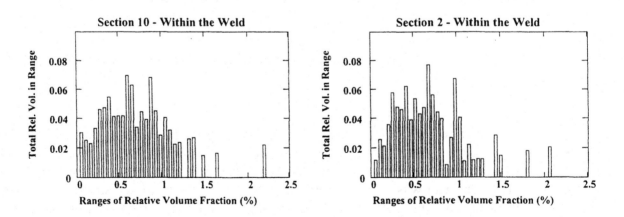

Figure 5: Histograms of full range of relative volume fractions.

Hardness Tests

The hardness of the aluminum alloy matrix material was measured using a Vicker's Hardness tester. Measurements were taken along the transverse face of the welded extrusion, at various distances from the weld centerline and at three levels from the top to bottom surface of the weld. The silicon-carbide particles and the extruded swirl of pure aluminum were avoided. The results, shown in Figure 8 indicate a sharp increase of hardness at roughly the outer edge of the weld, known as the **H**eat **A**ffected **Z**one (HAZ). At locations closer to the weld nugget, the hardness decreases from that of the HAZ, but remains higher than outside of the weld. Because the temperature processing history of the composite is unknown, it is difficult to assess the effect of the weld heating on the constitutive response of the matrix material, as the observed response will be due to the combined effects of fabrication, extrusion processing, and the welding process.

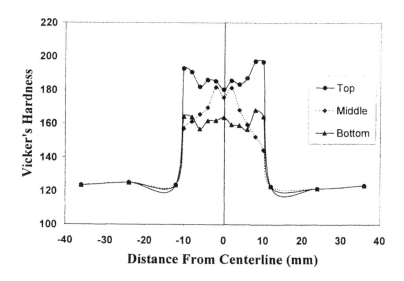

Figure 8: Hardness Tests of Aluminum matrix material. Tests were made near the top, center and bottom of the weld, across the transverse face of the welded sample. Higher hardnesses are observed within the weld.

Fracture surfaces

Fracture surfaces along the transverse axes, through the weld and through the as-extruded sample were produced by bending. Figure 9a shows an SEM (scanning electron micrograph) of the interface between the pure aluminum swirl and the reinforced aluminum matrix. The large size of the voids in the pure aluminum, lower left corner, is characteristic of fracture in very ductile materials. There will be fewer nucleation sites within the ductile pure aluminum, but each will grow to a relatively large void and the resulting fracture is high-energy. Figures 9b and 9c show higher magnification SEMs of the reinforced aluminum within and outside of the weld, respectively. The smooth dark areas are SiC particles. Within the weld, voids within the matrix aluminum are difficult to see, their small size the result of a more brittle matrix and numerous nucleation sites. Outside of the weld, the larger voids suggest the lowered hardness and somewhat less brittle nature of the matrix. These images illustrate the significant effect of the

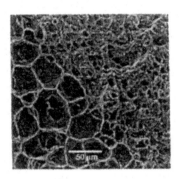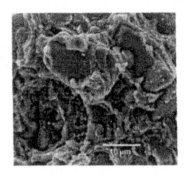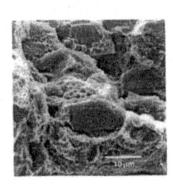

Figure 9: SEMs of fracture surfaces: a) Interface of pure aluminum and reinforced matrix within the weld, pure aluminum is in the lower left corner  b) matrix within the weld, smooth darker regions are SiC particles small voids not visible in the aluminum matrix c) matrix outside of the weld larger voids.

heat treatment on the matrix material during the welding process. While the change in particle size distribution may also have an effect, smaller SiC particles may potentially influence yield stress and subsequent hardening slope of stress strain curves, it is not the dominant factor in the fracture mechanism. The dominant effect on the matrix material within the weld probably arises from precipitation processes that occur simultaneously with the friction stir welding.

## Summary and Conclusions

Previous modeling and experimental work [3] have demonstrated the presence of high shear rates within the flow-influenced area of a friction stir weld. While the flow mechanism produces high shear rates and strains, it appears that material displacements are small. The current work shows that while the volume fraction of the included particulate phase remains relatively constant through the extrusion and weld, the number of 'small' particles increases significantly within the weld, likely due to the high shear strains encountered in the welding process and contact with the tool. Increased clustering of the particles within the weld is not apparent. The change in the distribution of relative percent volume fraction, contributed by large or small particles, in contrast to the distribution of particles sizes, is not significant, suggesting that corners of large particles are being rounded rather than large particles shattered.

Elastic properties in composites are dominated by the volume fraction of the phases. The microstructural geometry, i.e. the distribution of particle sizes and shapes, can potentially influence the yield stress and hardening slope in an inelastic regime [4], but this effect is expected to be relatively small. While this work suggests that the dominant influence in fracture is the heat treatment of the aluminum alloy matrix during the welding process, an increase in the number of small particles also has the potential to influence fracture mechanisms. This effect of friction stir welding on DRA properties might be more readily observed if post weld heat treatments are utilized to homogenize the matrix properties inside and outside of the weld region.

## Acknowledgements

This work was supported by an NSF Research Experience for Undergraduates supplement to NSF-CMS –9875067, Dr. Jorn Larsen-Basse program director and by USAF/AFRL, Dr. Kumar Jata, Technical Point of Contact, contract number TMC-96-5835-0083-01. John G. Crowley and Keith A. Blinn contributed to this work.

## References

[1] A.P. Reynolds, "Visualization of material flow in autogeneous friction stir welds", Science and Technology of Welding and Joining, 5(2): (2000) 120 -124.

[2] A.B. Pandey, B.S. Majumdar, and D.B. Miracle, "Effect of aluminum particles on the fracture toughness of a 7093/SiC/15p composite,". Materials Science & Engineering, A259 (1999) 296-307.

[3] S. Xu, X. Deng, A.P. Reynolds and T. Seidel (2000), "Finite Element Simulation of Material Flow in Friction Stir Welding", (2000) to appear in Science and Technology of Welding and Joining.

[4] A.C. Sankurathri, S.C. Baxter, and M-J. Pindera "The Effect of Fiber Architecture on the Inelastic Response of Metal Matrix Composites with Interfacial and Fiber Damage", in Damage and Interfacial Debonding in Composites, G. Z. Voyiadjis and D. H. Allen (Eds.), pp. 235-257, Elsevier Science B.V., the Netherlands 1996

# LIGHTWEIGHT ALLOYS FOR AEROSPACE APPLICATION

*Edited by:*
*Dr. Kumar Jata, Dr. Eui Whee Lee,*
*Dr. William Frazier and Dr. Nack J. Kim*

## COMPOSITES

## In-Situ Formation of AlN Reinforced Al Alloy Composites Using Ammonia

*Qingjun Zheng, Banqiu Wu and Ramana G. Reddy*

Pgs. 295-307

184 Thorn Hill Road
Warrendale, PA 15086-7514
(724) 776-9000

# *IN-SITU* FORMATION OF AlN REINFORCED Al ALLOY COMPOSITES USING AMMONIA

Qingjun Zheng[1], Banqiu Wu[2], and Ramana G. Reddy[3]

[1]Graduate Student, [2]Research Engineer, and [3]ACIPCO Professor
Department of Metallurgical and Materials Engineering
The University of Alabama
Tuscaloosa, AL 35487-0202

## Abstract

*In-situ* formation of Al alloy matrix composites was investigated. The synthesis of Al alloy matrix composites reinforced with *in-situ* formed AlN particles was achieved in the temperature range of 1273-1473K by directly bubbling ammonia ($NH_3$) through an Al alloy melt. Products were characterized using X-ray diffraction, optical microscopy, SEM, and EDS analyses. Thermodynamic analysis of Al-Si-N ternary system was carried out using Gibbs energy minimization method and its phase diagram at 1473K was calculated. The results showed that the AlN content in the composites is up to 27.4 wt% and AlN particles are uniformly dispersed in the Al-alloy matrix. Formed AlN particles are non-spherical and small in size. The results also indicated that AlN reinforcement is thermodynamically stable in Al and Al-Si alloy matrix over a wide temperature range.

Lightweight Alloys for Aerospace Applications
Edited by Kumar Jata, Eui Whee Lee,
William Frazier and Nack J. Kim
TMS (The Minerals, Metals & Materials Society), 2001

## Introduction

In addition to the applications in defense and aerospace industries, discontinuously reinforced Al-based composites (DRACs) can also be used in tribological, automotive, electronic packaging, and other industries. The increasing applications of DRACs depend on their quality/cost (Q/C) relationship [1]. So far a competitive Q/C relationship has been established for wear applications by means of traditional casting and solid-state techniques. However DRACs produced by these techniques are still unprofitable for structural applications, in comparison with cast iron and structural steel. One important reason is that the small-size reinforcing particles, necessary for achieving good mechanical properties, are currently very expensive. Additionally, the features of interface between reinforcement and matrix play an important role in determining the quality of composites, because a significant amount of failures in composites are through interface mechanisms such as debonding, stress damping, crack propagation and boundary pinning. Strong interfacial bond is effective in transferring stress from matrix to the stronger particles, while weak bond may cause debonding or micro-cracks at the inter-phase boundaries under loading. During service life, composites may experience phase transformation, changes in equilibrium composition, and chemical reaction between matrix and reinforcement, resulting in thermodynamic instability of the interface.

The inherent economy and the long-term superior performance are guaranteed by the economic processing techniques and the stable phases in the composites during their service. This scientific awareness has inspired a large amount of research on *in-situ* processing of composites [2-9]. *In-situ* processing flaws the traditional routes, because the equilibrium reinforcing particles are formed *in-situ* and dispersed in the matrix melts. This not only simplifies the production process, lowers the costs, but also minimizes the contamination of the reinforcing particles, giving thermodynamically stable interface.

In our previous research [3, 4], it was found that AlN is a thermodynamically equilibrium phase with low solubility in Al and Al-Mg alloy over a wide temperature range, covering the whole processing and application temperature range of Al-based composites. AlN-Al composites were *in-situ* synthesized by bubbling deoxidized $N_2$ through Al alloy melts at high temperatures. It was also found that the *in-situ* formed particles had small size and were well bonded with matrix. In addition, it was also showed that interfacial reaction is the rate-controlling step and $O_2$ and oxygen-bearing compounds are deleterious to the synthesis of AlN reinforcing phase. The efficiency of the *in-situ* formation of AlN reinforcing phase is low and may be improved by introducing reducing agents or surface-catalytic agents [4]. With this purpose, $NH_3$ is being investigated as the nitrogen precursor for substituting $N_2$. In this paper, thermodynamic modeling on this process and experimental research were carried out.

## Thermodynamic Modeling

Thermodynamic analysis is based on Gibbs energy minimization method, owing to the fact that the Gibbs energy of the system reaches its minimum value at equilibrium. The thermodynamic formulation is given as [9]:

$$G = \sum_{gas} n_i \left( g_i^o + RT \ln P_i \right) + \sum_{\substack{pure \\ condensed \\ phase}} n_i g_i^o + \sum_{solution-1} n_i \left( g_i^o + RT \ln x_i + RT \ln \gamma_i \right)$$

$$+ \sum_{solution-2} n_i \left( g_i^o + RT \ln x_i + RT \ln \gamma_i \right) \tag{1}$$

where, G - total Gibbs energy of the system, $g_i^\circ$ - the standard molar Gibbs energy of species i at P and T; $n_i$ – molar number of species i; $P_i$ – partial pressure of species i; $x_i$ – mole fraction of species i; and $\gamma_i$ – activity coefficient of species i. Using Gibbs energy minimization method, the stability of AlN in Al alloy and the condition of AlN formation using ammonia as the nitrogen precursor were analyzed.

Stability of AlN in Al-Si alloy

Figure 1 shows the ternary phase diagram of Al-Si-N system at 1473K calculated using Gibbs minimization method. In the calculation, Al-Si alloy melt was assumed to be an ideal solution at this temperature based on the model proposed by Reddy and Kocherginsky [7]. It is indicated that AlN is thermodynamically stable in Al-Si-N ternary system, and attains equilibrium with Al-Si alloy in the whole composition range at 1473K, while $Si_3N_4$ is stable and can be formed only when Al is consumed in the alloy melt.

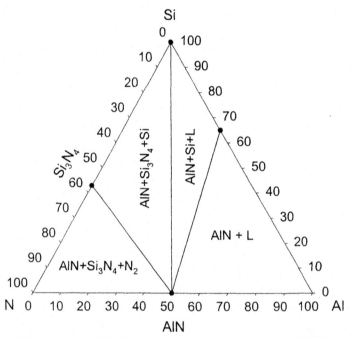

Figure 1: Al-Si-N ternary phase diagram at 1473K.

The equilibrium compositions in Al-NH3 and Al-Si-NH3 systems at different temperatures were calculated using Gibbs energy minimization method and the results are shown in Figures 2 and 3. AlN is in equilibrium with Al and Al-Si alloy from ambient temperature to 2000K, covering the whole temperature range for the application and synthesis of Al-based composites, while $Si_3N_4$ is not a stable phase in Al-Si-NH3 system. The possible reactions in Al-Si-NH3 systems are:

$$2Al(l) + N_2(g) = 2AlN(s) \tag{2}$$

$$3Si(l) + 2N_2(g) = Si_3N_4(s) \tag{3}$$

$$4Al(\text{in Al-Si alloy melt}) + Si_3N_4(s) \rightleftharpoons 4AlN(s) + 3Si(\text{in Al-Si alloy melt}) \tag{4}$$

The reaction expressed in Equation 4 shows the equilibrium between AlN and $Si_3N_4$ in Al-Si alloy melt. If the reaction occurs toward the right side, AlN is thermodynamically stable, and if inversely, $Si_3N_4$ is stable. The condition for the stability of AlN is expressed as Equation 5, where $a_{Si}$ and $a_{Al}$ are the activities of Si and Al in the Al-Si alloy melt respectively.

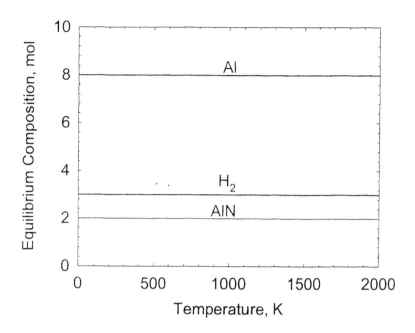

Figure 2: Equilibrium composition of Al-NH$_3$ system vs. temperature (K), initial input: Al: 10mol, NH$_3$: 2mol.

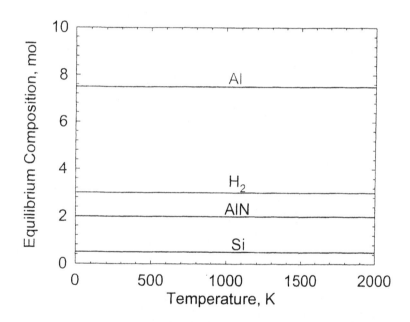

Figure 3: Equilibrium composition of Al-Si-NH$_3$ system vs. temperature (K), initial input: Al: 9.5mol, Si: 0.5mol, NH$_3$: 2mol.

$$\log \frac{a_{Si}^3}{a_{Al}^4} \leq \frac{22113}{T} - 2.69 \tag{5}$$

At temperatures higher than 973K, $a_{Al}$ approaches $x_{Al}$ based on the activity model of Al-Si alloy melt proposed by Reddy and Kocherginsky [7]. In the limiting case, $a_{Si}=1$, Equation 5 can be rewritten as:

$$\log x_{Al} \geq -\frac{5528.25}{T} + 0.6725 \qquad (6)$$

Figure 4 shows the stability regions of AlN and $Si_3N_4$, which is plotted based on Equation 6. Above the line, AlN is stable and below the line, $Si_3N_4$ is stable. $Si_3N_4$ is stable only at low concentration of Al in Al-Si alloy. AlN is thermodynamically stable and forms in Al-Si alloys with appreciable content of Al.

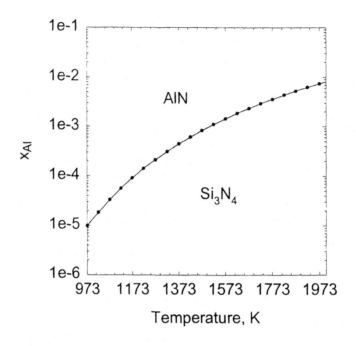

Figure 4: Stability regions of AlN and $Si_3N_4$ in Al-Si alloy melts.

### Stability of ammonia

In addition to $N_2$, $NH_3$ can also be used as nitrogen precursor. $NH_3$ is unstable and is dissociated into $N_2$ and $H_2$ at high temperatures according to the following reaction as:

$$NH_3(g) = 1/2N_2(g) + 3/2H_2(g) \qquad (7)$$

The equilibrium composition of N-H system at different temperatures was calculated using Gibbs energy minimization method and the results are shown in Figure 5. $NH_3$ is almost completely dissociated into $H_2$ and $N_2$ at temperatures exceeding 773K.

### Effect of $H_2$ in bubbling gas

$O_2$ and moisture in the bubbling gas are detrimental to the formation of AlN. They are more active than $N_2$, and have preference to react with Al to form an impermeable oxide skin at the interface between the gas bubble and the alloy melt through reaction expressed in Equation 8. Once the impermeable oxide skin is formed, it is necessary for nitrogen to diffuse through the skin to react with the fresh Al, the atomic diffusion through the dense $Al_2O_3$ skin is extremely slow, and then the formation of AlN is inhibited.

$$4Al(l) + 3O_2(g) = 2Al_2O_3(s) \qquad (8)$$

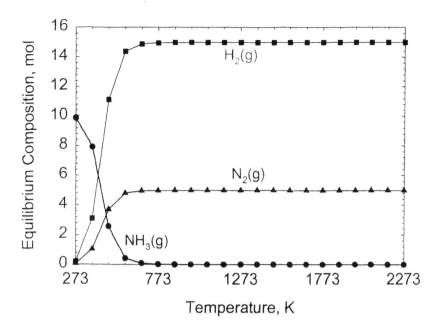

Figure 5: Equilibrium composition in N-H system, initial input $NH_3$ =10mol.

When $N_2$ is used as the nitrogen precursor, the maximum permissible partial pressure of $O_2$, $P_{O2}$, is given by:

$$\log P_{O_2} = -\frac{58941}{T} + 11.53 \tag{9}$$

where, the partial pressure of $N_2$ ($P_{N2}$) approaches the total pressure, $P_t$ (= 1atm) and so the permissible partial pressure of $O_2$ can be calculated. The results are shown in Figure 6, indicating that the partial pressure of $O_2$ in $N_2$ bubbling gas is required to be lower than the threshold value to form AlN. At 1423K, the permissible partial pressure of $O_2$ is about $10^{-29}$ atm.

The main advantage of substituting $N_2$ by $NH_3$ as nitrogen precursor is the introduction of the reducing agent, $H_2$, into the reaction system. In this case, the permissible partial pressure of $O_2$ can be obtained based on the reactions described by Equations 10 and 11, and is expressed in Equation 12.

$$2H_2(g) + O_2(g) = 2H_2O(g) \tag{10}$$

$$2Al(l) + 3H_2O(g) = Al_2O_3(s) + 3H_2(g) \tag{11}$$

$$\log P_{O_2} = \frac{1}{2}\log P_{H_2O} = \frac{-16420}{T} + \log P_{H_2} + 2.492 \tag{12}$$

Since the total pressure in the gas bubble approaches 1 atm in the bubbling process, $P_{H2}$ in Equations 12 can be considered to be 0.75 atm based on the reaction described by Equation 7. In this case, the original permissible partial pressure of $O_2$ in $NH_3$ bubbling gas can be calculated. The results are shown in Figure 6, indicating that the permissible content of $O_2$ is significantly increased. At 1473K, the permissible partial pressure of $O_2$ is $1.69 \times 10^{-9}$ atm in $NH_3$ bubbling gas, $10^{13}$ order higher than that in $N_2$ bubbling gas.

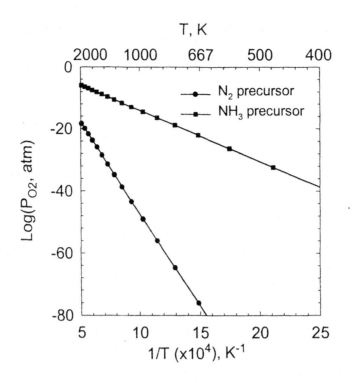

Figure 6: Maximum permissible partial pressure of $O_2$
in bubbling gas for different nitrogen precursor

## Experimental

The experimental setup is shown in Figure 7. A vertical Lindberg resistance furnace with
working temperature range of 773-1773K was used in the experiments. The *in-situ* reaction was
performed in an alumina crucible, located in a furnace tube with bottom end closed and upper
end sealed by a furnace cover. Pure Al of 99.99% and Al-5 at.% Si alloy with Si of 5 at. %
were used as the matrix. Before heating, the furnace tube was evacuated and flushed with pure
Ar gas for 2-3 times to remove $O_2$ and moisture. Then, pure Ar gas was introduced at a constant
rate from the top of the furnace throughout the experimental process to maintain an inert
atmosphere. An opening in the furnace cover serves as the exit for gas. However, the gas in the
environment cannot enter the system due to the small pressure difference maintained by the
flowing Ar gas. After the reaction temperature was attained, $NH_3$ of 99.999% was bubbled
through the alloy melts through an alumina tube with nozzle diameter of 1.5 mm. The length of
the bubbling tube in the furnace tube is long enough, ensuring that the bubbling gas is heated in
it to the temperature of melts before entering alloy melt. This design can avoid the fluctuation
of the melt temperature. In this case, $NH_3$ can be considered to have completely dissociated into
$N_2$ and $H_2$ before entering the matrix melt in this study, where the reaction temperature is not
less than 1173K. After bubbling for 2-5 hours, the furnace was turned off, allowing the
products cool in the furnace. Temperature in the reaction chamber was measured by a Type-S
thermocouple positioned directly above the melt. The whole experimental process was
monitored through an eyehole on the furnace cover. The experiments were done at temperatures
ranging from 1173K to 1473K.

The average AlN content in the products was analyzed based on materials balance before and
after the experiments. The products were characterized using X-ray diffraction (XRD),
scanning electron microscopy (SEM), and energy-disperse X-ray (EDS) analyses.

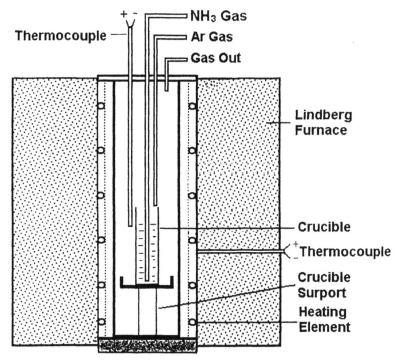

Figure 7: Experimental setup.

**Results and Discussion**

Yield efficiency of AlN

The reaction temperature has significant effect on the formation of AlN. After bubbling $NH_3$ through Al melt for 2 hours at 1473K, the content of AlN is about 27.4 wt % in the synthesized AlN-Al composites. However, even after bubbling for more than 4 hours at temperature lower than 1273K, the weight gain of the alloy is unappreciable, indicating low synthesis efficiency of AlN.

A comparison was made between the calculated and the experimental yield efficiency of AlN in Al-$NH_3$ system at different temperatures, and the results are shown in Figure 8. Based on the thermodynamic calculation, the yield efficiency of AlN can be 100% in Al-$NH_3$ system in a wide temperature range. However, the actual yield efficiency of AlN is sensitive to the temperature. AlN is not significantly formed at reaction temperatures lower than 1273K. At temperatures exceeding 1273K, the yield efficiency of AlN increases with the rising temperature, and reaches 96% at 1473K. There are two causes for the lower efficiency of AlN formation in practice and the efficiency temperature dependence. Firstly, the thermodynamic prediction does not consider the effects of trace $O_2$ and moisture in the bubbling gas. $O_2$ and $H_2O$ are more reactive than $N_2$ and may react with Al to form an impermeable skin at the interface between gas bubble and alloy melt, resulting in the difficulty in the formation of AlN. In addition, the equilibrium is not always achieved in the practical conditions due to the kinetic restraints. In our previous research [4], it was found that the activation energy for the dissociation of $N_2$ at the interface between gas bubble and alloy melt is very high, and long time is required for establishing the thermodynamic equilibrium. In bubbling process, the residence time of a gas bubble in the alloy melt may not be sufficiently long for establishing the thermodynamic equilibrium.

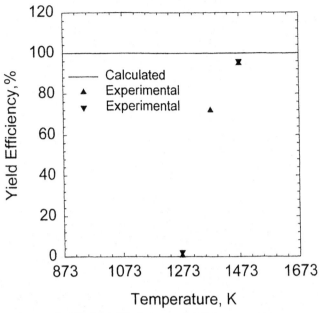

Figure 8: Yield efficiency of AlN vs. temperature.

## Microstructure of AlN-Al Composites

Figure 9 shows a typical secondary electron (SE) image of the sample from the AlN-Al composites formed *in-situ*. The composites are formed in the crucible by bubbling $NH_3$ through Al melt for 2 hours at 1473K. The corresponding EDX analysis shows that the bright domains are AlN with content of about 22 wt.% in the sample, and the dark region is the Al matrix. As shown in Figure 9, AlN particles are non-spherical, small in size, and uniformly dispersed in Al matrix. Additionally, the interface between the reinforcement and matrix is clean, where no other phase is found, showing that AlN is thermodynamically stable in Al matrix. The powder X-ray diffraction (XRD) pattern of the composites is shown in Figure 10. Except for Al and AlN, no other phases were identified, also indicating that the composites are mainly composed of AlN and Al.

Figure 9: Typical SE image of the sample from the AlN-Al composites formed as the bulk products in Al-$NH_3$ system.

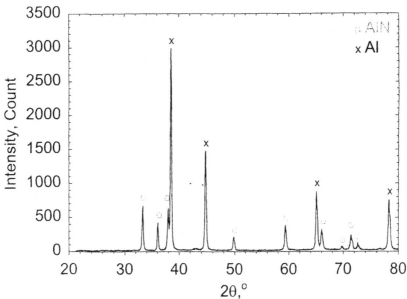

Figure 10: XRD spectrum of the sample from the AlN-Al
composites formed in Al–NH₃ system.

Effect of Alloy Composition

In this study, Al-5% Si alloy was also used as the matrix of the composites. Si does not exhibit obvious effects on the formation of AlN reinforcement. Materials balance calculation shows that the content of AlN in Al-Si alloy matrix is comparable with that for Al matrix under the same experimental conditions. Figure 11 is a typical XRD spectrum of the AlN reinforced Al-Si alloy composites formed after bubbling NH₃ through the Al-Si alloy for 2 hours at 1473K, manifesting that the composite is composed mainly of Al, AlN and Si phases. Figure 12 shows a typical topography of the composites, where bright domains are AlN particles, the dark massive phase is the eutectic Al-Si alloy, and the remaining region is primary Al. In the *in-situ* formed AlN-reinforced Al-Si alloy composites, AlN particles are uniformly dispersed in the matrix and have even smaller size, in comparison with the AlN-Al composites formed in Al-NH₃ system.

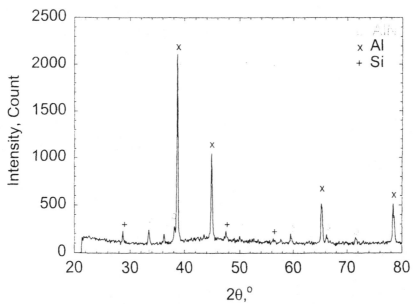

Figure 11: XRD spectrum of the sample from the AlN-Al
composites formed in Al-Si-NH₃ system.

305

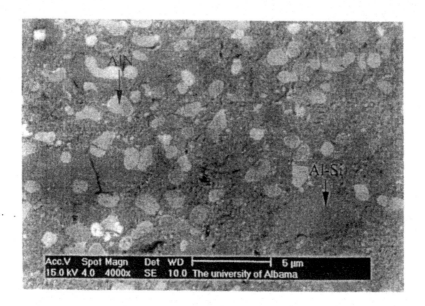

Figure 12: Typical SE image of the sample from the AlN-Al
composites formed in Al-Si-NH₃ system

## Conclusions

1) AlN is a thermodynamically stable phase in Al-Si alloy over a wide temperature range, covering the entire temperature range for the processing and application of Al alloy composites.

2) AlN reinforcing particles can be formed *in-situ* by using $N_2$ and $NH_3$ as the nitrogen precursor. When $N_2$ is used as the nitrogen precursor, the permissible partial pressure of $O_2$ in the bubbling gas is low. In practice, the trace $O_2$ in it is required to be removed before bubbling.

3) Using $NH_3$ as the nitrogen precursor introduces $H_2$, which is a reducing agent, hereby increasing significantly the permissible partial pressure of $O_2$ for the formation of AlN reinforcement.

4) The *in-situ* formed AlN reinforcing particles in Al alloys are uniformly dispersed in matrix and are small in size ($\leq 2\mu m$).

5) Si in alloy does not exhibit obvious effect on the formation of AlN reinforcing particles. The particles formed are also small in size ($\leq 2\mu m$) and are uniformly dispersed in Al-Si alloy matrix.

## Acknowledgements

The authors are pleased to acknowledge the financial support for this research by National Science Foundation (DMI 9714321).

## References

1.  Varuzan M. Kevorkijan, "Aluminum-base Composites," Advanced Materials and Processes, 155 (5) 1999, 27-31.

2.  M. J. Koczak and M. K. Premkumar, "Emerging Technology for *in-situ* Production of MMCs," Journal of Metals, 45 (1) (1993), 44-48.

3. Qingjun Zheng, R. G. Reddy and Banqiu Wu, "*In-situ* Processing of AlN-Al alloy composites," State of Art in Cast Metal Composites in the Next Millennium - 2000 TMS Fall Meeting, Edited by P. K. Rohatgi, St. Louis, Missouri, October 8-12, 2000, pp. 1-12.

4. Qingjun Zheng and R. G. Reddy, "*In-situ* Synthesis of AlN-Al Alloy composites," The 2001 NSF Design and Manufacturing Research Conference, Florida.

5. R. G. Reddy, "A Novel Manufacturing Process for Lightweight Alloy Composites," Composites in Manufacturing, 14 (3) (1998), 5-7

6. Banqiu Wu and R. G. Reddy, "*In–situ* Formation of SiC Reinforced Al-Si Alloy Composite From Methane," Metallurgical Transaction B, submitted.

7. D. M. Kocherginsky and R. G. Reddy, "*In-situ* Processing of Al/SiC Composites," Proc. *In-situ* Reactions for Synthesis of Composites, Ceramics, and Intermetallics, TMS, (1995), 159-167.

8. Ramana G Reddy and Banqiu Wu, "*In-situ* Formation of SiC Reinforced Al-Si Alloy Composites From Methane," Patent, 09/537412, (Pending).

9. Yucel Birol, "*In-situ* Processing of TiC$_p$-Al Composites by Reacting Graphite with Al-Ti Melts," J. Mat. Sci., 34 (7) (1999): 1653-1657.

10. N. A. Gokcen and R. G. Reddy, Thermodynamics, New York Plenum Press, 1986.

# AUTHOR INDEX

# SUBJECT INDEX